Probabilistic Methods
for
Financial and Marketing Informatics

Richard E. Neapolitan   Xia Jiang

Publisher	Diane D. Cerra
Publishing Services Manager	George Morrison
Project Manager	Kathryn Liston
Assistant Editor	Asma Palmeiro
Interior printer	The Maple-Vail Book Manufacturing Group
Cover printer	Phoenix Color

Morgan Kaufmann Publishers is an imprint of Elsevier.
500 Sansome Street, Suite 400, San Francisco, CA 94111

This book is printed on acid-free paper.

Library of Congress Cataloging-in-Publication Data

Application submitted.

ISBN 13: 978-0-12-370477-1
ISBN10: 0-12-370477-4

For information on all Morgan Kaufmann publications, visit our Web site at www.mkp.com or www.books.elsevier.com.

Printed and bound in the United Kingdom

Transferred to Digital Printing 2011

Preface

This book is based on a course I recently developed for computer science majors at Northeastern Illinois University (NEIU). The motivation for developing this course came from guidance I obtained from the NEIU Computer Science Department Advisory Board. One objective of this Board is to advise the Department concerning the maintenance of curricula that is relevant to the needs of companies in Chicagoland. The Board consists of individuals in IT departments from major companies such as Walgreen's, AON Company, United Airlines, Harris Bank, and Microsoft. After the dot.com bust and the introduction of outsourcing, it became evident that students, trained only in the fundamentals of computer science, programming, web design, etc., often did not have the skills to compete in the current U.S. job market. So I asked the Advisory Board what else the students should know. The board unanimously felt the students needed business skills such as knowledge of IT project management, marketing, and finance. As a result, our revised curriculum, for students who hoped to obtain employment immediately following graduation, contained a number of business courses. However, several members of the board said they'd like to see students equipped with knowledge of cutting edge applications of computer science to areas such as decision analysis, risk management, data mining, and market basket analysis. I realized that some of the best work in these areas was being done in my own field, namely Bayesian networks. After consulting with colleagues worldwide and checking on topics taught in similar programs at other universities, I decided it was time for a course on applying probabilistic reasoning to business problems. So my new course called "Informatics for MIS Students" and this book called *Probabilistic Methods for Financial and Marketing Informatics* were conceived.

Part I covers the basics of Bayesian networks and decision analysis. Much of this material is based on my 2004 book *Learning Bayesian Networks*. However, I've tried to make the material more accessible. Rather than dwelling on rigor, algorithms, and proofs of theorems, I concentrate on showing examples and using the software package Netica to represent and solve problems. The specific content of Part I is as follows: Chapter 1 provides a definition of informatics and probabilistic informatics. Chapter 2 reviews the probability and statistics needed to understand the remainder of the book. Chapter 3 presents Bayesian networks and inference in Bayesian networks. Chapter 4 concerns learning Bayesian networks from data. Chapter 5 introduces decision analysis and influence diagrams, and Chapter 6 covers further topics in decision analysis. There is overlap between the material in Part I and that which would be found in a book on decision analysis. However, I discuss Bayesian networks and learning Bayesian networks in more detail, whereas a decision analysis book would show more examples of solving problems using decision analysis. Sections and subsections in Part I that are marked with a star ($\bigstar$) contain material that either requires a background in continuous mathematics or that seems to be inherently more difficult than the material in the rest of the book. For the most part, these sections can be skipped without impacting one's mastery of the rest of the book. The only exception is that if Section 3.6 (which covers d-separation) is omitted, it will be necessary to briefly review the faithfulness condition in order to understand Sections 4.4.1 and 4.5.1, which concern the constraint-based method for learning faithful DAGs from data. I believe one can gain an intuition for this type of learning from a few simple examples, and one does not need a formal knowledge of d-separation to understand these examples. I've presented constraint-based learning in this fashion at several talks and workshops worldwide and found that the audience could always understand the material. Furthermore, this is how I present the material to my students.

Part II presents financial applications. Specifically, Chapter 7 presents the basics of investment science and develops a Bayesian network for portfolio risk analysis. Sections 7.2 and 7.3 are marked with a star ($\bigstar$) because the material in these sections seems inherently more difficult than most of the other material in the book. However, they do not require as background the material from Part I that is marked with a star ($\bigstar$). Chapter 8 discusses modeling real options, which concerns decisions a company must make as to what projects it should pursue. Chapter 9 covers venture capital decision making, which is the process of deciding whether to invest money in a start-up company. Chapter 10 discusses a model for bankruptcy prediction.

Part III contains chapters on two important areas of marketing. First, Chapter 11 shows methods for doing collaborative filtering and market basket analysis. These disciplines concern determining what products an individual might prefer based on how the individual feels about other products. Finally, Chapter 12 presents a technique for doing targeted advertising, which is the process of identifying those customers to whom advertisements should be sent.

There is too much material for me to cover the entire book in a one semester course at NEIU. Since the course requires discrete mathematics and business statistics as prerequisites, I only review most of the material in Chapter 2. However, I do discuss conditional independence in depth because ordinarily the students have not been exposed to this concept. I then cover the following sections from the remainder of the book:

Chapter 3: 3.1-3.5.1

Chapter 4: 4.1, 4.2, 4.4.1, 4.5.1, 4.6

Chapter 5: 5.1-5.3.2, 5.3.4

Chapters 6 - 12: All sections

The course is titled "Informatics for MIS Students," and is a required course in the MIS (Management Information Science) concentration of NEIU's Computer Science M.S. Degree Program. This book should be appropriate for any similar course in an MIS, computer science, business, or MBA program. It is intended for upper level undergraduate and graduate students. Besides having taken one or two courses covering basic probability and statistics, it would be useful but not necessary for the student to have studied data structures.

Part I of the book could also be used for the first part of any course involving probabilistic reasoning using Bayesian networks. That is, although many of the examples in Part I concern the stock market and applications to business problems, I've presented the material in a general way. Therefore, an instructor could use Part I to cover basic concepts and then provide papers relative to a particular domain of interest. For example, if the course is "Probabilistic Methods for Medical Informatics," the instructor could cover Part I of this book, and then provide papers concerning applications in the medical domain.

For the most part, the applications discussed in Part II were the results of research done at the School of Business of the University of Kansas, while the applications in Part III were the results of research done by the Machine Learning and Applied Statistics Group of Microsoft Research. The reason is not that I have any particular affiliations with either of this institutions. Rather, I did an extensive search for financial and marketing applications, and the ones I found that seemed to be most carefully designed and evaluated came from these institutions.

I thank Catherine Shenoy for reviewing the chapter on investment science and Dawn Homes, Francisco Javier Díez, and Padmini Jyotishmati for reviewing the entire book. They all offered many useful comments and criticisms. I thank Prakash Shenoy and Edwin Burmeister for correspondence concerning some of the content of the book. I thank my co-author, Xia Jiang, for giving me the idea to write this book in the first place, and for her efforts on the book itself. Finally, I thank Prentice Hall for granting me permission to reprint material from my 2004 book *Learning Bayesian Networks*.

Rich Neapolitan
RE-Neapolitan@neiu.edu

Contents

About the Authors

Richard E. Neapolitan is Professor and Chair of Computer Science at Northeastern Illinois University. He has previously written three books including the seminal 1990 Bayesian network text *Probabilistic Reasoning in Expert Systems*. More recently, he wrote the 2004 text *Learning Bayesian networks*, and *Foundations of Algorithms*, which has been translated to three languages and is one of the most widely-used algorithms texts world-wide. His books have the reputation of making difficult concepts easy to understand because of the logical flow of the material, the simplicity of the explanations, and the clear examples.

Xia Jiang received an M.S. in Mechanical Engineering from Rose Hulman University and is currently a Ph.D. candidate in the Biomedical Informatics Program at the University of Pittsburgh. She has published theoretical papers concerning Bayesian networks, along with applications of Bayesian networks to biosurveillance.

Part I

Bayesian Networks and Decision Analysis

Chapter 1

Probabilistic Informatics

Informatics programs in the United States go back at least to the 1980s when Stanford University offered a Ph.D. in medical informatics. Since that time, a number of informatics programs in other disciplines have emerged at universities throughout the United States. These programs go by various names, including bioinformatics, medical informatics, chemical informatics, music informatics, marketing informatics, etc. What do these programs have in common? To answer that question we must articulate what we mean by the term "informatics." Since other disciplines are usually referenced when we discuss informatics, some define informatics as the application of information technology in the context of another field. However, such a definition does not really tell us the focus of informatics itself. First, we explain what we mean by the term informatics. Then we discuss why we have chosen to concentrate on the probabilistic approach in this book. Finally, we provide an outline of the material that will be covered in the rest of the book.

1.1   What Is Informatics?

In much of western Europe, informatics has come to mean the rough translation of the English "computer science," which is the discipline that studies computable processes. Certainly, there is overlap between computer science programs and informatics programs, but they are not the same. Informatics programs ordinarily investigate subjects such as biology and medicine, whereas computer science programs do not. So the European definition does not suffice for the way the word is currently used in the United States.

To gain insight into the meaning of informatics, let us consider the suffix "-ics," which means the science, art, or study of some entity. For example, "linguistics" is the study of the nature of language, "economics" is the study of the production and distribution of goods, and "photonics" is the study of electromagnetic energy whose basic unit is the photon. Given this, informatics should be the study of information. Indeed, WordNet 2.1 defines informatics as "the science concerned with gathering, manipulating, storing, retrieving and classifying recorded information." To proceed from this definition we need to define the word "information." Most dictionary definitions do not help as far as giving us anything concrete. That is, they define information either as knowledge or as a collection of data, which means we are left with the situation of determining the meaning of knowledge and data. To arrive at a concrete definition of informatics, let's define data, information, and knowledge first.

By **datum** we mean a character string that can be recognized as a unit. For example, the nucleotide G in the nucleotide sequence GATC is a datum, the field "cancer" in a record in a medical data base is a datum, and the field "Gone with the Wind" in a movie data base is a datum. Note that a single character, a word, or a group of words can be a datum depending on the particular application. **Data** then are more than one datum. By **information** we mean the meaning given to data. For example, in a medical data base the data "Joe Smith" and "cancer" in the same record mean that Joe Smith has cancer. By **knowledge** we mean dicta which enable us to infer new information from existing information. For example, suppose we have the following item of knowledge (dictum):[1]

IF the stem of the plant is woody

AND the position is upright

AND there is one main trunk

THEN the plant is a tree.

Suppose further that I am looking at a plant in my backyard and I observe that its stem is woody, its position is upright, and it has one main trunk. Then using the above knowledge item, we can deduce the new information that the plant in my backyard is a tree.

Finally, we define **informatics** as the discipline that applies the methodologies of science and engineering to information. It concerns organizing data

---

[1]Such an item of knowledge would be part of a rule-based expert system.

into information, learning knowledge from information, learning new information from existing information and knowledge, and making decisions based on the knowledge and information learned. We use engineering to develop the algorithms that learn knowledge from information and that learn information from information and knowledge. We use science to test the accuracy of these algorithms.

Next, we show several examples that illustrate how informatics pertains to other disciplines.

Example 1.1 (*medical informatics*) *Suppose we have a large data file of patients records as follows:*

Patient	Smoking History	Bronchitis	Lung Cancer	Fatigue	Positive Chest X-Ray
1	yes	yes	yes	no	yes
2	no	no	no	no	no
3	no	no	yes	yes	no
⋮	⋮	⋮	⋮	⋮	⋮
10,000	yes	no	no	no	no

From the **information** *in this data file we can use the methodologies of informatics to obtain* **knowledge** *such as "25% of people with smoking history have bronchitis" and "60% of people with lung cancer have positive chest X-rays." Then from this knowledge and the information that "Joe Smith has a smoking history and a positive chest X-ray" we can use the methodologies of informatics to obtain the new* **information** *that "there is a 5% chance Joe Smith also has lung cancer."*

Example 1.2 (*bioinformatics*) *Suppose we have long homologous DNA sequences from the human, the chimpanzee, the gorilla, the orangutan, and the rhesus monkey. From this* **information** *we can use the methodologies of informatics to obtain the new* **information** *that it is most probable that the human and the chimpanzee are the most closely related of the five species.*

Example 1.3 (*marketing informatics*) *Suppose we have a large data file of movie ratings as follows:*

Person	Aviator	Shall We Dance	Dirty Dancing	Vanity Fair
1	1	5	5	4
2	5	1	1	2
3	4	1	2	1
4	2	5	4	5
⋮	⋮	⋮	⋮	⋮
10,000	1	4	5	5

This means, for example, that Person 1 rated Aviator the lowest (1) and Shall We Dance the highest (5). From the information in this data file, we can develop a knowledge system that will enable us to estimate how an individual will rate a particular movie. For example, suppose Kathy Black rates Aviator as 1, Shall We Dance as 5, and Dirty Dancing as 5. The system could estimate how Kathy will rate Vanity Fair. Just by eyeballing the data in the five records shown, we see that Kathy's ratings on the first three movies are similar to those of Persons 1, 4, and 5. Since they all rated Vanity Fair high, based on these five records, we would suspect Kathy would rate it high. An informatics algorithm can formalize a way to make these predictions. This task of predicting the utility of an item to a particular user based on the utilities assigned by other users is called **collaborative filtering***.*

In this book we concentrate on two related areas of informatics, namely financial informatics and marketing informatics. **Financial informatics** involves applying the methods of informatics to the management of money and other assets. In particular, it concerns determining the risk involved in some financial venture. As an example, we might develop a tool to improve portfolio risk analysis. **Marketing informatics** involves applying the methods of informatics to promoting and selling products or services. For example, we might determine which advertisements should be presented to a given Web user based on that user's navigation pattern.

Before ending this section, let's discuss the relationship between informatics and the relatively new expression "'data mining." The term data mining can be traced back to the *First International Conference on Knowledge Discovery and Data Mining* (KDD-95) in 1995. Briefly, **data mining** is the process of extrapolating unknown knowledge from a large amount of observational data. Recall that we said informatics concerns (1) organizing data into information, (2) learning knowledge from information, (3) learning new information from existing information and knowledge, and (4) making decisions based on the knowledge and information learned. So, technically speaking, data mining is a subfield of informatics that includes only the first two of these procedures. However, both terms are still evolving, and some individuals use data mining to refer to all four procedures.

1.2   Probabilistic Informatics

As can be seen in Examples 1.1, 1.2, and 1.3, the knowledge we use to process information often does not consist of IF-THEN rules, such as the one concerning plants discussed earlier. Rather, we only know relationships such as "smoking makes lung cancer more likely." Similarly, our conclusions are uncertain. For example, we feel it is most likely that the closest living relative of the human is the chimpanzee, but we are not certain of this. So ordinarily we must reason under uncertainty when handling information and knowledge. In the 1960s and 1970s a number of new formalisms for handling uncertainty were developed, including certainty factors, the Dempster-Shafer Theory of Evidence, fuzzy logic, and fuzzy set theory. Probability theory has a long history of representing uncertainty in a formal axiomatic way. Neapolitan [1990] contrasts the various approaches and argues for the use of probability theory.[2] We will not present that argument here. Rather, we accept probability theory as being the way to handle uncertainty and explain why we choose to describe informatics algorithms that use the model-based probabilistic approach.

A **heuristic algorithm** uses a commonsense rule to solve a problem. Ordinarily, heuristic algorithms have no theoretical basis and therefore do not enable us to prove results based on assumptions concerning a system. An example of a heuristic algorithm is the one developed for collaborative filtering in Chapter 11, Section 11.1.

An **abstract model** is a theoretical construct that represents a physical process with a set of variables and a set of quantitative relationships (axioms) among them. We use models so we can reason within an idealized framework and thereby make predictions/determinations about a system. We can mathematically prove these predictions/determinations are "correct," but they are correct only to the extent that the model accurately represents the system. A **model-based algorithm** therefore makes predictions/determinations within the framework of some model. Algorithms that make predictions/determinations within the framework of probability theory are model-based algorithms. We can prove results concerning these algorithms based on the axioms of probability theory, which are discussed in Chapter 2. We concentrate on such algorithms in this book. In particular, we present algorithms that use Bayesian networks to reason within the framework of probability theory.

1.3 Outline of This Book

In Part I we cover the basics of Bayesian networks and decision analysis. Chapter 2 reviews the probability and statistics necessary to understanding the remainder of the book. In Chapter 3 we present Bayesian networks, which are graphical structures that represent the probabilistic relationships among many related variables. Bayesian networks have become one of the most prominent architectures for representing multivariate probability distributions and enabling probabilistic inference using such distributions. Chapter 4 shows how we can learn Bayesian networks from data. A Bayesian network augmented with a value node and decision nodes is called an influence diagram. We can use an influence diagram to recommend a decision based on the uncertain relationships among the variables and the preferences of the user. The field that investigates such decisions is called decision analysis. Chapter 5 introduces decision analysis, while Chapter 6 covers further topics in decision analysis. Once you have completed Part I, you should have a basic understanding of how Bayesian networks and decision analysis can be used to represent and solve real-world problems. Parts II and III then cover applications to specific problems. Part II covers financial applications. Specifically, Chapter 7 presents the basics of investment science and develops a Bayesian network for portfolio risk analysis. In Chapter

---

[2]Fuzzy set theory and fuzzy logic model a different class of problems than probability theory and therefore complement probability theory rather than compete with it. See [Zadeh, 1995] or [Neapolitan, 1992].

8 we discuss the modeling of real options, which concerns decisions a company must make as to what projects it should pursue. Chapter 9 covers venture capital decision making, which is the process of deciding whether to invest money in a start-up company. In Chapter 10 we show an application to bankruptcy prediction. Finally, Part III contains chapters on two of the most important areas of marketing. First, Chapter 11 shows an application to collaborative filtering/market basket analysis. These disciplines concern determining what products an individual might prefer based on how the user feels about other products. Second, Chapter 12 presents an application to targeted advertising, which is the process of identifying those customers to whom advertisements should be sent.

Chapter 2

Probability and Statistics

This chapter reviews the probability and statistics you need to read the remainder of this book. In Section 2.1 we present the basics of probability theory, while in Section 2.2 we review random variables. Section 2.3 briefly discusses the meaning of probability. In Section 2.4 we show how random variables are used in practice. Finally, Section 2.5 reviews concepts in statistics, such as expected value, variance, and covariance.

2.1 Probability Basics

After defining probability spaces, we discuss conditional probability, independence and conditional independence, and Bayes' Theorem.

2.1.1 Probability Spaces

You may recall using probability in situations such as drawing the top card from a deck of playing cards, tossing a coin, or drawing a ball from an urn. We call the process of drawing the top card or tossing a coin an **experiment**. Probability theory has to do with experiments that have a set of distinct **outcomes**. The set of all outcomes is called the **sample space** or **population**. Mathematicians ordinarily say "sample space," while social scientists ordinarily say "population." We will say sample space. In this simple review we assume the sample space is finite. Any subset of a sample space is called an **event**. A subset containing exactly one element is called an **elementary event**.

Example 2.1 *Suppose we have the experiment of drawing the top card from an ordinary deck of cards. Then the set*

$$E = \{jack \ of \ hearts, \ jack \ of \ clubs, \ jack \ of \ spades, \ jack \ of \ diamonds\}$$

is an event, and the set

$$F = \{jack \ of \ hearts\}$$

is an elementary event.

The meaning of an event is that one of the elements of the subset is the outcome of the experiment. In the previous example, the meaning of the event E is that the card drawn is one of the four jacks, and the meaning of the elementary event F is that the card is the jack of hearts.

We articulate our certainty that an event contains the outcome of the experiment with a real number between 0 and 1. This number is called the probability of the event. In the case of a finite sample space, a probability of 0 means we are certain the event does not contain the outcome, while a probability of 1 means we are certain it does. Values in between represent varying degrees of belief. The following definition formally defines probability when the sample space is finite.

Definition 2.1 *Suppose we have a sample space Ω containing n distinct elements: that is,*

$$\Omega = \{e_1, e_2, \ldots, e_n\}.$$

*A function that assigns a real number $P(\mathsf{E})$ to each event $\mathsf{E} \subseteq \Omega$ is called a **probability function** on the set of subsets of Ω if it satisfies the following conditions:*

1. *$0 \leq P(e_i) \leq 1$ for $1 \leq i \leq n$.*

2. *$P(e_1) + P(e_2) + \ldots + P(e_n) = 1$.*

3. *For each event that is not an elementary event, $P(E)$ is the sum of the probabilities of the elementary events whose outcomes are in E. For example, if*

$$E = \{e_3, e_6, e_8\}$$

then

$$P(E) = P(e_3) + P(e_6) + P(e_8).$$

*The pair (Ω, P) is called a **probability space**.*

Because probability is defined as a function whose domain is a set of sets, we should write $P(\{e_i\})$ instead of $P(e_i)$ when denoting the probability of an elementary event. However, for the sake of simplicity, we do not do this. In the same way, we write $P(e_3, e_6, e_8)$ instead of $P(\{e_3, e_6, e_8\})$.

The most straightforward way to assign probabilities is to use the **Principle of Indifference**, which says that outcomes are to be considered equiprobable if we have no reason to expect one over the other. According to this principle, when there are n elementary events, each has probability equal to $1/n$.

Example 2.2 *Let the experiment be tossing a coin. Then the sample space is*

$$\Omega = \{heads, tails\},$$

and, according to the Principle of Indifference, we assign

$$P(heads) = P(tails) = .5.$$

We stress that there is nothing in the definition of a probability space that says we must assign the value of .5 to the probabilities of heads and tails. We could assign $P(heads) = .7$ and $P(tails) = .3$. However, if we have no reason to expect one outcome over the other, we give them the same probability.

Example 2.3 *Let the experiment be drawing the top card from a deck of 52 cards. Then Ω contains the faces of the 52 cards, and, according to the Principle of Indifference, we assign $P(e) = 1/52$ for each $e \in \Omega$. For example,*

$$P(\text{jack of hearts}) = \frac{1}{52}.$$

The event

$$E = \{\text{jack of hearts, jack of clubs, jack of spades, jack of diamonds}\}$$

means that the card drawn is a jack. Its probability is

$$
\begin{aligned}
P(E) &= P(\text{jack of hearts}) + P(\text{jack of clubs}) + \\
&\quad P(\text{jack of spades}) + P(\text{jack of diamonds}) \\
&= \frac{1}{52} + \frac{1}{52} + \frac{1}{52} + \frac{1}{52} = \frac{1}{13}.
\end{aligned}
$$

We have Theorem 2.1 concerning probability spaces. Its proof is left as an exercise.

Theorem 2.1 *Let* (Ω, P) *be a probability space. Then*

   1. $P(\Omega) = 1$.

   2. $0 \le P(\mathsf{E}) \le 1$ *for every* $\mathsf{E} \subseteq \Omega$.

   3. For every two subsets E *and* F *of* Ω *such that* $\mathsf{E} \cap \mathsf{F} = \emptyset$,

$$P(\mathsf{E} \cup \mathsf{F}) = P(\mathsf{E}) + P(\mathsf{F}),$$

where $\emptyset$ *denotes the empty set.*

Example 2.4 *Suppose we draw the top card from a deck of cards. Denote by* Queen *the set containing the 4 queens and by* King *the set containing the 4 kings. Then*

$$P(\mathsf{Queen} \cup \mathsf{King}) = P(\mathsf{Queen}) + P(\mathsf{King}) = \frac{1}{13} + \frac{1}{13} = \frac{2}{13}$$

because Queen $\cap$ King $= \emptyset$. *Next denote by* Spade *the set containing the 13 spades. The sets* Queen *and* Spade *are not disjoint; so their probabilities are not additive. However, it is not hard to prove that, in general,*

$$P(\mathsf{E} \cup \mathsf{F}) = P(\mathsf{E}) + P(\mathsf{F}) - P(\mathsf{E} \cap \mathsf{F}).$$

So

$$
\begin{aligned}
P(\mathsf{Queen} \cup \mathsf{Spade}) &= P(\mathsf{Queen}) + P(\mathsf{Spade}) - P(\mathsf{Queen} \cap \mathsf{Spade}) \\
&= \frac{1}{13} + \frac{1}{4} - \frac{1}{52} = \frac{4}{13}.
\end{aligned}
$$

2.1.2 Conditional Probability and Independence

We start with a definition.

Definition 2.2 *Let* E *and* F *be events such that* $P(\mathsf{F}) \ne 0$. *Then the* **conditional probability** *of* E *given* F, *denoted* $P(\mathsf{E}|\mathsf{F})$, *is given by*

$$P(\mathsf{E}|\mathsf{F}) = \frac{P(\mathsf{E} \cap \mathsf{F})}{P(\mathsf{F})}.$$

We can gain intuition for this definition by considering probabilities that are assigned using the Principle of Indifference. In this case, $P(\mathsf{E}|\mathsf{F})$, as defined above, is the ratio of the number of items in $\mathsf{E} \cap \mathsf{F}$ to the number of items in F. We show this as follows. Let n be the number of items in the sample space, n_F be the number of items in F, and n_EF be the number of items in $\mathsf{E} \cap \mathsf{F}$. Then

$$\frac{P(\mathsf{E} \cap \mathsf{F})}{P(\mathsf{F})} = \frac{n_\mathsf{EF}/n}{n_\mathsf{F}/n} = \frac{n_\mathsf{EF}}{n_\mathsf{F}},$$

which is the ratio of the number of items in $\mathsf{E} \cap \mathsf{F}$ to the number of items in F. As far as the meaning is concerned, $P(\mathsf{E}|\mathsf{F})$ is our belief that E contains the outcome given that we know F contains the outcome.

Example 2.5 *Again, consider drawing the top card from a deck of cards. Let* Jack *be the set of the 4 jacks,* RedRoyalCard *be the set of the 6 red royal cards,*[1] *and* Club *be the set of the 13 clubs. Then*

$$P(\mathsf{Jack}) = \frac{4}{52} = \frac{1}{13}$$

$$P(\mathsf{Jack}|\mathsf{RedRoyalCard}) = \frac{P(\mathsf{Jack} \cap \mathsf{RedRoyalCard})}{P(\mathsf{RedRoyalCard})} = \frac{2/52}{6/52} = \frac{1}{3}$$

$$P(\mathsf{Jack}|\mathsf{Club}) = \frac{P(\mathsf{Jack} \cap \mathsf{Club})}{P(\mathsf{Club})} = \frac{1/52}{13/52} = \frac{1}{13}.$$

Notice in the previous example that $P(\mathsf{Jack}|\mathsf{Club}) = P(\mathsf{Jack})$. This means that finding out the card is a club does not change the likelihood that it is a jack. We say that the two events are independent in this case, which is formalized in the following definition:

Definition 2.3 *Two events* E *and* F *are **independent** if one of the following holds:*

 1. $P(\mathsf{E}|\mathsf{F}) = P(\mathsf{E})$ and $P(\mathsf{E}) \neq 0, P(\mathsf{F}) \neq 0.$

 2. $P(\mathsf{E}) = 0$ *or* $P(\mathsf{F}) = 0.$

Notice that the definition states that the two events are independent even though it is in terms of the conditional probability of E given F. The reason is that independence is symmetric. That is, if $P(\mathsf{E}) \neq 0$ and $P(\mathsf{F}) \neq 0$, then $P(\mathsf{E}|\mathsf{F}) = P(\mathsf{E})$ if and only if $P(\mathsf{F}|\mathsf{E}) = P(\mathsf{F})$. It is straightforward to prove that E and F are independent if and only if $P(\mathsf{E} \cap \mathsf{F}) = P(\mathsf{E})P(\mathsf{F})$.

If you've previously studied probability, you should have already been introduced to the concept of independence. However, a generalization of independence, called conditional independence, is not covered in many introductory texts. This concept is important to the applications discussed in this book. We discuss it next.

Definition 2.4 *Two events* E *and* F *are **conditionally independent** given* G *if* $P(\mathsf{G}) \neq 0$ *and one of the following holds:*

 1. $P(\mathsf{E}|\mathsf{F} \cap \mathsf{G}) = P(\mathsf{E}|\mathsf{G})$ and $P(\mathsf{E}|\mathsf{G}) \neq 0, P(\mathsf{F}|\mathsf{G}) \neq 0.$

 2. $P(\mathsf{E}|\mathsf{G}) = 0$ *or* $P(\mathsf{F}|\mathsf{G}) = 0.$

Notice that this definition is identical to the definition of independence except that everything is conditional on G. The definition entails that E and F are independent once we know that the outcome is in G. The next example illustrates this.

---

[1] A royal card is a jack, queen, or king.

Figure 2.1: Using the Principle of Indifference, we assign a probability of 1/13 to each object.

Example 2.6 *Let* Ω *be the set of all objects in Figure 2.1. Using the Principle of Indifference, we assign a probability of 1/13 to each object. Let* Black *be the set of all black objects,* White *be the set of all white objects,* Square *be the set of all square objects, and* A *be the set of all objects containing an "A." We then have*

$$P(\mathsf{A}) = \frac{5}{13}$$
$$P(\mathsf{A}|\mathsf{Square}) = \frac{3}{8}.$$

So A *and* Square *are not independent. However,*

$$P(\mathsf{A}|\mathsf{Black}) = \frac{3}{9} = \frac{1}{3}$$
$$P(\mathsf{A}|\mathsf{Square} \cap \mathsf{Black}) = \frac{2}{6} = \frac{1}{3}.$$

We see that A *and* Square *are conditionally independent given* Black. *Furthermore,*

$$P(\mathsf{A}|\mathsf{White}) = \frac{2}{4} = \frac{1}{2}$$
$$P(\mathsf{A}|\mathsf{Square} \cap \mathsf{White}) = \frac{1}{2}.$$

So A *and* Square *are also conditionally independent given* White.

Next, we discuss an important rule involving conditional probabilities. Suppose we have n events $\mathsf{E}_1, \mathsf{E}_2, \ldots, \mathsf{E}_n$ such that

$$\mathsf{E}_i \cap \mathsf{E}_j = \varnothing \qquad \text{for } i \neq j$$

and

$$\mathsf{E}_1 \cup \mathsf{E}_2 \cup \ldots \cup \mathsf{E}_n = \Omega.$$

Such events are called **mutually exclusive and exhaustive**. Then the **law of total probability** says for any other event F,

$$P(\mathsf{F}) = P(\mathsf{F} \cap \mathsf{E}_1) + P(\mathsf{F} \cap \mathsf{E}_2) + \cdots + P(\mathsf{F} \cap \mathsf{E}_n). \tag{2.1}$$

You are asked to prove this rule in the exercises. If $P(\mathsf{E}_i) \neq 0$, then $P(\mathsf{F} \cap \mathsf{E}_i) = P(\mathsf{F}|\mathsf{E}_i)P(\mathsf{E}_i)$. Therefore, if $P(\mathsf{E}_i) \neq 0$ for all i, the law is often applied in the following form:

$$P(\mathsf{F}) = P(\mathsf{F}|\mathsf{E}_1)P(\mathsf{E}_1) + P(\mathsf{F}|\mathsf{E}_2)P(\mathsf{E}_2) + \cdots + P(\mathsf{F}|\mathsf{E}_n)P(\mathsf{E}_n). \qquad (2.2)$$

Example 2.7 *Suppose we have the objects discussed in Example 2.6. Then according to the law of total probability*

$$
\begin{aligned}
P(\mathsf{A}) &= P(\mathsf{A}|\mathsf{Black})P(\mathsf{Black}) + P(\mathsf{A}|\mathsf{White})P(\mathsf{White}) \\
&= \left(\frac{1}{3}\right)\left(\frac{9}{13}\right) + \left(\frac{1}{2}\right)\left(\frac{4}{13}\right) = \frac{5}{13}.
\end{aligned}
$$

2.1.3 Bayes' Theorem

We can compute conditional probabilities of events of interest from known probabilities using the following theorem:

Theorem 2.2 (Bayes) *Given two events* E *and* F *such that* $P(\mathsf{E}) \neq 0$ *and* $P(\mathsf{F}) \neq 0$, *we have*

$$P(\mathsf{E}|\mathsf{F}) = \frac{P(\mathsf{F}|\mathsf{E})P(\mathsf{E})}{P(\mathsf{F})}. \qquad (2.3)$$

Furthermore, given n *mutually exclusive and exhaustive events* $\mathsf{E}_1, \mathsf{E}_2, \ldots, \mathsf{E}_n$ *such that* $P(\mathsf{E}_i) \neq 0$ *for all* i, *we have for* $1 \leq i \leq n$,

$$P(\mathsf{E}_i|\mathsf{F}) = \frac{P(\mathsf{F}|\mathsf{E}_i)P(\mathsf{E}_i)}{P(\mathsf{F}|\mathsf{E}_1)P(\mathsf{E}_1) + P(\mathsf{F}|\mathsf{E}_2)P(\mathsf{E}_2) + \cdots P(\mathsf{F}|\mathsf{E}_n)P(\mathsf{E}_n)}. \qquad (2.4)$$

Proof. *To obtain Equality 2.3, we first use the definition of conditional probability as follows:*

$$P(\mathsf{E}|\mathsf{F}) = \frac{P(\mathsf{E} \cap \mathsf{F})}{P(\mathsf{F})} \qquad \text{and} \qquad P(\mathsf{F}|\mathsf{E}) = \frac{P(\mathsf{F} \cap \mathsf{E})}{P(\mathsf{E})}.$$

Next we multiply each of these equalities by the denominator on its right side to show that

$$P(\mathsf{E}|\mathsf{F})P(\mathsf{F}) = P(\mathsf{F}|\mathsf{E})P(\mathsf{E})$$

because they both equal $P(\mathsf{E} \cap \mathsf{F})$. *Finally, we divide this last equality by* $P(\mathsf{F})$ *to obtain our result.*

To obtain Equality 2.4, we place the expression for F, *obtained using the rule of total probability (Equality 2.2), in the denominator of Equality 2.3.* ∎

Both of the formulas in the preceding theorem are called **Bayes' Theorem** because the original version was developed by Thomas Bayes (published in 1763). The first enables us to compute $P(\mathsf{E}|\mathsf{F})$ if we know $P(\mathsf{F}|\mathsf{E})$, $P(\mathsf{E})$, and $P(\mathsf{F})$, while the second enables us to compute $P(\mathsf{E}_i|\mathsf{F})$ if we know $P(\mathsf{F}|\mathsf{E}_j)$ and $P(\mathsf{E}_j)$ for $1 \leq j \leq n$. The next example illustrates the use of Bayes' Theorem.

Example 2.8 *Let* Ω *be the set of all objects in Figure 2.1, and assign each object a probability of* $1/13$. *Let* A *be the set of all objects containing an "A," * B *be the set of all objects containing a "B," and* Black *be the set of all black objects. Then according to Bayes' Theorem,*

$$
\begin{aligned}
P(\text{Black}|\text{A}) &= \frac{P(\text{A}|\text{Black})P(\text{Black})}{P(\text{A}|\text{Black})P(\text{Black}) + P(\text{A}|\text{White})P(\text{White})} \\
&= \frac{\left(\frac{1}{3}\right)\left(\frac{9}{13}\right)}{\left(\frac{1}{3}\right)\left(\frac{9}{13}\right) + \left(\frac{1}{2}\right)\left(\frac{4}{13}\right)} = \frac{3}{5},
\end{aligned}
$$

which is the same value we get by computing $P(\text{Black}|\text{A})$ *directly.*

In the previous example we can just as easily compute $P(\text{Black}|\text{A})$ directly. We will see a useful application of Bayes' Theorem in Section 2.4.

2.2   Random Variables

In this section we present the formal definition and mathematical properties of a random variable. In Section 2.4 we show how they are developed in practice.

2.2.1   Probability Distributions of Random Variables

Definition 2.5 *Given a probability space* (Ω, P), *a* **random variable** X *is a function whose domain is* Ω.

The range of X is called the **space** of X.

Example 2.9 *Let* Ω *contain all outcomes of a throw of a pair of six-sided dice, and let* P *assign* $1/36$ *to each outcome. Then* Ω *is the following set of ordered pairs:*

$$\Omega = \{(1,1), (1,2), (1,3), (1,4), (1,5), (1,6), (2,1), (2,2), \ldots (6,5), (6,6)\}.$$

Let the random variable X assign the sum of each ordered pair to that pair, and let the random variable Y assign "odd" to each pair of odd numbers and "even" to a pair if at least one number in that pair is an even number. The following table shows some of the values of X and Y:

e	$X(e)$	$Y(e)$
$(1,1)$	2	odd
$(1,2)$	3	even
$\ldots$	$\ldots$	$\ldots$
$(2,1)$	3	even
$\ldots$	$\ldots$	$\ldots$
$(6,6)$	12	even

The space of X is $\{2, 3, 4, 5, 6, 7, 8, 9, 10, 11, 12\}$, and that of Y is $\{\text{odd}, \text{even}\}$.

For a random variable X, we use $X = x$ to denote the subset containing all elements $e \in \Omega$ that X maps to the value of x. That is,

$$X = x \quad \text{represents the event} \quad \{e \text{ such that } X(e) = x\}.$$

Note the difference between X and x. Small x denotes any element in the space of X, while X is a function.

Example 2.10 *Let Ω, P, and X be as in Example 2.9. Then*

$$X = 3 \quad \text{represents the event} \quad \{(1,2),(2,1)\} \text{ and}$$

$$P(X = 3) = \frac{1}{18}.$$

Notice that

$$\sum_{x \in space(X)} P(X = x) = 1.$$

Example 2.11 *Let Ω, P, and Y be as in Example 2.9. Then*

$$\sum_{y \in space(Y)} P(Y = y) \quad = \quad P(Y = odd) + P(Y = even)$$

$$= \quad \frac{9}{36} + \frac{27}{36} = 1.$$

We call $P(X = x)$ the **probability distribution** of the random variable X. We often use x alone to represent the event $X = x$, and so we write $P(x)$ instead of $P(X = x)$.

Let Ω, P, and X be as in Example 2.9. Then if $x = 3$,

$$P(x) = P(X = x) = \frac{1}{18}.$$

Given two random variables X and Y, defined on the same sample space Ω, we use $X = x, Y = y$ to denote the subset containing all elements $e \in \Omega$ that are mapped both by X to x and by Y to y. That is,

$$X = x, Y = y \quad \text{represents the event}$$

$$\{e \text{ such that } X(e) = x\} \cap \{e \text{ such that } Y(e) = y\}.$$

Example 2.12 *Let Ω, P, X, and Y be as in Example 2.9. Then*

$$X = 4, Y = odd \quad \text{represents the event} \quad \{(1,3),(3,1)\},$$

and so

$$P(X = 4, Y = odd) = 1/18.$$

We call $P(X = x, Y = y)$ the **joint probability distribution** of X and Y. If $A = \{X, Y\}$, we also call this the **joint probability distribution** of A. Furthermore, we often just say "joint distribution" or "probability distribution."

For brevity, we often use x, y to represent the event $X = x, Y = y$, and so we write $P(x, y)$ instead of $P(X = x, Y = y)$. This concept extends to three or more random variables. For example, $P(X = x, Y = y, Z = z)$ is the joint probability distribution function of the random variables X, Y, and Z, and we often write $P(x, y, z)$.

Example 2.13 Let Ω, P, X, and Y be as in Example 2.9. Then if $x = 4$ and $y = odd$,

$$P(x, y) = P(X = x, Y = y) = 1/18.$$

If we want to refer to all values of, for example, the random variables X and Y, we sometimes write $P(X, Y)$ instead of $P(X = x, Y = y)$ or $P(x, y)$.

Example 2.14 Let Ω, P, X, and Y be as in Example 2.9. It is left as an exercise to show that for all values of x and y we have

$$P(X = x, Y = y) < 1/2.$$

For example, as shown in Example 2.12

$$P(X = 4, Y = odd) = 1/18 < 1/2.$$

We can restate this fact as follows: for all values of X and Y we have that

$$P(X, Y) < 1/2.$$

If, for example, we let $A = \{X, Y\}$ and $a = \{x, y\}$, we use

$$A = a \qquad \text{to represent} \qquad X = x, Y = y,$$

and we often write $P(a)$ instead of $P(A = a)$.

Example 2.15 Let Ω, P, X, and Y be as in Example 2.9. If $A = \{X, Y\}$, $a = \{x, y\}$, $x = 4$, and $y = odd$, then

$$P(A = a) = P(X = x, Y = y) = 1/18.$$

Recall the law of total probability (Equalities 2.1 and 2.2). For two random variables X and Y, these equalities are as follows:

$$P(X = x) = \sum_y P(X = x, Y = y). \tag{2.5}$$

$$P(X = x) = \sum_y P(X = x | Y = y) P(Y = y). \tag{2.6}$$

It is left as an exercise to show this.

Example 2.16 Let Ω, P, X, and Y be as in Example 2.9. Then owing to Equality 2.5,

$$
\begin{aligned}
P(X = 4) &= \sum_y P(X = 4, Y = y) \\
&= P(X = 4, Y = odd) + P(X = 4, Y = even) = \frac{1}{18} + \frac{1}{36} = \frac{1}{12}.
\end{aligned}
$$

Example 2.17 Again, let Ω, P, X, and Y be as in Example 2.9. Then due to Equality 2.6,

$$
\begin{aligned}
P(X = 4) &= \sum_y P(X = x|Y = y)P(Y = y) \\
&= P(X = 4|Y = odd)P(Y = odd) + \\
&\quad P(X = 4|Y = even)P(Y = even) \\
&= \frac{2}{9} \times \frac{9}{36} + \frac{1}{27} \times \frac{27}{36} = \frac{1}{12}.
\end{aligned}
$$

In Equality 2.5 the probability distribution $P(X = x)$ is called the **marginal probability distribution** of X relative to the joint distribution $P(X = x, Y = y)$ because it is obtained using a process similar to adding across a row or column in a table of numbers. This concept also extends in a straightforward way to three or more random variables. For example, if we have a joint distribution $P(X = x, Y = y, Z = z)$ of X, Y, and Z, the marginal distribution $P(X = x, Y = y)$ of X and Y is obtained by summing over all values of Z. If A = $\{X, Y\}$, we also call this the **marginal probability distribution** of A.

The next example reviews the concepts covered so far concerning random variables.

Example 2.18 Let Ω be a set of 12 individuals, and let P assign $1/12$ to each individual. Suppose the sexes, heights, and wages of the individuals are as follows:

Case	Sex	Height (inches)	Wage ($)
1	female	64	30,000
2	female	64	30,000
3	female	64	40,000
4	female	64	40,000
5	female	68	30,000
6	female	68	40,000
7	male	64	40,000
8	male	64	50,000
9	male	68	40,000
10	male	68	50,000
11	male	70	40,000
12	male	70	50,000

Let the random variables S, H, and W, respectively, assign the sex, height, and wage of an individual to that individual. Then the probability distributions of the three random variables are as follows (recall that, for example, $P(s)$ represents $P(S = s)$):

s	$P(s)$
female	1/2
male	1/2

h	$P(h)$
64	1/2
68	1/3
70	1/6

w	$P(w)$
30,000	1/4
40,000	1/2
50,000	1/4

The joint distribution of S and H is as follows:

s	h	$P(s, h)$
female	64	1/3
female	68	1/6
female	70	0
male	64	1/6
male	68	1/6
male	70	1/6

The following table also shows the joint distribution of S and H and illustrates that the individual distributions can be obtained by summing the joint distribution over all values of the other variable:

h s	64	68	70	Distribution of S
female	1/3	1/6	0	1/2
male	1/6	1/6	1/6	1/2
Distribution of H	1/2	1/3	1/6	

The table that follows shows the first few values in the joint distribution of S, H, and W. There are 18 values in all, many of which are 0.

s	h	w	$P(s, h, w)$
female	64	30,000	1/6
female	64	40,000	1/6
female	64	50,000	0
female	68	30,000	1/12
...	...	...	...

We close with the **chain rule** for random variables, which says that given n random variables $X_1, X_2, \ldots, X_n$, defined on the same sample space Ω,

$$P(x_1, x_2, \ldots, x_n) = P(x_n | x_{n-1}, x_{n-2}, \ldots, x_1) \cdots \times P(x_2 | x_1) \times P(x_1)$$

whenever $P(x_1, x_2, \ldots, x_n) \neq 0$. It is straightforward to prove this rule using the rule for conditional probability.

Example 2.19 *Suppose we have the random variables in Example 2.18. Then according to the chain rule for all values s, h, and w of S, H, and W,*

$$P(s, h, w) = P(w|h, s)P(h|s)P(s).$$

There are eight combinations of values of the three random variables. The table that follows shows that the equality holds for two of the combinations:

s	h	w	$P(s, h, w)$	$P(w|h, s)P(h|s)P(s)$
$female$	64	30,000	$\frac{1}{6}$	$\left(\frac{1}{2}\right)\left(\frac{2}{3}\right)\left(\frac{1}{2}\right) = \frac{1}{6}$
$female$	64	40,000	$\frac{1}{12}$	$\left(\frac{1}{2}\right)\left(\frac{1}{3}\right)\left(\frac{1}{2}\right) = \frac{1}{12}$

It is left as an exercise to show that the equality holds for the other six combinations.

2.2.2 Independence of Random Variables

The notion of independence extends naturally to random variables.

Definition 2.6 *Suppose we have a probability space (Ω, P) and two random variables X and Y defined on Ω. Then X and Y are* **independent** *if, for all values x of X and y of Y, the events $X = x$ and $Y = y$ are independent. When this is the case, we write*

$$I_P(X, Y),$$

where I_P stands for independent in P.

Example 2.20 *Let Ω be the set of all cards in an ordinary deck, and let P assign $1/52$ to each card. Define random variables as follows:*

Variable	Value	Outcomes Mapped to This Value
R	r_1	All royal cards
	r_2	All nonroyal cards
S	s_1	All spades
	s_2	All nonspades

Then the random variables R and S are independent. That is,

$$I_P(R, S).$$

To show this, we need show for all values of r and s that

$$P(r|s) = P(r).$$

The following table shows this is the case:

s	r	$P(r)$	$P(r\|s)$
s_1	r_1	$\frac{12}{52} = \frac{3}{13}$	$\frac{3}{13}$
s_1	r_2	$\frac{40}{52} = \frac{10}{13}$	$\frac{10}{13}$
s_2	r_1	$\frac{12}{52} = \frac{3}{13}$	$\frac{9}{39} = \frac{3}{13}$
s_2	r_2	$\frac{40}{52} = \frac{10}{13}$	$\frac{30}{39} = \frac{10}{13}$

The concept of conditional independence also extends naturally to random variables.

Definition 2.7 *Suppose we have a probability space (Ω, P), and three random variables X, Y, and Z defined on Ω. Then X and Y are **conditionally independent given** Z if for all values x of X, y of Y, and z of Z, whenever $P(z) \neq 0$, the events $X = x$ and $Y = y$ are conditionally independent given the even $Z = z$. When this is the case, we write*

$$I_P(X, Y | Z).$$

Example 2.21 *Let Ω be the set of all objects in Figure 2.1, and let P assign $1/13$ to each object. Define random variables S (for shape), L (for letter), and C (for color) as follows:*

Variable	Value	Outcomes Mapped to This Value
L	l_1	All objects containing an "A"
	l_2	All objects containing a "B"
S	s_1	All square objects
	s_2	All circular objects
C	c_1	All black objects
	c_2	All white objects

Then L and S are conditionally independent given C. That is,

$$I_P(L, S | C).$$

To show this, we need to show for all values of l, s, and c that

$$P(l|s, c) = P(l|c).$$

There are a total of eight combinations of the three variables. The table that follows shows that the equality holds for two of the combinations:

c	s	l	$P(l\|s, c)$	$P(l\|c)$
c_1	s_1	l_1	$\frac{2}{6} = \frac{1}{3}$	$\frac{3}{9} = \frac{1}{3}$
c_1	s_1	l_2	$\frac{4}{6} = \frac{2}{3}$	$\frac{6}{9} = \frac{2}{3}$

It is left as an exercise to show that it holds for the other combinations.

Independence and conditional independence can also be defined for sets of random variables.

Definition 2.8 *Suppose we have a probability space* (Ω, P) *and two sets* A *and* B *containing random variables defined on* Ω. *Let* a *and* b *be sets of values of the random variables in* A *and* B, *respectively. The sets* A *and* B *are said to be* **independent** *if, for all values of the variables in the sets* a *and* b, *the events* A = a *and* B = b *are independent. When this is the case, we write*

$$I_P(\mathsf{A}, \mathsf{B}),$$

where I_P *stands for independent in* P.

Example 2.22 *Let* Ω *be the set of all cards in an ordinary deck, and let* P *assign* $1/52$ *to each card. Define random variables as follows:*

Variable	Value	Outcomes Mapped to This Value
R	r_1	All royal cards
	r_2	All nonroyal cards
T	t_1	All tens and jacks
	t_2	All cards that are neither tens nor jacks
S	s_1	All spades
	s_2	All nonspades

Then the sets $\{R, T\}$ *and* $\{S\}$ *are independent. That is,*

$$I_P(\{R, T\}, \{S\}). \tag{2.7}$$

To show this, we need to show for all values of r, t, *and* s *that*

$$P(r, t | s) = P(r, t).$$

There are eight combinations of values of the three random variables. The table that follows shows that the equality holds for two of the combinations:

s	r	t	$P(r, t\|s)$	$P(r, t)$
s_1	r_1	t_1	$\frac{1}{13}$	$\frac{4}{52} = \frac{1}{13}$
s_1	r_1	t_2	$\frac{2}{13}$	$\frac{8}{52} = \frac{2}{13}$

It is left as an exercise to show that it holds for the other combinations.

When a set contains a single variable, we do not ordinarily show the braces. For example, we write Independency 2.7 as

$$I_P(\{R, T\}, S).$$

Definition 2.9 *Suppose we have a probability space* (Ω, P) *and three sets* A, B, *and* C *containing random variables defined on* Ω. *Let* a, b, *and* c *be sets of values of the random variables in* A, B, *and* C, *respectively. Then the sets* A *and* B *are said to be* **conditionally independent given the set** C *if, for all values of the variables in the sets* a, b, *and* c, *whenever* $P(\mathsf{c}) \neq 0$, *the events* A = a *and* B = b *are conditionally independent given the event* C = c. *When this is the case, we write*

$$I_P(\mathsf{A}, \mathsf{B} | \mathsf{C}).$$

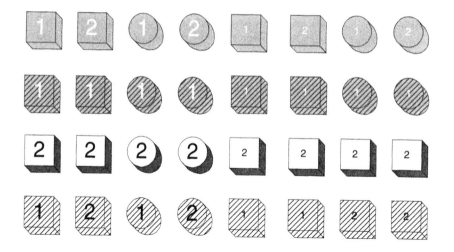

Figure 2.2: Objects with five properties.

Example 2.23 *Suppose we use the Principle of Indifference to assign proba-bilities to the objects in Figure 2.2, and define random variables as follows:*

Variable	Value	Outcomes Mapped to This Value
V	v_1	All objects containing a "1"
	v_2	All objects containing a "2"
L	l_1	All objects covered with lines
	l_2	All objects not covered with lines
C	c_1	All grey objects
	c_2	All white objects
S	s_1	All square objects
	s_2	All circular objects
F	f_1	All objects containing a number in a large font
	f_2	All objects containing a number in a small font

It is left as an exercise to show for all values of v, l, c, s, and f that

$$P(v, l|s, f, c) = P(v, l|c).$$

So we have

$$I_P(\{V, L\}, \{S, F\}|C).$$

2.3   The Meaning of Probability

When one does not have the opportunity to study probability theory in depth, one is often left with the impression that all probabilities are computed using ratios. Next, we discuss the meaning of probability in more depth and show that this is not how probabilities are ordinarily determined.

2.3.1 Relative Frequency Approach to Probability

A classic textbook example of probability concerns tossing a coin. Because the coin is symmetrical, we use the Principle of Indifference to assign

$$P(\text{Heads}) = P(\text{Tails}) = .5.$$

Suppose instead we toss a thumbtack. It can also land one of two ways. That is, it could land on its flat end, which we will call "heads," or it could land with the edge of the flat end and the point touching the ground, which we will call "tails." Because the thumbtack is not symmetrical, we have no reason to apply the Principle of Indifference and assign probabilities of .5 to both outcomes. How then should we assign the probabilities? In the case of the coin, when we assign $P(heads) = .5$, we are implicitly assuming that if we tossed the coin a large number of times it would land heads about half the time. That is, if we tossed the coin 1000 times, we would expect it to land heads about 500 times. This notion of repeatedly performing the experiment gives us a method for computing (or at least estimating) the probability. That is, if we repeat an experiment many times, we are fairly certain that the probability of an outcome is about equal to the fraction of times the outcome occurs. For example, a student tossed a thumbtack 10,000 times and it landed heads 3761 times. So

$$P(\text{Heads}) \approx \frac{3761}{10,000} = .3761.$$

Indeed, in 1919 Richard von Mises used the limit of this fraction as the definition of probability. That is, if n is the number of tosses and S_n is the number of times the thumbtack lands heads, then

$$P(\text{Heads}) \equiv \lim_{n \to \infty} \frac{S_n}{n}.$$

This definition assumes that a limit actually is approached. That is, it assumes the ratio does not fluctuate. For example, there is no reason *a priori* to assume that the ratio is not .5 after 100 tosses, .1 after 1000 tosses, .5 after 10,000 tosses, .1 after 100,000 tosses, and so on. Only experiments in the real world can substantiate that a limit is approached. In 1946 J. E. Kerrich conducted many such experiments using games of chance in which the Principle of Indifference seemed to apply (e.g., drawing a card from a deck). His results indicated that the relative frequency does appear to approach a limit and that this limit is the value obtained using the Principle of Indifference.

This approach to probability is called the **relative frequency approach to probability**, and probabilities obtained using this approach are called **relative frequencies**. A **frequentist** is someone who feels this is the only way we can obtain probabilities. Note that, according to this approach, we can never know a probability for certain. For example, if we tossed a coin 10,000 times and it landed heads 4991 times, we would estimate

$$P(\text{Heads}) \approx \frac{4991}{10,000} = .4991.$$

On the other hand, if we used the Principle of Indifference, we would assign $P(\text{Heads}) = .5$. In the case of the coin, the probability may not actually be .5 because the coin may not be perfectly symmetrical. For example, Kerrich [1946] found that the six came up the most in the toss of a die and that one came up the least. This makes sense because, at that time, the spots on the die were hollowed out of the die. So the die was lightest on the side with a six. On the other hand, in experiments involving cards or urns, it seems we can be certain of probabilities obtained using the Principle of Indifference.

Example 2.24 *Suppose we toss an asymmetrical six-sided die, and in 1000 tosses we observe the six sides coming up the following number of times:*

Side	Number of Times
1	250
2	150
3	200
4	70
5	280
6	50

So we estimate $P(1) \approx .25$, $P(2) \approx .15$, $P(3) \approx .2$, $P(4) \approx .07$, $P(5) \approx .28$, *and* $P(6) \approx .05$.

Repeatedly performing an experiment (so as to estimate a relative frequency) is called **sampling**, and the set of outcomes is called a **random sample** (or simply sample), while the set from which we sample is called a **population**.

Example 2.25 *Suppose our population is all males in the United States between the ages of 31 and 85, and we are interested in the probability of such males having high blood pressure. Then if we sample 10,000 males, this set of males is our sample. Furthermore, if 3210 have high blood pressure, we estimate*

$$P(\text{High Blood Pressure}) \approx \frac{3210}{10,000} = .321.$$

Technically, we should not call the set of all current males in this age group the population. Rather, the theory says that there is a propensity for a male in this group to have high blood pressure, and that this propensity is the probability. This propensity might not be equal to the fraction of current males in the group that have high blood pressure. In theory, we would have to have an infinite number of males to determine the probability exactly. The current set of males in this age group is called a **finite population**. The fraction of them with high blood pressure is the probability of obtaining a male with high blood pressure when we sample him from the set of all males in the age group. This latter probability is simply the ratio of males with high blood pressure. When doing statistical inference, we sometimes want to estimate the ratio in a finite population from a sample of the population, and other times we want to estimate a propensity from a finite sequence of observations. For example, TV raters ordinarily want to estimate the actual fraction of people in a nation watching a show from a sample of those people. On the other hand, medical scientists want to estimate the propensity with which males tend to have high blood pressure from a finite sequence of males. One can create an infinite sequence from a finite population by returning a sampled item back to the population before sampling the next item. This is called "**sampling with replacement**." In practice, it is rarely done, but ordinarily the finite population is so large that statisticians make the simplifying assumption that it is done. That is, they do not replace the item, but still assume the ratio is unchanged for the next item sampled.

When sampling, the observed relative frequency is called the **maximum likelihood estimate** of the probability (limit of the relative frequency) because it is the estimate of the probability that makes the observed sequence most probable when we assume the trials (repetitions of the experiment) are probabilistically independent.

Example 2.26 *Suppose we toss a thumbtack four times and we observe the sequence* [*heads, tails, heads, heads*]. *Then the maximum likelihood estimate of the probability of heads is* 3/4. *Let* T_1, T_2, T_3, *and* T_4 *be random variables such that the value of* T_i *is the outcome on the ith toss. If we estimate* $P(T_i = heads)$ *to be* .75 *for all i and assume the trials are independent, then*

$$
\begin{aligned}
P(T_1 &= heads, T_2 = tails, T_3 = heads, T_4 = heads) \\
&= P(T_1 = heads)P(T_2 = tails)P(T_3 = heads)P(T_4 = heads) \\
&= \frac{3}{4} \times \frac{1}{4} \times \frac{3}{4} \times \frac{3}{4} = .1055.
\end{aligned}
$$

It is possible to show that any other estimate will make this sequence less probable. For example, if we estimate $P(T_i = heads)$ *to be* 1/2, *we have*

$$
\begin{aligned}
P(T_1 &= heads, T_2 = tails, T_3 = heads, T_4 = heads) \\
&= P(T_1 = heads)P(T_2 = tails)P(T_3 = heads)P(T_4 = heads) \\
&= \frac{1}{2} \times \frac{1}{2} \times \frac{1}{2} \times \frac{1}{2} = .0625.
\end{aligned}
$$

Another facet of von Mises' relative frequency approach is that a random process is generating the sequence of outcomes. According to the von Mises theory, a **random process** is defined to be a repeatable experiment for which the infinite sequence of outcomes is assumed to be a random sequence. Intuitively, a **random sequence** is one which shows no regularity or pattern. For example, the finite binary sequence "1011101100" appears random, whereas the sequence "1010101010" does not because it has the pattern "10" repeated five times. There is evidence that experiments such as coin tossing and dice throwing are indeed random processes. In 1971 Iversen et al. ran many experiments with dice indicating the sequence of outcomes is random. It is believed that unbiased sampling also yields a random sequence and is therefore a random process. See [van Lambalgen, 1987] for a formal treatment of random sequences.

2.3.2   Subjective Approach to Probability

If we tossed a thumbtack 10,000 times and it landed heads 6000 times, we would estimate $P(heads)$ to be .6. Exactly what does this number approximate? Is there some probability, accurate to an arbitrary number of digits, of the thumbtack landing heads? It seems not. Indeed, as we toss the thumbtack, its shape will slowly be altered, changing any propensity for landing heads. As another example, is there really an exact propensity for a male in a certain age group to have high blood pressure? Again, it seems not. So it seems that, outside of games of chance involving cards and urns, the relative frequency notion of probability is only an idealization. Regardless, we obtain useful insights concerning our beliefs from this notion. For example, after the thumbtack lands heads 6000 times out of 10,000 tosses, we believe it has about a .6 chance of landing heads on the next toss and bet accordingly. That is, we would consider it fair to win \$.40 if the thumbtack landed heads and to lose $\$1 - \$.60 = \$.60$ if the thumbtack landed tails. Since the bet is considered fair, the opposite position, namely, to lose \$.40 if the thumbtack landed heads and to win \$.60 if it landed tails, would also be considered fair. Hence, we would take either side of the bet. This notion of a probability as a value that determines a fair bet is called a **subjective approach to probability**, and probabilities assigned within this frame are called **subjective probabilities** or **beliefs**. A **subjectivist** is someone who feels we can assign probabilities within this framework. More concretely, in this approach the **subjective probability** of an uncertain event is the fraction p of units of money we would agree it is fair to give (lose) if the event does not occur in exchange for the promise to receive (win) $1 - p$ units if it does occur.

Example 2.27 *Suppose we estimate that* $P(\text{Heads}) = .6$. *This means we would agree it is fair to give \$.60 if heads does not occur for the promise to receive \$.40 if it does occur. Notice that if we repeated the experiment 100 times and heads did occur 60% of the time (as we expected), we would win* $60(\$.40) = \$24$ *and lose* $40(\$.60) = \$24$. *That is, we would break even.*

Unlike the relative frequency approach to probability, the subjective approach allows us to compute probabilities of events that are not repeatable. A classic example concerns betting at the racetrack. In order to decide on how to bet, we must first determine how likely we feel it is that each horse will win. A particular race has never run before and never will be run again. So we cannot look at previous occurrences of the race to obtain our belief. Rather, we obtain this belief from a careful analysis of the horses' overall previous performance, of track conditions, of jockeys, etc. Clearly, not everyone will arrive at the same probabilities based on their analyses. This is why these probabilities are called subjective. They are particular to individuals. In general, they do not have objective values in nature upon which we all must agree. Of course, if we did do an experiment such as tossing a thumbtack 10,000 times and it landed heads 6000 times, most would agree the probability of heads is about .6. Indeed, de Finetti [1937] showed that if we make certain reasonable assumptions about your beliefs, this would have to be your probability.

Before pursuing this matter further, we discuss a concept related to probability, namely, odds. Mathematically, if $P(E)$ is the probability of event E, then the **odds** $O(E)$ is defined by

$$O(E) = \frac{P(E)}{1 - P(E)}.$$

As far as betting, $O(E)$ is the amount of money we would consider it fair to lose if E did not occur in return for gaining \$1 if E did occur.

Example 2.28 *Let E be the event that the horse Oldnag wins the Kentucky Derby. If we feel $P(E) = .2$, then*

$$O(E) = \frac{P(E)}{1 - P(E)} = \frac{.2}{1 - .2} = .25.$$

This means we would consider it fair to lose \$.25 if the horse did not win in return for gaining \$1 if it did win.

If we state the fair bet in terms of probability (as discussed above), we would consider it fair to lose \$.20 if the horse did not win in return for gaining \$.80 if it did win. Notice that with both methods the ratio of the amount won to the amount lost is 4; so they are consistent in how they determine betting behavior.

At the racetrack, the betting odds shown are the odds against the event. That is, they are the odds of the event not occurring. If $P(E) = .2$ and $\neg E$ denotes that E does not occur, then

$$O(\neg E) = \frac{P(\neg E)}{1 - P(\neg E)} = \frac{.8}{1 - .8} = 4,$$

and the odds shown at the racetrack are 4 to 1 against E. If you bet on E, you will lose \$1 if E does not occur and win \$4 if E does occur. Note that these are the track odds based on the betting behavior of all participants. If you believe $P(E) = .5$, for you the odds against E are 1 to 1 (even money), and you should jump at the chance to get 4 to 1.

Some individuals are uncomfortable at being forced to consider wagering in order to assess a subjective probability. There are other methods for ascertaining these probabilities. One of the most popular is the following, which was suggested by Lindley in 1985. This method says an individual should liken the uncertain outcome to a game of chance by considering an urn containing white and black balls. The individual should determine for what fraction of white balls the individual would be indifferent between receiving a small prize if the uncertain event E happened (or turned out to be true) and receiving the same small prize if a white ball was drawn from the urn. That fraction is the individual's probability of the outcome. Such a probability can be constructed using binary cuts. If, for example, you were indifferent when the fraction was .8, for you $P(E) = 8$. If someone else were indifferent when the fraction was .6, for that individual $P(E) = .6$. Again, neither individual is right or wrong.

It would be a mistake to assume that subjective probabilities are only important in gambling situations. Actually, they are important in all the applications discussed in this book. In the next section we illustrate interesting uses of subjective probabilities.

See [Neapolitan, 1990] for more on the two approaches to probability presented here.

2.4   Random Variables in Applications

Notice that in Example 2.26 we defined random variables without first articulating a sample space Ω. Although it is mathematically elegant to first specify a sample space and then define random variables on the space, in practice this is not what we ordinarily do. In practice there is some single entity, or set of entities, which has features, the states of which we wish to determine, but which we cannot determine for certain. So we settle for determining how likely it is that a particular feature is in a particular state. An example of a single entity is a jurisdiction in which we are considering introducing an economically beneficial chemical which might be carcinogenic. We would want to determine the relative risk of the chemical versus its benefits. An example of a set of entities is a set of patients with similar diseases and symptoms. In this case, we would want to diagnose diseases based on symptoms. As mentioned in Section 2.3.1, this set of entities is called a **population**, and technically it is usually not the set of all currently existing entities, but rather is in theory an infinite set of entities.

In these applications, a random variable represents some feature of the entity being modeled, and we are uncertain as to the value of this feature. In the case of a single entity, we are uncertain as to the value of the feature for that entity, while in the case of a set of entities, we are uncertain as to the value of the feature for some members of the set. To help resolve this uncertainty, we develop probabilistic relationships among the variables. When there is a set of entities, we assume the entities in the set all have the same probabilistic relationships concerning the variables used in the model. When this is not the case, our analysis is not applicable. In the case of the scenario concerning introducing a chemical, features may include the amount of human exposure and the carcinogenic potential. If these are our features of interest, we identify the random variables *HumanExposure* and *CarcinogenicPotential* (for simplicity, our illustrations include only a few variables. An actual application ordinarily includes many more than this). In the case of a set of patients, features of interest might include whether or not diseases such as lung cancer are present, whether or not manifestations of diseases such as a chest X-ray are present, and whether or not causes of diseases such as smoking are present. Given these features, we would identify the random variables *ChestXray*, *LungCancer*, and *SmokingHistory*. After identifying the random variables, we distinguish a set of mutually exclusive and exhaustive values for each of them. The possible values of a random variable are the different states that the feature can take. For example, the state of *LungCancer* could be *present* or *absent*, the state of

ChestXray could be *positive* or *negative*, and the state of *SmokingHistory* could be *yes* or *no*, where *yes* might mean the patient has smoked one or more packs of cigarettes every day during the past 10 years.

After distinguishing the possible values of the random variables (i.e., their spaces), we judge the probabilities of the random variables having their values. However, in general, we do not directly determine values in a joint probability distribution of the random variables. Rather, we ascertain probabilities, concerning relationships among random variables, which are accessible to us. We can then reason with these variables using Bayes' Theorem to obtain probabilities of events of interest. The next example illustrates this.

Example 2.29 *Suppose Sam plans to marry, and to obtain a marriage licence in the state in which he resides one must take the blood test ELISA (enzyme linked immunosorbent assay) which tests for the presence of HIV (human immunodeficiency virus). Sam takes the test and it comes back positive for HIV. How likely is it that Sam is infected with HIV? Without knowing the accuracy of the test, Sam really has no way of knowing how probable it is that he is infected with HIV. The data we ordinarily have on such tests are the true positive rate (sensitivity) and the true negative rate (specificity). The true positive rate is the fraction of people who have the infection that test positive. For example, to obtain this number for ELISA 10,000 people who were known to be infected with HIV were identified. This was done using the Western Blot, which is the gold standard test for HIV. These people were then tested with ELISA and 9990 tested positive. Therefore, the true positive rate is .999. The true negative rate is the fraction of people who do not have the infection that test negative. To obtain this number for ELISA 10,000 nuns who denied risk factors for HIV infection were tested. Of these 9980 tested negative using the ELISA test. Furthermore, the 20 positive testing nuns tested negative using the Western Blot test. So, the true negative rate is .998, which means the false negative rate is .002. We therefore formulate the following random variables and subjective probabilities:*

$$P(ELISA = positive | HIV = present) = .999 \qquad (2.8)$$

$$P(ELISA = positive | HIV = absent) = .002. \qquad (2.9)$$

You may wonder why we called these subjective probabilities when we obtained them from data. Recall that the frequentist approach says we can never know the actual relative frequencies (objective probabilities). We can only estimate them from data. However, within the subjective approach, we can make our beliefs (subjective probabilities) equal to the fractions obtained from the data.

It may seem that Sam almost certainly is infected with HIV since the test is so accurate. However, notice that neither the probability in Equality 2.8 nor the one in Equality 2.9 is the probability of Sam being infected with HIV. Since we know that Sam tested positive on ELISA, that probability is

$$P(HIV = present | ELISA = positive).$$

We can compute this probability using Bayes' Theorem if we know $P(HIV = present)$. Recall that Sam took the blood test simply because the state required it. He did not take it because he thought for any reason he was infected with HIV. So the only other information we have about Sam is that he is a male in the state in which he resides. So if 1 in 100,000 men in Sam's state is infected with HIV, we assign the following subjective probability:

$$P(HIV = present) = .00001.$$

We now employ Bayes' Theorem to compute

$P(present|positive)$

$$= \frac{P(positive|present)P(present)}{P(positive|present)P(present) + P(positive|absent)P(absent)}$$

$$= \frac{(.999)(.00001)}{(.999)(.00001) + (.002)(.99999)}$$

$$= .00497.$$

Surprisingly, we are fairly confident that Sam is not infected with HIV.

A probability such as $P(HIV = present)$ is called a **prior probability** because, in a particular model, it is the probability of some event prior to updating the probability of that event, within the framework of that model, using new information. Do not mistakenly think it means a probability prior to any information. A probability like $P(HIV = present|ELISA = positive)$ is called a **posterior probability** because it is the probability of an event after its prior probability has been updated, within the framework of some model, based on new information. In the previous example the reason the posterior probability is small, even though the test is fairly accurate, is that the prior probability is extremely low. The next example shows how dramatically a different prior probability can change things.

Example 2.30 Suppose Mary and her husband have been trying to have a baby and she suspects she is pregnant. She takes a pregnancy test which has a true positive rate of .99 and a false positive rate of .02. Suppose further that 20% of all women who take this pregnancy test are indeed pregnant. Using Bayes' Theorem we then have

$P(present|positive)$

$$= \frac{P(positive|present)P(present)}{P(positive|present)P(present) + P(positive|absent)P(absent)}$$

$$= \frac{(.99)(.2)}{(.99)(.2) + (.02)(.8)}$$

$$= .92523.$$

Even though Mary's test was far less accurate than Sam's test, she probably is pregnant while he probably is not infected with HIV. This is due to the prior information. There was a significant prior probability (.2) that Mary was pregnant because only women who suspect they are pregnant on other grounds take pregnancy tests. Sam, on the other hand, took his test simply because he wanted to get married. We had no previous information indicating he could be infected with HIV.

In the previous two examples we obtained our beliefs (subjective probabilities) directly from the observed fractions in the data. Although this is often done, it is not necessary. In general, we obtain our beliefs from our information about the past, which means these beliefs are a composite of all our experience rather than merely observed relative frequencies. We will see examples of this throughout the book. The following example illustrates such a case.

Example 2.31 *Suppose I feel there is a .4 probability the NASDAQ will go up at least 1% today. This is based on my knowledge that, after trading closed yesterday, excellent earnings were reported by several big companies in the technology sector and it was learned that U.S. crude oil supplies unexpectedly increased. Furthermore, if the NASDAQ does go up at least 1% today, I feel there is a .1 probability that my favorite stock NTPA will go up at least 10% today. If the NASDAQ does not go up at least 1% today, I feel there is only a .02 probability NTPA will go up at least 10% today. I have these beliefs because I know from the past that NTPA's performance is linked to overall performance in the technology sector. I checked NTPA after the close of trading, and I noticed it went up over 10%. What is the probability that the NASDAQ went up at least 1%? Using Bayes' Theorem we have*

$$P(Nasdaq = up\ 1\%|NTPA = up\ 10\%)$$

$$= \frac{P(up\ 10\%|up\ 1\%)P(up\ 1\%)}{P(up\ 10\%|up\ 1\%)P(up\ 1\%) + P(up\ 10\%|not\ up\ 1\%)P(not\ up\ 1\%)}$$

$$= \frac{(.1)\,(.4)}{(.1)\,(.4) + (.02)(.6)} = .769.$$

In the previous three examples we used Bayes' Theorem to compute posterior subjective probabilities from known subjective probabilities. In a rigorous sense, we can only do this within the subjective framework. That is, since strict frequentists say we can never know probabilities for certain, they cannot use Bayes' Theorem. They can only do analyses such as the computation of a confidence interval for the value of an unknown probability based on the data. These techniques are discussed in any classic statistics text such as [Hogg and Craig, 1972]. Since subjectivists are the ones who use Bayes' Theorem, they are often called **Bayesians**.

2.5   Statistical Concepts

Next, we review basic concepts in statistics. Although we present the material for discrete variables, the concepts extend in a straightforward way to continuous variables.

2.5.1   Expected Value

Definition 2.10 *Suppose we have a discrete numeric random variable X, whose space is*

$$\{x_1, x_2, \ldots, x_n\}.$$

*Then the **expected value** $E(X)$ is given by*

$$E(X) = x_1 P(x_1) + x_2 P(x_2) + \cdots + x_n P(x_n).$$

Example 2.32 *Suppose we have a symmetric die such that the probability of each side showing is $1/6$. Let X be a random variable whose value is the number that shows when we toss the die. Then*

$$
\begin{aligned}
E(X) &= 1P(1) + 2P(2) + 3P(3) + 4P(4) + 5P(5) + 6P(6) \\
&= 1\left(\frac{1}{6}\right) + 2\left(\frac{1}{6}\right) + 3\left(\frac{1}{6}\right) + 4\left(\frac{1}{6}\right) + 5\left(\frac{1}{6}\right) + 6\left(\frac{1}{6}\right) \\
&= 3.5.
\end{aligned}
$$

If we threw the die many times, we would expect the average of the numbers showing to equal about 3.5.

Example 2.33 *Suppose we have the random variables H (height) and W (wage) discussed in Example 2.18. Recall that they were based on sex, height, and wage being distributed according to the following table:*

Case	Sex	Height (inches)	Wage ($)
1	*female*	64	30,000
2	*female*	64	30,000
3	*female*	64	40,000
4	*female*	64	40,000
5	*female*	68	30,000
6	*female*	68	40,000
7	*male*	64	40,000
8	*male*	64	50,000
9	*male*	68	40,000
10	*male*	68	50,000
11	*male*	70	40,000
12	*male*	70	50,000

Then

$$E(H) = 64\left(\frac{6}{12}\right) + 68\left(\frac{4}{12}\right) + 70\left(\frac{2}{12}\right) = 66.33$$

$$E(W) = 30{,}000 \left(\frac{3}{12}\right) + 40{,}000 \left(\frac{6}{12}\right) + 50{,}000 \left(\frac{3}{12}\right) = 40{,}000.$$

The expected value of X is also called the **mean** and is sometimes denoted $\bar{X}$. In situations where probabilities are assigned according to how many times a value appears in a population (as in the previous example), it is simply the average of the values taken over all members of the population.

We have the following theorem, whose proof is left as an exercise, concerning two independent random variables:

Theorem 2.3 *If X and Y are independent random variables, then*

$$E(XY) = E(X)E(Y).$$

Example 2.34 *Let X be a random variable whose value is the number that shows when we toss a symmetric die, and let Y be a variable whose value is 1 if heads is the result of tossing a fair coin and 0 if tails is the result. Then X and Y are independent. We show next that $E(XY) = E(X)E(Y)$ as the previous theorem entails. To that end,*

$$
\begin{aligned}
E(XY) &= (1 \times 1)P(1,1) + (2 \times 1)P(2,1) + (3 \times 1)P(3,1) + \\
&\quad (4 \times 1)P(4,1) + (5 \times 1)P(5,1) + (6,1)P(6,1) + \\
&\quad (1 \times 0)P(1,0) + (2 \times 1)P(2,0) + (3 \times 0)P(3,1) + \\
&\quad (4 \times 0)P(4,0) + (5 \times 0)P(5,0) + (6,1)P(6,0) \\
&= 1 \left(\frac{1}{6}\right)\left(\frac{1}{2}\right) + 2 \left(\frac{1}{6}\right)\left(\frac{1}{2}\right) + 3 \left(\frac{1}{6}\right)\left(\frac{1}{2}\right) + \\
&\quad 4 \left(\frac{1}{6}\right)\left(\frac{1}{2}\right) + 5 \left(\frac{1}{6}\right)\left(\frac{1}{2}\right) + 6 \left(\frac{1}{6}\right)\left(\frac{1}{2}\right) \\
&= 1.75.
\end{aligned}
$$

Furthermore,

$$E(X)E(Y) = 3.5 \times .5 = 1.75.$$

2.5.2 Variance and Covariance

While the expected value is a summary statistic that often tells us approximately what values occur in a population, the variance tells us about how far the actual values are from the expected value. We discuss the variance next.

Definition 2.11 *Suppose we have a discrete numeric random variable X, whose space is*

$$\{x_1, x_2, \ldots, x_n\}.$$

*Then the **variance** $Var(X)$ is given by*

$$Var(X) = E\left([X - E(X)]^2\right).$$

The next theorem often enables us to compute the variance more easily. Its proof is left as an exercise.

Theorem 2.4 *We have*

$$Var(X) = E\left(X^2\right) - (E(X))^2.$$

Example 2.35 *Suppose again we have the die discussed in Example 2.32. Then*

$$
\begin{aligned}
E\left(X^2\right) &= 1^2 P(1) + 2^2 P(2) + 3^2 P(3) + 4^2 P(4) + 5^2 P(5) + 6^2 P(6) \\
&= 1^2\left(\frac{1}{6}\right) + 2^2\left(\frac{1}{6}\right) + 3^2\left(\frac{1}{6}\right) + 4^2\left(\frac{1}{6}\right) + 5^2\left(\frac{1}{6}\right) + 6^2\left(\frac{1}{6}\right) \\
&= 15.167.
\end{aligned}
$$

So

$$
\begin{aligned}
Var(X) &= E\left(X^2\right) - (E(X))^2 \\
&= 15.167 - (3.5)^2 = 2.917.
\end{aligned}
$$

In order to compare quantities of the same order of magnitude, the variance is often converted to the standard deviation σ_X, which is given by

$$\sigma_X = \sqrt{Var(X)}.$$

Example 2.36 *For the die discussed in the previous example,*

$$\sigma_X = \sqrt{2.917} = 1.708.$$

Example 2.37 *Suppose there are two neighboring communities, one community consists of "amazons," while the other consists of "pygmies." So in the first community everyone is around 7 feet tall, while in the second community everyone is around 4 feet tall. Suppose further that there are the same number of people in both communities. Let H be a random variable representing the height of individuals in the two communities combined. Then it is left as an exercise to show*

$$
\begin{aligned}
E(H) &= 5.5 \\
\sigma_H &= 1.5.
\end{aligned}
$$

Example 2.38 *Suppose now there is a community in which everyone is about 5.5 feet tall. Then*

$$
\begin{aligned}
E(H) &= 5.5 \\
\sigma_H &= 0.
\end{aligned}
$$

The previous two examples illustrate that the expected value alone cannot tell us much about what we can expect concerning the actual values in the population. In the first example, we would never see individuals 5.5 feet tall, while in the second example that is all we would see. The standard deviation gives us an idea as to how far the values actually deviate from the expected value. However, it too is only a summary statistic, and the expected value and standard deviation together do not, in general, contain all information in the probability distribution. In the second example above they do indeed do this, while in the first example there could be many other distributions which yield the same expected value and standard deviation.

The next definition concerns the relationship between two random variables.

Definition 2.12 *Suppose we have two discrete numeric random variables X and Y. Then the **covariance** $Cov(X, Y)$ of X and Y is given by*

$$Cov(X, Y) = E\left([X - E(X)][Y - E(Y)]\right).$$

$Cov(X, Y)$ is often denoted σ_{XY}. It is left as an exercise to prove the following theorem.

Theorem 2.5 *We have*

$$Cov(X, Y) = E\left(XY\right) - E(X)E(Y).$$

The previous theorem makes it obvious that

$$Cov(X, X) = Var(X),$$

and

$$Cov(X, Y) = Cov(Y, X).$$

Example 2.39 *Suppose we have the random variables in Example 2.33. Then*

$E(HW)$

$$
\begin{aligned}
= \ & (64 \times 30{,}000)P(64, 30{,}000) + (64 \times 40{,}000)P(64, 40{,}000) + \\
& (64 \times 50{,}000)P(64, 50{,}000) + (68 \times 30{,}000)P(68, 30{,}000) + \\
& (68 \times 40{,}000)P(68, 40{,}000) + (68 \times 50{,}000)P(68, 40{,}000) + \\
& (70 \times 40{,}000)P(70, 40{,}000) + (70 \times 50{,}000)P(70, 50{,}000) \\
= \ & (64 \times 30{,}000)\left(\frac{2}{12}\right) + (64 \times 40{,}000)\left(\frac{3}{12}\right) + (64 \times 50{,}000)\left(\frac{1}{12}\right) + \\
& (68 \times 30{,}000)\left(\frac{1}{12}\right) + (68 \times 40{,}000)\left(\frac{2}{12}\right) + (68 \times 50{,}000)\left(\frac{1}{12}\right) + \\
& (70 \times 40{,}000)\left(\frac{1}{12}\right) + (70 \times 50{,}000)\left(\frac{1}{12}\right) \\
= \ & 2{,}658{,}333.
\end{aligned}
$$

Therefore,

$$
\begin{aligned}
Cov(H,W) &= E(HW) - E(H)E(W) \\
&= 2{,}658{,}333 - 66.33 \times 40{,}000 \\
&= 5133.
\end{aligned}
$$

The covariance itself does not convey much meaning concerning the relationship between X and Y. To accomplish this we compute the correlation coefficient from it.

Suppose we have two discrete numeric random variables X and Y. Then the **correlation coefficient** $\rho(X,Y)$ of X and Y is given by

$$
\rho(X,Y) = \frac{Cov(X,Y)}{\sigma_X \sigma_Y}.
$$

Example 2.40 *Suppose we have the random variables in Example 2.33. It is left as an exercise to show*

$$
\begin{aligned}
\sigma_H &= 2.516 \\
\sigma_W &= 7071.
\end{aligned}
$$

We then have

$$
\begin{aligned}
\rho(H,W) &= \frac{Cov(H,W)}{\sigma_H \sigma_W} \\
&= \frac{5133}{2.516 \times 7071} \\
&= .2885.
\end{aligned}
$$

Notice the $\rho(H,W)$ is positive and less than 1. The correlation coefficient is between -1 and $+1$. A value greater than 0 indicates the variables are positively correlated, and a value less than 0 indicates they are negatively correlated. By positively correlated we mean as one increases the other increases, while by negatively correlated we mean as one increases the other decreases. In the example above as height increases wage tends to increase. That is, tall people tend to have larger wages. The next theorem states these results concretely.

Theorem 2.6 *For any two numeric random variables X and Y,*

$$
|\rho(X,Y)| \le 1.
$$

Furthermore,

$$
\rho(X,Y) = 1
$$

if and only if there exists constants $a > 0$ and b such that

$$
Y = aX + b,
$$

and

$$\rho(X, Y) = -1$$

if and only if there exists constants $a < 0$ and b such that

$$Y = aX + b.$$

Finally, if X and Y are independent, then

$$\rho(X, Y) = 0.$$

Proof. *The proof can be found in [Feller, 1968].* ∎

Example 2.41 *Let X be a random variable whose value is the number that shows when we toss a symmetric die, and let Y be a random variable whose value is two times the number that shows. Then*

$$Y = 2X.$$

According to the previous theorem, $\rho(X, Y) = 1$. We show this directly.

$$
\begin{aligned}
Cov(XY) &= E(XY) - E(X)E(Y) \\
&\quad (1 \times 2)\left(\frac{1}{6}\right) + (2 \times 4)\left(\frac{1}{6}\right) + (3 \times 6)\left(\frac{1}{6}\right) + \\
&\quad (4 \times 8)\left(\frac{1}{6}\right) + (5 \times 10)\left(\frac{1}{6}\right) + (6 \times 12)\left(\frac{1}{6}\right) - (3.5)(7.0) \\
&= 5.833.
\end{aligned}
$$

Therefore,

$$
\begin{aligned}
\rho(X, Y) &= \frac{Cov(X, Y)}{\sigma_X \sigma_Y} \\
&= \frac{5.833}{1.708 \times 3.416} = 1.
\end{aligned}
$$

Example 2.42 *Let X be a random variable whose value is the number that shows when we toss a symmetric die, and let Y be a random variable whose value is the negative of the number that shows. Then*

$$Y = -X.$$

According to the previous theorem, $\rho(X, Y) = -1$. It is left as an exercise to show this directly.

Example 2.43 *As in Example 2.34, let X be a random variable whose value is the number that shows when we toss a symmetric die, and let Y be a variable whose value is 1 if heads is the result of tossing a fair coin and 0 if tails is the result. According to the previous theorem, $\rho(X, Y) = 0$. It is left as an exercise to show this directly.*

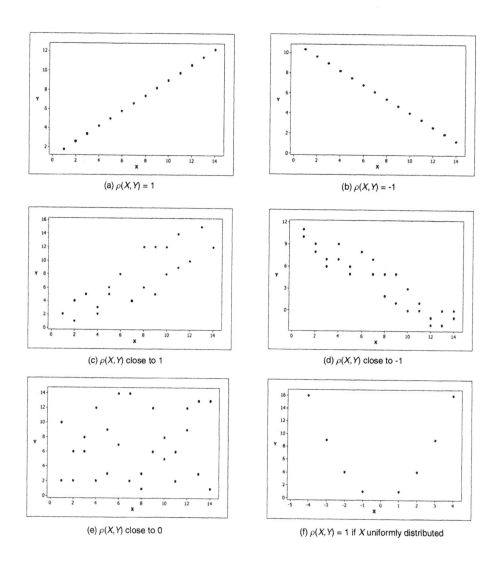

Figure 2.3: Several correlation coefficients of two random variables.

Notice that the theorem does not say that $\rho(X, Y) = 0$ if and only if X and Y are independent. Indeed, X and Y can be deterministically related and still have a zero correlation coefficient. The next example illustrates this.

Example 2.44 *Let X's space be $\{-1, -2, 1, 2\}$, let each value have probability .25, let Y's space be $\{1, 4\}$, and let $Y = X^2$. Then $E(XY)$ is as follows:*

$E(XY)$

$$
\begin{aligned}
= \quad & (-1)(1)P(X = -1, Y = 1) + (-1)(4)P(X = -1, Y = 4) + \\
& (-2)(1)P(X = -2, Y = 1) + (-2)(4)P(X = -2, Y = 4) + \\
& (1)(1)P(X = 1, Y = 1) + (1)(4)P(X = 1, Y = 4) + \\
& (2)(1)P(X = 2, Y = 1) + (2)(4)P(X = 2, Y = 4)
\end{aligned}
$$

$$
= \quad -1\left(\frac{1}{4}\right) - 4(0) - 2(0) - 8\left(\frac{1}{4}\right) + 1\left(\frac{1}{4}\right) + 4(0) + 2(0) + 8\left(\frac{1}{4}\right)
$$

$$
= \quad 0.
$$

Since clearly $E(X) = 0$, we conclude that $\rho(X, Y) = 0$.

We see that the correlation coefficient is not a general measure of the dependency between X and Y. Rather, it is a measure of the degree of linear dependence. Figure 2.3 shows scatterplots illustrating correlation coefficients associated with some distributions. In each scatterplot, each point represents one (x, y) pair occurring in the joint distribution of X and Y.

We close with a theorem that gives us the variance of the sum of several random variables. Its proof is left as an exercise.

Theorem 2.7 *For random variables $X_1, X_2, \ldots, X_n$, we have*

$$
Var(X_1 + X_2 + \cdots + X_n) = \sum_{i=1}^{n} Var(X_i) + 2 \sum_{i \neq j} Cov(X_i, X_j).
$$

For two random variables X and Y, this equality is as follows:

$$
Var(X + Y) = Var(X) + Var(Y) + 2Cov(X, Y).
$$

Note that if X and Y are independent, then the variance of their sum is just the sum of their variances.

2.5.3 Linear Regression

Often we want to assume a linear relationship exists between two random variables, and then try to learn that relationship from a random sample taken from the population over which the random variables are defined. The purpose is to be able to predict the value of one of the variables from the value of the other. Linear regression is used to accomplish this. Next, we briefly review linear regression. We do not develop the theory here, but rather only review the technique. An applied statistics text such as [Aczel and Sounderpandian, 2002] develops the theory.

Simple Linear Regression

In **simple linear regression**, we assume we have an independent variable X and a dependent variable Y such that

$$y = \beta_0 + \beta_1 x + \epsilon_x, \qquad (2.10)$$

where ϵ_x is a random variable, which depends on the value x of X, with the following properties:

1. For every value x of X, ϵ_x is normally distributed with 0 mean.

2. For every value x of X, ϵ_x has the same standard deviation σ.

3. The random variables ϵ_x for all x are mutually independent.

Note that these assumptions entail that the expected value of Y given a value x of X is given by

$$E(Y|X = x) = \beta_0 + \beta_1 x.$$

The idea is that the expected value of Y is a deterministic linear function of x. However, the actual value y of Y is not uniquely determined by the value of X because of a random error term ϵ_x.

Once we make these assumptions about two random variables, we use simple linear regression to try to discover the linear relationship shown in Equality 2.10 from a random sample of values of X and Y. An example follows.

Example 2.45 *This example is taken from [Aczel and Sounderpandian, 2002]. American Express suspected that charges on American Express cards increased with the number of miles traveled by the card holder. To investigate this matter, a research firm randomly selected 25 card holders and obtained the data shown in Table 2.1. Figure 2.4 shows a scatterplot of the data. We see that it appears a linear relationship holds between Dollars (Y) and Miles (X).*

To estimate the values of β_0 and β_1, we find the values of b_0 and b_1 that minimize the expression

$$\sum_{i=1}^{n}[y_i - (b_0 + b_1 x_i)]^2,$$

where n is the size of the sample and x_i and y_i are the values of X and Y for the ith item in the sample. These values are our estimates. We do not review how to find the minimizing values here. Any statistics package such as MINITAB, SAS, or SPSS has a linear regression module. If we use one such package to do a linear regression analysis based on the data in Table 2.1, we obtain

$$b_0 = 274.8 \qquad b_1 = 1.26.$$

So the linear relationship is estimated to be

$$y = 274.8 + 1.26x + \epsilon_x.$$

Passenger	Miles (X)	Dollars (Y)
1	1211	1802
2	1345	2405
3	1422	2005
4	1687	2511
5	1849	2332
6	2026	2305
7	2133	3016
8	2253	3385
9	2400	3090
10	2468	3694
11	2699	3371
12	2806	3998
13	3082	3555
14	3209	4692
15	3466	4244
16	3643	5298
17	3852	4801
18	4033	5147
19	4267	5738
20	4498	6420
21	4533	6059
22	4804	6426
23	5090	6321
24	5233	7026
25	5439	6964

Table 2.1: Miles and dollars for 25 American Express card holders.

Other information provided by a statistics package, when used to do the linear regression analysis in Example 2.45, includes the following:

Predictor	Coefficient	SE Coefficient	T	P
Constant	$b_0 = 274.8$	170.3	1.61	.12
x	$b_1 = 1.255$	.0497	25.25	0

$$R^2 = 96.5\%.$$

Briefly, we discuss what each of these quantities means.

1. SE coefficient: This quantity, called the **standard error of the coefficient**, enables us to compute our confidence in how close our approximations are to the true values of β_0 and β_1 (assuming a linear relationship exists). Recall that for the normal density function, 95% of the mass falls in an interval whose endpoints are the mean $\pm$ 1.96 times the standard deviation. So if σ_0 is the SE coefficient for b_0, we can be 95% confident that

$$\beta_0 \in (b_0 - 1.96\sigma_0, b_0 + 1.96\sigma_0).$$

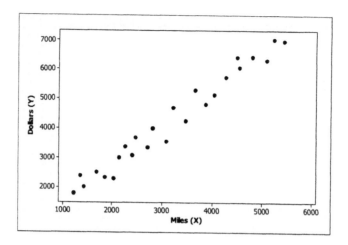

Figure 2.4: Scatterplot of dollars verses miles for 25 American Express card holders

In Example 2.45, we can be 95% confident that

$$\beta_0 \in (274.8 - 1.96 \times 170.3, 274.8 + 1.96 \times 170.3)$$
$$= (-58.988, 608.588)$$

$$\beta_1 \in (1.255 - 1.96 \times .0497, 1.255 + 1.96 \times .0497)$$
$$= (1.158, 1.352).$$

Note that we can be much more confident in the estimate of β_1.

2. T: We have

$$T = \frac{\text{coefficient}}{\text{SE coefficient}}$$

The larger the value of T, the more confident we can be that the estimate is close to the true value. Notice in Example 2.45 that T is quite large for b_1 and not very large for b_0.

3. P: If the true value of the parameter (b_0 or b_1) is 0, then this is the probability of obtaining data more unlikely than the result. In Example 2.45, if $b_0 = 0$, then the probability of obtaining data more unlikely than the result is .12, while if $b_1 = 0$, the probability of obtaining data more unlikely than the result is 0. So we can be very confident $b_1 \neq 0$, while we cannot be so confident that $b_0 \neq 0$. This means that the data strongly implies a linear dependence of Y on X, but it is not improbable that the constant is 0.

4. R^2: This value, called the **coefficient of determination**, is the square of the sample correlation coefficient, written as a percent. So its value

Driver	Miles (X_1)	Deliveries (X_2)	Travel Time (Y)
1	100	4	9.3
2	50	3	4.8
3	100	4	8.9
4	100	2	6.5
5	50	2	4.2
6	80	2	6.2
7	75	3	7.4
8	65	4	6.0
9	90	3	7.6
10	90	2	6.1

Table 2.2: Data on miles traveled, number of deliveries, and travel time in hours.

is between 0% and 100%. A value of 0% means that there is no linear dependence between the sample values of X and Y, while a value of 100% means there is a perfect linear dependence. Clearly, the larger the value of R^2, the more confidence we can have that there really is a linear relationship between X and Y.

Multiple Linear Regression

Multiple linear regression is just like simple linear regression except there is more than one independent variable. That is, we assume we have n independent variables $X_1, X_2, \ldots, X_n$ and a dependent variable Y such that

$$y = b_0 + b_1 x_1 + b_2 x_2 + \ldots + b_n x_n + \epsilon_{x_1, x_2, \ldots, x_n},$$

where $\epsilon_{x_1, x_2, \ldots, x_n}$ is as described at the beginning of Section 2.5.3.

Example 2.46 *This example is taken from [Anderson et al., 2005]. Suppose a delivery company wants to investigate the dependence of drivers' travel time on miles traveled and number of deliveries. Assume we have the data in Table 2.2. We obtain the following results from a regression analysis based on these data:*

$$
\begin{aligned}
b_0 &= -.869 \\
b_1 &= .062 \\
b_2 &= .923.
\end{aligned}
$$

So the linear relationship is estimated to be

$$y = -0.8689 + 0.061 x_1 + 0.923 x_2 + \epsilon_{x_1, x_2}.$$

Furthermore, we have the following:

Predictor	Coefficient	SE Coefficient	T	P
Constant	$b_0 = -.8687$	.9515	$-.91$	.392
Miles (X_1)	$b_1 = .061135$	.009888	6.18	0
Deliveries (X_2)	$b_2 = .9234$	.2211	4.18	.004

$$R^2 = 90.4\%$$

We see that we can be fairly confident a linear relationship exists in terms of the two variables, but the constant term could well be 0.

EXERCISES

Section 2.1

Exercise 2.1 Let the experiment be drawing the top card from a deck of 52 cards. Let Heart be the event a heart is drawn, and RoyalCard be the event a royal card is drawn.

1. Compute $P(\mathsf{Heart})$.

2. Compute $P(\mathsf{RoyalCard})$.

3. Compute $P(\mathsf{Heart} \cup \mathsf{RoyalCard})$.

Exercise 2.2 Prove Theorem 2.1.

Exercise 2.3 Example 2.5 showed that, in the draw of the top card from a deck, the event Jack is independent of the event Club. That is, it showed $P(\mathsf{Jack}|\ \mathsf{Club}) = P(\mathsf{Jack})$.

1. Show directly that the event Club is independent of the event Jack. That is, show $P(\mathsf{Club}|\mathsf{Jack}) = P(\mathsf{Club})$. Show also that $P(\mathsf{Jack} \cap \mathsf{Club}) = P(\mathsf{Jack})P(\mathsf{Club})$.

2. Show, in general, that if $P(\mathsf{E}) \neq 0$ and $P(\mathsf{F}) \neq 0$, then $P(\mathsf{E}|\mathsf{F}) = P(\mathsf{E})$ if and only if $P(\mathsf{F}|\mathsf{E}) = P(\mathsf{F})$, and each of these holds if and only if $P(\mathsf{E} \cap \mathsf{F}) = P(\mathsf{E})P(\mathsf{F})$.

Exercise 2.4 The complement of a set E consists of all the elements in Ω that are not in E and is denoted by $\overline{\mathsf{E}}$.

1. Show that E is independent of F if and only if $\overline{\mathsf{E}}$ is independent of F, which is true if and only if $\overline{\mathsf{E}}$ is independent of $\overline{\mathsf{F}}$.

2. Example 2.6 showed that, for the objects in Figure 2.1, A and Square are conditionally independent given Black and given White. Let B be the set of all objects containing a "B," and Circle be the set of all circular objects. Use the result just obtained to conclude that A and Circle, B and Square, and B and Circle are each conditionally independent given either Black or White.

Exercise 2.5 Show that in the draw of the top card from a deck the event $E = \{kh, ks, qh\}$ and the event $F = \{kh, kc, qh\}$ are conditionally independent given the event $G = \{kh, ks, kc, kd\}$. Determine whether E and F are conditionally independent given $\overline{G}$.

Exercise 2.6 Prove the rule of total probability, which says if we have n mutually exclusive and exhaustive events $E_1, E_2, \ldots, E_n$, then for any other event F,

$$P(F) = P(F \cap E_1) + P(F \cap E_2) + \cdots + P(F \cap E_n).$$

Exercise 2.7 Let Ω be the set of all objects in Figure 2.1, and assign each object a probability of $1/13$. Let A be the set of all objects containing an "A," and Square be the set of all square objects. Compute $P(A|\text{Square})$ directly and using Bayes' Theorem.

Section 2.2

Exercise 2.8 Consider the probability space and random variables given in Example 2.18.

1. Determine the joint distribution of S and W, the joint distribution of W and H, and the remaining values in the joint distribution of S, H, and W.

2. Show that the joint distribution of S and H can be obtained by summing the joint distribution of S, H, and W over all values of W.

Exercise 2.9 Let a joint probability distribution be given. Using the law of total probability, show that, in general, the probability distribution of any one of the random variables is obtained by summing over all values of the other variables.

Exercise 2.10 The chain rule says that for n random variables $X_1, X_2, \ldots, X_n$, defined on the same sample space Ω,

$$P(x_1, x_2, \ldots, x_n) = P(x_n | x_{n-1}, x_{n-2}, \ldots x_1) \cdots \times P(x_2 | x_1) \times P(x_1)$$

whenever $P(x_1, x_2, \ldots, x_n) \neq 0$. Prove this rule.

Exercise 2.11 Use the results in Exercise 2.4 (1) to conclude that it was only necessary in Example 2.20 to show that $P(r,t) = P(r,t|s_1)$ for all values of r and t.

Exercise 2.12 Suppose we have two random variables X and Y with spaces $\{x_1, x_2\}$ and $\{y_1, y_2\}$, respectively.

1. Use the results in Exercise 2.4 (1) to conclude that we need only show $P(y_1|x_1) = P(y_1)$ to conclude $I_P(X, Y)$.

2. Develop an example showing that if X and Y both have spaces containing more than two values, then we need to check whether $P(y|x) = P(y)$ for all values of x and y to conclude $I_P(X, Y)$.

Exercise 2.13 Consider the probability space and random variables given in Example 2.18.

1. Are H and W independent?

2. Are H and W conditionally independent given S?

3. If this small sample is indicative of the probabilistic relationships among the variables in some population, what causal relationships might account for this dependency and conditional independency?

Exercise 2.14 In Example 2.23, it was left as an exercise to show for all values v of V, l of L, c of C, s of S, and f of F that

$$P(v, l|s, f, c) = P(v, l|c).$$

Show this.

Section 2.3

Exercise 2.15 Kerrich [1946] performed experiments such as tossing a coin many times, and he found that the relative frequency did appear to approach a limit. That is, for example, he found that after 100 tosses the relative frequency may have been .51, after 1000 tosses it may have been .508, after 10,000 tosses it may have been .5003, and after 100,000 tosses it may have been .50006. The pattern is that the 5 in the first place to the right of the decimal point remains in all relative frequencies after the first 100 tosses, the 0 in the second place remains in all relative frequencies after the first 1000 tosses, etc. Toss a thumbtack at least 1000 times and see if you obtain similar results.

Exercise 2.16 Pick some upcoming event. It could be a sporting event or it could be the event that you will get an "A" in this course. Determine your probability of the event using Lindley's [1985] method of comparing the uncertain event to a draw of a ball from an urn (see the discussion following Example 2.28).

Section 2.4

Exercise 2.17 A forgetful nurse is supposed to give Mr. Nguyen a pill each day. The probability that she will forget to give the pill on a given day is .3. If he receives the pill, the probability he will die is .1. If he does not receive the pill, the probability he will die is .8. Mr. Nguyen died today. Use Bayes' Theorem to compute the probability the nurse forgot to give him the pill.

Exercise 2.18 An oil well may be drilled on Professor Neapolitan's farm in Texas. Based on what has happened on similar farms, we judge the probability of oil being present to be .5, the probability of only natural gas being present to be .2, and the probability of neither being present to be .3. If oil is present, a geological test will give a positive result with probability .9; if only natural gas is present, it will give a positive result with probability .3; and if neither are present, the test will be positive with probability .1. Suppose the test comes back positive. Use Bayes' Theorem to compute the probability that oil is present.

Section 2.5

Exercise 2.19 Suppose a fair six-sided die is to be rolled. The player will receive a number of dollars equal to the number of dots that show, except when five or six dots show, in which case the player will lose $5 or $6 respectively.

1. Compute the expected value of the amount of money the player will win or lose.

2. Compute the variance and standard deviation of the amount of money the player will win or lose.

3. If the game is repeated 100 times, compute the most the player could win, the most the player could lose, and the amount the player can expect to win or lose.

Exercise 2.20 Prove Theorem 2.3.

Exercise 2.21 Prove Theorem 2.4.

Exercise 2.22 Prove Theorem 2.5.

Exercise 2.23 Prove Theorem 2.7.

Exercise 2.24 Suppose we have the following data on 10 college students concerning their grade point averages (GPA) and scores on the graduate record exam (GRE):

Student	GPA	GRE
1	2.2	1400
2	2.4	1300
3	2.8	1550
4	3.1	1600
5	3.3	1400
6	3.3	1700
7	3.4	1650
8	3.7	1800
9	3.9	1700
10	4.0	1800

1. Develop a scatterplot of GRE versus GPA.

2. Determine the sample covariance.

3. Determine the sample correlation coefficient. What does this value tell us about the relationship between the two variables?

Exercise 2.25 Consider again the data in Exercise 2.24. Using some statistical package such as MINITAB, do a linear regression for GRE in terms of GPA. Show the values of b_0 and b_1 and the values of the SE coefficient, T, and P for each of them. Show R^2. Do these results indicate a linear relationship between GRE and GPA? Does it seem that the constant term is significant?

Exercise 2.26 Suppose we have the same data as in Exercise 2.24, except we also have data on family income as follows:

Student	GPA	GRE	Income ($1000)
1	2.2	1400	44
2	2.4	1300	40
3	2.8	1550	46
4	3.1	1600	50
5	3.3	1400	40
6	3.3	1700	54
7	3.4	1650	52
8	3.7	1800	56
9	3.9	1700	50
10	4.0	1800	56

Using some statistical package such as MINITAB, do a linear regression for GRE in terms of GPA and income. Show the values of b_0, b_1, and b_2 and the values of the SE coefficient, T, and P for each of them. Show R^2. Does it seem that we have improved our predictive accuracy for the GRE by also considering income?

Exercise 2.27 Suppose we have the same data as in Exercise 2.24, except that we also have data on the students' American College Test (ACT) scores as follows:

Student	GPA	GRE	ACT
1	2.2	1400	22
2	2.4	1300	23
3	2.8	1550	25
4	3.1	1600	26
5	3.3	1400	27
6	3.3	1700	27
7	3.4	1650	28
8	3.7	1800	29
9	3.9	1700	30
10	4.0	1800	31

Using some statistical package such as MINITAB, do a linear regression for GRE in terms of GPA and ACT. Show the values of b_0, b_1, and b_2 and the values of the SE coefficient, T, and P for each of them. Show R^2. Does it seem that we have improved our predictive accuracy for GRE by also considering ACT score? If not, what do you think would be an explanation for this?

Exercise 2.28 Using the data in Exercise 2.27 and some statistical package such as MINITAB, do a linear regression for GPA in terms of ACT. Show the values of b_0 and b_1, and the values of the SE coefficient, T, and P for each of them. Show R^2. Relate this result to that obtained in Exercise 2.27.

Chapter 3

Bayesian Networks

The Reverend Thomas Bayes (1702-1761) developed Bayes' Theorem in the 18th century. Since that time the theorem has had a great impact on statistical inference because it enables us to infer the probability of a cause when its effect is observed. In the 1980s, the method was extended to model the probabilistic relationships among many causally related variables. The graphical structures that describe these relationships have come to be known as Bayesian networks. We introduce these networks next. Applications of Bayesian networks to finance and marketing appear in Parts II and III. In Sections 3.1 and 3.2 we define Bayesian networks and discuss their properties. Section 3.3 shows how causal graphs often yield Bayesian networks. In Section 3.4 we discuss doing probabilistic inference using Bayesian networks. Section 3.5 concerns obtaining the conditional probabilities necessary to a Bayesian network. Finally, Section 3.6 shows the conditional independencies entailed by a Bayesian network.

3.1   What Is a Bayesian Network?

Recall that in Chapter 2, Example 2.29, we computed the probability of Joe having the HIV virus given that he tested positive for it using the ELISA test. Specifically, we knew that

$$P(ELISA = positive|HIV = present) = .999$$

$$P(ELISA = positive|HIV = absent) = .002,$$

and

$$P(HIV = present) = .00001,$$

and we then employed Bayes' Theorem to compute

$P(present|positive)$

$$= \frac{P(positive|present)P(present)}{P(positive|present)P(present) + P(positive|absent)P(absent)}$$

$$= \frac{(.999)(.00001)}{(.999)(.00001) + (.002)(.99999)}$$

$$= .00497.$$

We summarize the information used in this computation in Figure 3.1, which is a two-node/variable Bayesian network. Notice that it represents the random variables HIV and $ELISA$ by nodes in a directed acyclic graph (DAG) and the causal relationship between these variables with an edge from HIV to $ELISA$. That is, the presence of HIV has a causal effect on whether the test result is positive; so there is an edge from HIV to $ELISA$. Besides showing a DAG representing the causal relationships, Figure 3.1 shows the prior probability distribution of HIV, and the conditional probability distribution of $ELISA$ given each value of its parent HIV. In general, Bayesian networks consist of a DAG, whose edges represent relationships among random variables that are often (but not always) causal; the prior probability distribution of every variable that is a root in the DAG; and the conditional probability distribution of every non-root variable given each set of values of its parents. We use the terms "node" and "variable" interchangeably when discussing Bayesian networks.

Let's illustrate a more complex Bayesian network by considering the problem of detecting credit card fraud (taken from [Heckerman, 1996]). Let's say that we have identified the following variables as being relevant to the problem:

Variable	What the Variable Represents
Fraud (F)	Whether the current purchase is fraudulent
Gas (G)	Whether gas has been purchased in the last 24 hours
Jewelry (J)	Whether jewelry has been purchased in the last 24 hours
Age (A)	Age of the card holder
Sex (S)	Sex of the card holder

These variables are all causally related. That is, a credit card thief is likely to buy gas and jewelry, and middle-aged women are most likely to buy jewelry,

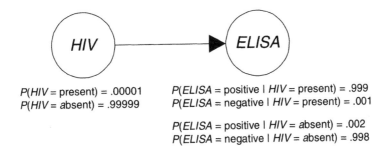

Figure 3.1: A two-node Bayesian network.

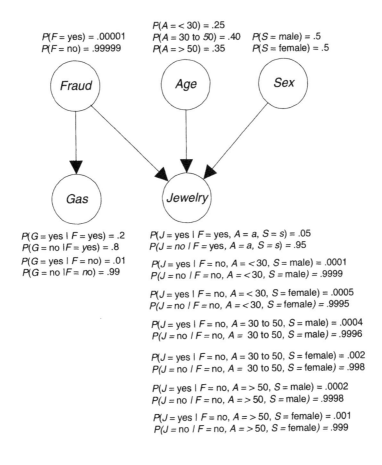

Figure 3.2: A Bayesian network for detecting credit card fraud.

while young men are least likely to buy jewelry. Figure 3.2 shows a DAG representing these causal relationships. Notice that it also shows the conditional probability distribution of every non-root variable given each set of values of its parents. The *Jewelry* variable has three parents, and there is a conditional probability distribution for every combination of values of those parents. The DAG and the conditional distributions together constitute a Bayesian network.

You may have a few questions concerning this Bayesian network. First, you may ask "What value does it have?" That is, what useful information can we obtain from it? Recall how we used Bayes' Theorem to compute $P(HIV = present|ELISA = positive)$ from the information in the Bayesian network in Figure 3.1. Similarly, we can compute the probability of credit card fraud given values of the other variables in this Bayesian network. For example, we can compute $P(F = yes|G = yes, J = yes, A = < 30, S = female)$. If this probability is sufficiently high we can deny the current purchase or require additional identification. The computation is not a simple application of Bayes' Theorem as was the case for the two-node Bayesian network in Figure 3.1. Rather it is done using sophisticated algorithms. Second, you may ask how we obtained the probabilities in the network. They can either be obtained from the subjective judgements of an expert in the area or be learned from data. In Chapter 4 we discuss techniques for learning them from data, while in Parts II and III we show examples of obtaining them from experts and learning them from data. Finally, you may ask why we are bothering to include the variables for age and sex in the network when the age and sex of the card holder has nothing to do with whether the card has been stolen (fraud). That is, fraud has no causal effect on the card holder's age or sex, and vice versa. The reason we include these variables is quite subtle. It is because fraud, age, and sex all have a common effect, namely the purchasing of jewelry. So, when we know jewelry has been purchased, the three variables are rendered probabilistically dependent owing to what psychologists call **discounting**. For example, if jewelry has been purchased in the last 24 hours, it increases the likelihood of fraud. However, if the card holder is a middle-aged woman, the likelihood of fraud is lessened (discounted) because such women are prone to buy jewelry. That is, the fact that the card holder is a middle-aged woman explains away the jewelry purchase. On the other hand, if the card holder is a young man, the likelihood of fraud is increased because such men are unlikely to purchase jewelry.

We have informally introduced Bayesian networks, their properties, and their usefulness. Next, we formally develop their mathematical properties.

3.2 Properties of Bayesian Networks

After defining Bayesian networks, we show how they are ordinarily represented.

3.2.1 Definition of a Bayesian Network

First, let's review the definition of a DAG. A **directed graph** is a pair (V, E), where V is a finite, nonempty set whose elements are called **nodes** (or vertices),

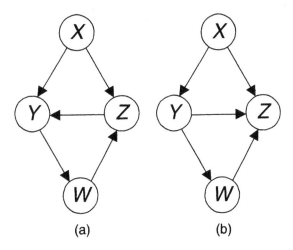

(a)   (b)

Figure 3.3: Both graphs are directed graphs; only the one in (b) is a directed acyclic graph.

and E is a set of ordered pairs of distinct elements of V. Elements of E are called **directed edges**, and if $(X, Y) \in \mathsf{E}$, we say there is an edge from X to Y. Figure 3.3 (a) is a directed graph. The set of nodes in that figure is

$$\mathsf{V} = \{X, Y, Z, W\},$$

and the set of edges is

$$\mathsf{E} = \{(X, Y), (X, Z), (Y, W), (W, Z), (Z, Y)\}.$$

A **path** in a directed graph is a sequence of nodes $[X_1, X_2, \ldots, X_k]$ such that $(X_{i-1}, X_i) \in \mathsf{E}$ for $2 \leq i \leq k$. For example, $[X, Y, W, Z]$ is a path in the directed graph in Figure 3.3 (a). A **chain** in a directed graph is a sequence of nodes $[X_1, X_2, \ldots, X_k]$ such that $(X_{i-1}, X_i) \in \mathsf{E}$ or $(X_i, X_{i-1}) \in \mathsf{E}$ for $2 \leq i \leq k$. For example, $[Y, W, Z, X]$ is a chain in the directed graph in Figure 3.3 (b), but it is not a path. A **cycle** in a directed graph is a path from a node to itself. In Figure 3.3 (a) $[Y, W, Z, Y]$ is a cycle from Y to Y. However, in Figure 3.3 (b) $[Y, W, Z, Y]$ is not a cycle because it is not a path. A directed graph $\mathbb{G}$ is called a **directed acyclic graph** (DAG) if it contains no cycles. The directed graph in Figure 3.3 (b) is a DAG, while the one in Figure 3.3 (a) is not.

Given a DAG $\mathbb{G} = (\mathsf{V}, \mathsf{E})$ and nodes X and Y in V, Y is called a **parent** of X if there is an edge from Y to X, Y is called a **descendent** of X and X is called an **ancestor** of Y if there is a path from X to Y, and Y is called a **nondescendent** of X if Y is not a descendent of X and Y is not a parent of X.[1]

---

[1] It is not standard to exclude a node's parents from its nondescendents, but this definition better serves our purposes.

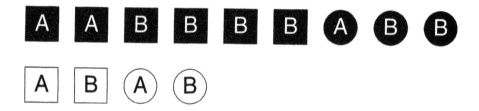

Figure 3.4: The random variables L and S are not independent, but they are conditionally independent given C.

We can now state the following definition:

Definition 3.1 *Suppose we have a joint probability distribution P of the random variables in some set V and a DAG $\mathbb{G} = (\mathsf{V}, \mathsf{E})$. We say that $(\mathbb{G}, P)$ satisfies the **Markov condition** if for each variable $X \in \mathsf{V}$, X is conditionally independent of the set of all its nondescendents given the set of all its parents. Using the notation established in Chapter 2, Section 2.2.2, this means that if we denote the sets of parents and nondescendents of X by PA and ND, respectively, then*

$$I_P(X, \mathsf{ND}|\mathsf{PA}).$$

*If $(\mathbb{G}, P)$ satisfies the Markov condition, we call $(\mathbb{G}, P)$ a **Bayesian network**.*

Example 3.1 *Recall Chapter 2, Figure 2.1, which appears again as Figure 3.4. In Chapter 2, Example 2.21 we let P assign $1/13$ to each object in the figure, and we defined these random variables on the set containing the objects:*

Variable	Value	Outcomes Mapped to This Value
L	l_1	All objects containing an "A"
	l_2	All objects containing a "B"
S	s_1	All square objects
	s_2	All circular objects
C	c_1	All black objects
	c_2	All white objects

We then showed that L and S are conditionally independent given C. That is, using the notation established in Chapter 2, Section 2.2.2, we showed

$$I_P(L, S|C).$$

Consider the DAG $\mathbb{G}$ in Figure 3.5. For that DAG we have the following:

Node	Parents	Nondescendents
L	C	S
S	C	L
C	$\varnothing$	$\varnothing$

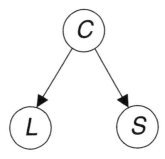

Figure 3.5: The joint probability distribution of L, S, and C constitutes a Bayesian network with this DAG.

For $(\mathbb{G}, P)$ to satisfy the Markov condition, we need to have

$$I_P(L, S|C)$$
$$I_P(S, L|C).$$

Note that since C has no nondescendents, we do not have a conditional independency for C. Since independence is symmetric, $I_P(L, S|C)$ implies $I_P(L, S|C)$. Therefore, all the conditional independencies required by the Markov condition are satisfied, and $(\mathbb{G}, P)$ is a Bayesian network.

Next, we further illustrate the Markov condition with a more complex DAG.

Example 3.2 Consider the DAG $\mathbb{G}$ in Figure 3.6. If $(\mathbb{G}, P)$ satisfied the Markov condition with some probability distribution P of X, Y, Z, W, and V, we would have the following conditional independencies:

Node	Parents	Nondescendents	Conditional Independency
X	$\varnothing$	$\varnothing$	None
Y	X	Z, V	$I_P(Y, \{Z, V\}|X)$
Z	X	Y	$I_P(Z, Y|X)$
W	Y, Z	X, W	$I_P(W, \{X, W\}|\{Y, Z\})$
V	Z	X, Y, W	$I_P(V, \{X, Y, W\}|Z)$

3.2.2 Representation of a Bayesian Network

A Bayesian network $(\mathbb{G}, P)$ by definition is a DAG $\mathbb{G}$ and joint probability distribution P that together satisfy the Markov condition. Then why in Figures 3.1 and 3.2 do we show a Bayesian network as a DAG and a set of conditional probability distributions? The reason is that $(\mathbb{G}, P)$ satisfies the Markov condition if and only if P is equal to the product of its conditional distributions in $\mathbb{G}$. Specifically, we have this theorem.

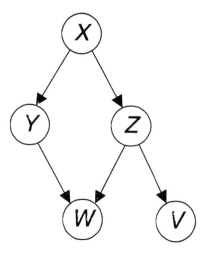

Figure 3.6: A DAG illustrating the Markov condition.

Theorem 3.1 $(\mathbb{G}, P)$ *satisfies the Markov condition (and thus is a Bayesian network) if and only if P is equal to the product of its conditional distributions of all nodes given their parents in G, whenever these conditional distributions exist.*

Proof. *The proof can be found in [Neapolitan, 2004].* ∎

Example 3.3 *We showed that the joint probability distribution P of the random variables L, S, and C defined on the set of objects in Figure 3.4 constitutes a Bayesian network with the DAG $\mathbb{G}$ in Figure 3.5. Next we illustrate that the preceding theorem is correct by showing that P is equal to the product of its conditional distributions in $\mathbb{G}$. Figure 3.7 shows those conditional distributions. We computed them directly from Figure 3.4. For example, since there are 9 black objects (c_1) and 6 of them are squares (s_1), we compute*

$$P(s_1|c_1) = \frac{6}{9} = \frac{2}{3}.$$

The other conditional distributions are computed in the same way. To show that the joint distribution is the product of the conditional distributions, we need to show for all values of l, s, and c that

$$P(s, l, c) = P(s|c)P(l|c)P(c).$$

There are a total of eight combinations of the three variables. We show that the equality holds for one of them. It is left as an exercise to show that it holds for the others. To that end, we have directly from Figure 3.4 that

$$P(s_1, l_1, c_1) = \frac{2}{13}.$$

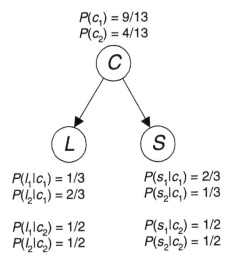

Figure 3.7: A Bayesian network representing the probability distribution P of the random variables L, S, and C defined on the set of objects in Figure 3.4.

From Figure 3.7 we have

$$P(s_1|c_1)P(l_1|c_1)P(c_1) = \frac{2}{3} \times \frac{1}{3} \times \frac{9}{13} = \frac{2}{13}.$$

Owing to Theorem 3.1, we can represent a Bayesian network $(\mathbb{G}, P)$ using the DAG $\mathbb{G}$ and the conditional distributions. We don't need to show every value in the joint distributions. These values can all be computed from the conditional distributions. So we always show a Bayesian network as the DAG and the conditional distributions as is done in Figures 3.1, 3.2, and 3.7. Herein lies the representational power of Bayesian networks. If there are a large number of variables, there are many values in the joint distribution. However, if the DAG is sparse, there are relatively few values in the conditional distributions. For example, suppose all variables are binary, and a joint distribution satisfies the Markov condition with the DAG in Figure 3.8. Then there are $2^{10} = 1024$ values in the joint distribution, but only $2 + 2 + 8 \times 8 = 68$ values in the conditional distributions. Note that we are not even including redundant parameters in this count. For example, in the Bayesian network in Figure 3.7 it is not necessary to show $P(c_2) = 4/13$ because $P(c_2) = 1 - P(c_1)$. So we need only show $P(c_1) = 9/13$. If we eliminate redundant parameters, there are only 34 values in the conditional distributions for the DAG in Figure 3.8, but still 1023 in the joint distribution. We see then that a Bayesian network is a structure for representing a joint probability distribution succinctly.

It is important to realize that we can't take just any DAG and expect a joint distribution to equal the product of its conditional distributions in the DAG. This is only true if the Markov condition is satisfied. The next example illustrates this.

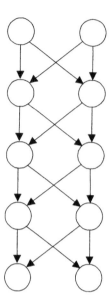

Figure 3.8: If all variables are binary, and a joint distribution satisifes the Markov condition with this DAG, there are 1024 values in the joint distribution, but only 68 values in the conditional distributions.

Example 3.4 *Consider again the joint probability distribution P of the random variables L, S, and C defined on the set of objects in Figure 3.4. Figure 3.9 shows its conditional distributions for the DAG in that figure. Note that we no longer show redundant parameters in our figures. If P satisfied the Markov condition with this DAG, we would have to have $I_P(L, S)$ because L has no parents, and S is the sole nondescendent of L. It is left as an exercise to show that this independency does not hold. Furthermore, P is not equal to the product of its conditional distributions in this DAG. For example, we have directly from Figure 3.4 that*

$$P(s_1, l_1, c_1) = \frac{2}{13} = .15385.$$

From Figure 3.9 we have

$$P(c_1|l_1, s_1)P(l_1)P(s_1) = \frac{2}{3} \times \frac{5}{13} \times \frac{8}{13} = .15779.$$

It seems we are left with a conundrum. That is, our goal is to represent a joint probability distribution succinctly using a DAG and conditional distributions for the DAG (a Bayesian network) rather than enumerating every value in the joint distribution. However, we don't know which DAG to use until we check whether the Markov condition is satisfied, and, in general, we would need to have the joint distribution to check this. A common way out of this

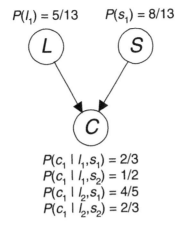

$$P(l_1) = 5/13 \qquad P(s_1) = 8/13$$

$$P(c_1 \mid l_1, s_1) = 2/3$$
$$P(c_1 \mid l_1, s_2) = 1/2$$
$$P(c_1 \mid l_2, s_1) = 4/5$$
$$P(c_1 \mid l_2, s_2) = 2/3$$

Figure 3.9: The joint probability distribution P of the random variables L, S, and C defined on the set of objects in Figure 3.4 does not satisfy the Markov condition with this DAG.

predicament is to construct a causal DAG, which is a DAG in which there is an edge from X to Y if X causes Y. The DAGs in Figures 3.1 and 3.2 are causal, while other DAGs shown so far in this chapter are not causal. Next, we discuss why a causal DAG should satisfy the Markov condition with the probability distribution of the variables in the DAG. A second way of obtaining the DAG is to learn it from data. This second way is discussed in Chapter 4.

3.3 Causal Networks as Bayesian Networks

Before discussing why a causal DAG should often satisfy the Markov condition with the probability distribution of the variables in the DAG, we formalize the notion of causality.

3.3.1 Causality

After providing an operational definition of a cause, we show a comprehensive example of identifying a cause according to this definition.

An Operational Definition of a Cause

One dictionary definition of a cause is "the one, such as a person, an event, or a condition, that is responsible for an action or a result." Although this definition is useful, it is certainly not the last word on the concept of causation, which has been investigated for centuries (see, e.g., [Hume, 1748], [Piaget, 1966], [Eells, 1991], [Salmon, 1997], [Spirtes et al., 1993, 2000], [Pearl, 2000]). This definition does, however, shed light on an operational method for identifying causal relationships. That is, if the action of making variable X take some value sometimes changes the value taken by variable Y, then we assume X is responsible for sometimes changing Y's value, and we conclude X is a cause of Y. More formally, we say we **manipulate** X when we force X to take some value, and we say X **causes** Y if there is some manipulation of X that leads to a change in the probability distribution of Y. We assume that if manipulating X leads to a change in the probability distribution of Y, then X obtaining a value by any means whatsoever also leads to a change in the probability distribution of Y. So we assume that causes and their effects are statistically correlated. However, as we shall discuss soon, variables can be correlated without one causing the other. A manipulation consists of a **randomized controlled experiment (RCE)** using some specific population of entities (e.g., individuals with chest pain) in some specific context (e.g., they currently receive no chest pain medication and they live in a particular geographical area). The causal relationship discovered is then relative to this population and this context.

Let's discuss how the manipulation proceeds. We first identify the population of entities we wish to consider. Our random variables are features of these entities. Next, we ascertain the causal relationship we wish to investigate. Suppose we are trying to determine if variable X is a cause of variable Y. We then sample a number of entities from the population. For every entity selected, we manipulate the value of X so that each of its possible values is given to the same number of entities (if X is continuous, we choose the values of X according to a uniform distribution). After the value of X is set for a given entity, we measure the value of Y for that entity. The more the resultant data show a dependency between X and Y, the more the data support that X causes Y. The manipulation of X can be represented by a variable M that is external to the system being studied. There is one value m_i of M for each value x_i of X; the probabilities of all values of M are the same; and when M equals m_i, X equals x_i. That is, the relationship between M and X is deterministic. The data support that X causes Y to the extent that the data indicate $P(y_i|m_j) \neq P(y_i|m_k)$ for $j \neq k$. Manipulation is actually a special kind of causal relationship that we assume exists primordially and is within our control so that we can define and discover other causal relationships.

An Illustration of Manipulation

We demonstrate these ideas with a comprehensive example concerning recent headline news. The pharmaceutical company Merck had been marketing its drug finasteride as medication for men with benign prostatic hyperplasia (BPH). Based on anecdotal evidence, it seemed that there was a correlation between use of the drug and regrowth of scalp hair. Let's assume that Merck took a random sample from the population of interest and, based on that sample, determined there is a correlation between finasteride use and hair regrowth. Should they conclude finasteride causes hair regrowth and therefore market it as a cure for baldness? Not necessarily. There are quite a few causal explanations for the correlation of two variables. We discuss these next.

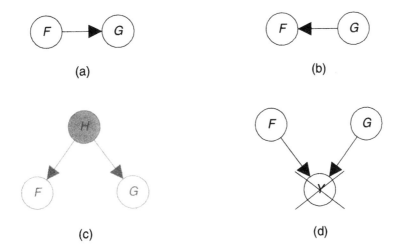

Figure 3.10: The edges in this graphs represent causal influences. All four causal relationships could account for F and G being correlated.

Possible Causal Relationships Let F be a variable representing finasteride use and G be a variable representing scalp hair growth. The actual values of F and G are unimportant to the present discussion. We could use either continuous or discrete values. If F caused G, then indeed they would be statistically correlated, but this would also be the case if G caused F or if they had some hidden common cause H. If we represent a causal influence by a directed edge, Figure 3.10 shows these three possibilities plus one more. Figure 3.10 (a) shows the conjecture that F causes G, which we already suspect might be the case. However, it could be that G causes F (Figure 3.10 (b)). You may argue that, based on domain knowledge, this does not seem reasonable. However, we do not, in general, have domain knowledge when doing a statistical analysis. So from the correlation alone, the causal relationships in Figure 3.10 (a) and (b) are equally reasonable. Even in this domain, G causing F seems possible. A man may have used some other hair regrowth product such as minoxidil which caused him to regrow hair, became excited about the regrowth, and decided to try other products such as finasteride which he heard might cause regrowth. A third possibility, shown in Figure 3.10 (c), is that F and G have some hidden common cause H which accounts for their statistical correlation. For example, a man concerned about hair loss might try both finasteride and minoxidil in his effort to regrow hair. The minoxidil may cause hair regrowth, while the finasteride may not. In this case the man's concern is a cause of finasteride use and hair regrowth (indirectly through minoxidil use), while the latter two are not causally related. A fourth possibility is that our sample (or even our entire population) consists of individuals who have some (possibly hidden) effect of both F and G. For example, suppose finasteride and apprehension about lack of hair regrowth are both causes of hypertension,[2] and our sample consists of individuals who have hypertension Y. We say a node is **instantiated** when we know its value for the entity currently being modeled. So we are saying the variable Y is instantiated to the same value for every entity in our sample. This situation is depicted in Figure 3.10 (d), where the cross through Y means the variable is instantiated. Ordinarily, the instantiation of a common effect creates a dependency between its causes because each cause explains away the occurrence of the effect, thereby making the other cause less likely. As noted earlier, psychologists call this **discounting**. So, if this were the case, discounting would explain the correlation between F and G. This type of dependency is called **selection bias**.[3] A final possibility (not depicted in Figure 3.10) is that F and G are not causally related at all. The most notable example of this situation is when our entities are points in time, and our random variables are values of properties at these different points in time. Such random variables are often correlated without having any apparent causal connection. For example, if our population consists of points in time, J is the Dow Jones Average at a given time, and L is Professor Neapolitan's hairline at a given time, then J and L are correlated.[4] Yet they do not seem to be causally connected. Some argue that there are hidden common causes beyond our ability to measure. We will not discuss this issue further here. We only wish to note the difficulty with such correlations. In light of all of the above, we see then that we cannot deduce the causal relationship between two variables from the mere fact that they are statistically correlated.

Note that any of the four causal relationships shown in Figure 3.10 could occur in combination, resulting in F and G being correlated. For example, it could be both that finasteride causes hair regrowth and that excitement about regrowth may cause use of finasteride, meaning we could have a causal loop or feedback. Therefore, we would have the causal relationships in both Figure 3.10 (a) and Figure 3.10 (b).

It may not be obvious why two variables with a common cause would be correlated. Consider the present example. Suppose H is a common cause of F and G and neither F nor G caused the other. Then H and F are correlated because H causes F, and H and G are correlated because H causes G, which implies F and G are correlated transitively through H. Here is a more detailed explanation. For the sake of example, suppose h_1 is a value of H that has a causal influence on F taking value f_1 and on G taking value g_1. Then if F had value f_1, each of its causes would become more probable because one of them should be responsible. So $P(h_1|f_1) > P(h_1)$. Now since the probability of h_1 has gone up, the probability of g_1 would also go up because h_1 causes g_1.

---

[2] There is no evidence that either finasteride or apprehension about the lack of hair regrowth causes hypertension. This is only for the sake of illustration.

[3] This could happen if our sample is a **convenience sample**, which is a sample where the participants are selected at the convenience of the researcher. The researcher makes no attempt to insure that the sample is an accurate representation of the larger population. In the context of the current example, this might be the case if it is convenient for the researcher to observe males hospitalized for hypertension.

[4] Unfortunately, his hairline did not go back down in 2003.

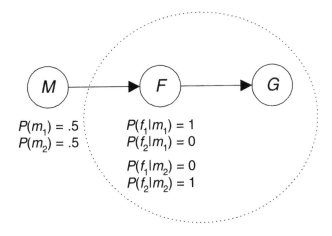

Figure 3.11: An RCE investigating whether F causes G.

Therefore, $P(g_1|f_1) > P(g_1)$, which means F and G are correlated.

Merck's Manipulation Study Since Merck could not conclude finasteride causes hair regrowth from their mere correlation alone, they did a manipulation study to test this conjecture. The study was done on 1879 men aged 18 to 41 with mild to moderate hair loss of the vertex and anterior mid-scalp areas. Half of the men were given 1 mg of finasteride, while the other half were given 1 mg of a placebo. The following table shows the possible values of the variables in the study, including the manipulation variable M:

Variable	Value	When the Variable Takes This Value
F	f_1	Subject takes 1 mg of finasteride.
	f_2	Subject takes 1 mg of a placebo.
G	g_1	Subject has significant hair regrowth.
	g_2	Subject does not have significant hair regrowth.
M	m_1	Subject is chosen to take 1mg of finasteride.
	m_2	Subject is chosen to take 1mg of a placebo.

An RCE used to test the conjecture that F causes G is shown in Figure 3.11. There is an oval around the system being studied (F and G and their possible causal relationship) to indicate that the manipulation comes from outside the system. The edges in Figure 3.11 represent causal influences. The RCE supports the conjecture that F causes G to the extent that the data support $P(g_1|m_1) \neq P(g_1|m_2)$. Merck decided that "significant hair regrowth" would be judged according to the opinion of independent dermatologists. A panel of independent dermatologists evaluated photos of the men after 24 months of treatment. The panel judged that significant hair regrowth was demonstrated in 66% of men treated with finasteride compared to 7% of men treated with placebo. Basing our probability on these results, we have $P(g_1|m_1) \approx .67$ and

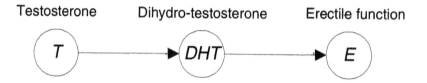

Figure 3.12: A causal DAG.

$P(g_1|m_2) \approx .07$. In a more analytical analysis, only 17% of men treated with finasteride demonstrated hair loss (defined as any decrease in hair count from baseline). In contrast, 72% of the placebo group lost hair, as measured by hair count. Merck concluded that finasteride does indeed cause hair regrowth and on Dec. 22, 1997, announced that the U.S. Food and Drug Administration granted marketing clearance to Propecia(TM) (finasteride 1 mg) for treatment of male pattern hair loss (androgenetic alopecia), for use in men only (see [McClennan and Markham, 1999] for more on this).

3.3.2   Causality and the Markov Condition

First, we more rigorously define a causal DAG. After that we state the causal Markov assumption and argue why it should be satisfied.

Causal DAGs

A **causal graph** is a directed graph containing a set of causally related random variables V such that for every $X, Y \in$ V there is an edge from X to Y if and only if X is a direct cause of Y. By a **direct cause** we mean a manipulation of X results in a change in the probability distribution of Y, and there is no subset of variables W of the set of variables in the graph such that if we knew the values of the variables in W, a manipulation of X would no longer change the probability distribution of Y. A causal graph is a **causal DAG** if the directed graph is acyclic (i.e., there are no causal feedback loops).

Example 3.5 *Testosterone (T) is known to convert to dihydro-testosterone (DHT), and DHT is believed to be the hormone necessary for erectile function (E). A study in [Lugg et al., 1995] tested the causal relationship among these variables in rats. They manipulated testosterone to low levels and found that both DHT and erectile function declined. They then held DHT fixed at low levels and found that erectile function was low regardless of the manipulated value of testosterone. Finally, they held DHT fixed at high levels and found that erectile function was high regardless of the manipulated value of testosterone. So they learned that, in a causal graph containing only the variables T, DHT, and E, T is a direct cause of DHT, DHT is a direct cause of E, but, although T is a cause of E, it is not a direct cause. So the causal graph (DAG) is the one in Figure 3.12.*

Notice that if the variable DHT were not in the DAG in Figure 3.12, there would be an edge from T directly into E instead of the directed path through DHT. In general, our edges always represent only the relationships among the identified variables. It seems we can usually conceive of intermediate, unidentified variables along each edge. Consider the following example taken from [Spirtes et al., 1993, 2000]; [p. 42]:

> If C is the event of striking a match, and A is the event of the match catching on fire, and no other events are considered, then C is a direct cause of A. If, however, we added B; the sulfur on the match tip achieved sufficient heat to combine with the oxygen, then we could no longer say that C directly caused A, but rather C directly caused B and B directly caused A. Accordingly, we say that B is a causal mediary between C and A if C causes B and B causes A.

Note that, in this intuitive explanation, a variable name is used to stand also for a value of the variable. For example, A is a variable whose value is *on-fire* or *not-on-fire*, and A is also used to represent that the match is on fire. Clearly, we can add more causal mediaries. For example, we could add the variable D, representing whether the match tip is abraded by a rough surface. C would then cause D, which would cause B, etc. We could go much further and describe the chemical reaction that occurs when sulfur combines with oxygen. Indeed, it seems we can conceive of a continuum of events in any causal description of a process. We see then that the set of observable variables is observer dependent. Apparently, an individual, given a myriad of sensory input, selectively records discernible events and develops cause/effect relationships among them. Therefore, rather than assuming that there is a set of causally related variables out there, it seems more appropriate to only assume that, in a given context or application, we identify certain variables and develop a set of causal relationships among them.

The Causal Markov Assumption

If we assume the observed probability distribution P of a set of random variables V satisfies the Markov condition with the causal DAG $\mathbb{G}$ containing the variables, we say we are making the **causal Markov assumption**, and we call $(\mathbb{G}, P)$ a **causal network**. Why should we make the causal Markov assumption? To answer this question we show several examples.

Example 3.6 *Consider again the situation involving testosterone (T), DHT, and erectile function (E). The manipulation study in [Lugg et al., 1995] showed that if we instantiate DHT, the value of E is independent of the value of T. So there is experimental evidence that the Markov condition is satisfied for a three-variable causal chain.*

Example 3.7 *A history of smoking (H) is known to cause both bronchitis (B) and lung cancer (L). Lung cancer and bronchitis both cause fatigue (F),*

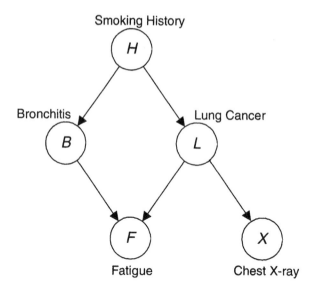

Figure 3.13: A causal DAG.

while only lung cancer can cause a chest X-ray (X) to be positive. There are no other causal relationships among the variables. Figure 3.13 shows a causal DAG containing these variables. The causal Markov assumption for that DAG entails the following conditional independencies:

Node	Parents	Nondescendents	Conditional Independency
H	$\varnothing$	$\varnothing$	None
B	H	L, X	$I_P(B, \{L, X\}|H)$
L	H	B	$I_P(L, B|H)$
F	B, L	H, X	$I_P(F, \{H, X\}|\{B, L\})$
X	L	H, B, F	$I_P(X, \{H, B, F\}|L)$

Given the causal relationship in Figure 3.13, we would not expect bronchitis and lung cancer to be independent because if someone had lung cancer it would make it more probable that they smoked (since smoking can cause lung cancer), which would make it more probable that another effect of smoking, namely bronchitis, was present. However, if we knew someone smoked, it would already be more probable that the person had bronchitis. Learning that they had lung cancer could no longer increase the probability of smoking (which is now 1), which means it cannot change the probability of bronchitis. That is, the variable H shields B from the influence of L, which is what the causal Markov condition says. Similarly, a positive chest X-ray increases the probability of lung cancer, which in turn increases the probability of smoking, which in turn increases the probability of bronchitis. So a chest X-ray and bronchitis are not independent. However, if we knew the person had lung cancer, the chest X-ray could not change the probability of lung cancer and thereby change the probability of bronchitis. So B is independent of X conditional on L, which is what the causal Markov condition says.

In summary, if we create a causal graph containing the variables X and Y, if X and Y do not have a hidden common cause (i.e. a cause that is not in our graph), if there are no causal paths from Y back to X (i.e. our graph is a DAG), and if we do not have selection bias (i.e. our probability distribution is not obtained from a population in which a common effect is instantiated to the same value for all members of the population), then we feel X and Y are independent if we condition on a set of variables including at least one variable in each of the causal paths from X to Y. Since the set of all parents of Y is such a set, we feel that the Markov condition holds relative to X and Y. So we conclude that the causal Markov assumption is justified for a causal graph if the following conditions are satisfied:

1. There are no hidden common causes. That is, all common causes are represented in the graph.

2. There are no causal feedback loops. That is, our graph is a DAG.

3. Selection bias is not present.

Note that, for the Markov condition to hold, there must an edge from X to Y whenever there is a causal path from X to Y besides the ones containing variables in our graph. However, we need not stipulate this requirement because it is entailed by the definition of a causal graph. Recall that in a causal graph there is an edge from X to Y if X is a direct cause of Y.

Perhaps the condition that is most frequently violated is that there can be no hidden common causes. We discuss this condition further with a final example.

Example 3.8 *Suppose we wanted to create a causal DAG containing the variables cold (C), sneezing (S), and runny nose (R). Since a cold can cause both sneezing and a runny nose and neither of these conditions can cause each other, we would create the DAG in Figure 3.14 (a). The causal Markov condition for that DAG would entail $I_P(S, R|C)$. However, if there were a hidden common cause of S and R as depicted in Figure 3.14 (b), this conditional independency would not hold because even if the value of C were known, S would change the probability of H, which in turn would change the probability of R. Indeed, there is at least one other cause of sneezing and runny nose, namely hay fever. So when making the causal Markov assumption, we must be certain that we have identified all common causes.*

3.3.3 The Markov Condition without Causality

We have argued that a causal DAG often satisfies the Markov condition with the joint probability distribution of the random variables in the DAG. This does not mean that the edges in a DAG in a Bayesian network must be causal. That

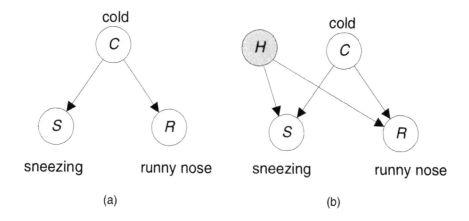

Figure 3.14: The causal Markov assumption would not hold for the DAG in (a) if there is a hidden common cause as depicted in (b).

is, a DAG can satisfy the Markov condition with the probability distribution of the variables in the DAG without the edges being causal. For example, we showed that the joint probability distribution P of the random variables L, S, and C defined on the set of objects in Figure 3.4 satisfies the Markov condition with the DAG $\mathbb{G}$ in Figure 3.5. However, we would not argue that the color of the objects causes their shape or the letter that is on them. As another example, if we reversed the edges in the DAG in Figure 3.12 to obtain the DAG $E \rightarrow DHT \rightarrow T$, the new DAG would also satisfy the Markov condition with the probability distribution of the variables, yet the edges would not be causal.

3.4   Inference in Bayesian Networks

As noted previously, a standard application of Bayes' Theorem is inference in a two-node Bayesian network. Larger Bayesian networks address the problem of representing the joint probability distribution of a large number of variables. For example, Figure 3.2, which appears again as Figure 3.15, represents the joint probability distribution of variables related to credit card fraud. Inference in this network consists of computing the conditional probability of some variable (or set of variables) given that other variables are instantiated to certain values. For example, we may want to compute the probability of credit card fraud given gas has been purchased, jewelry has been purchased, and the card holder is male. To accomplish this inference we need sophisticated algorithms. First, we show simple examples illustrating how one of these algorithms uses the Markov condition and Bayes' Theorem to do inference. Then we reference papers describing some of the algorithms. Finally, we show examples of using the algorithms to do inference.

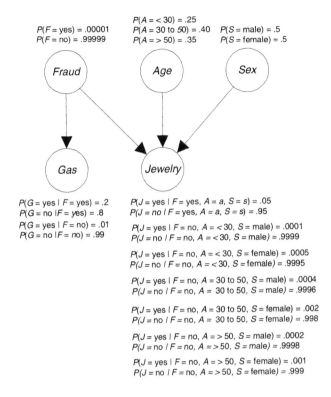

Figure 3.15: A Bayesian network for detecting credit card fraud.

3.4.1 Examples of Inference

Next we present some examples illustrating how the conditional independencies entailed by the Markov condition can be exploited to accomplish inference in a Bayesian network.

Example 3.9 *Consider the Bayesian network in Figure 3.16 (a). The prior probabilities of all variables can be computed using the law of total probability:*

$$P(y_1) = P(y_1|x_1)P(x_1) + P(y_1|x_2)P(x_2) = (.9)(.4) + (.8)(.6) = .84$$

$$P(z_1) = P(z_1|y_1)P(y_1) + P(z_1|y_2)P(y_2) = (.7)(.84) + (.4)(.16) = .652$$

$$P(w_1) = P(w_1|z_1)P(z_1) + P(w_1|z_2)P(z_2) = (.5)(.652) + (.6)(.348) = .5348.$$

These probabilities are shown in Figure 3.16 (b). Note that the computation for each variable requires information determined for its parent. We can therefore consider this method a message-passing algorithm in which each node passes its child a message needed to compute the child's probabilities. Clearly, this algorithm applies to an arbitrarily long linked list and to trees.

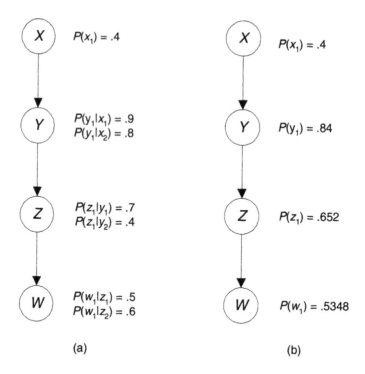

Figure 3.16: A Bayesian network appears in (a), and the prior probabilities of the variables in that network are shown in (b). Each variable only has two values, so only the probability of one is shown in (a).

Example 3.10 *Suppose now that X is instantiated for x_1. Since the Markov condition entails that each variable is conditionally independent of X given its parent, we can compute the conditional probabilities of the remaining variables by again using the law of total probability (however, now with the background information that $X = x_1$) and passing messages down as follows:*

$$P(y_1|x_1) \quad = \quad .9$$

$$
\begin{aligned}
P(z_1|x_1) \quad &= \quad P(z_1|y_1, x_1)P(y_1|x_1) + P(z_1|y_2, x_1)P(y_2|x_1) \\
&= \quad P(z_1|y_1)P(y_1|x_1) + P(z_1|y_2)P(y_2|x_1) \qquad // \text{ Markov condition} \\
&= \quad (.7)(.9) + (.4)(.1) = .67
\end{aligned}
$$

$$
\begin{aligned}
P(w_1|x_1) \quad &= \quad P(w_1|z_1, x_1)P(z_1|x_1) + P(w_1|z_2, x_1)P(z_2|x_1) \\
&= \quad P(w_1|z_1)P(z_1|x_1) + P(w_1|z_2)P(z_2|x_1) \\
&= \quad (.5)(.67) + (.6)(1 - .67) = .533.
\end{aligned}
$$

Clearly, this algorithm also applies to an arbitrarily long linked list and to trees.

The preceding example shows how we can use downward propagation of messages to compute the conditional probabilities of variables below the instantiated variable. Next, we illustrate how to compute conditional probabilities of variables above the instantiated variable.

Example 3.11 *Suppose W is instantiated for w_1 (and no other variable is instantiated). We can use upward propagation of messages to compute the conditional probabilities of the remaining variables. First, we use Bayes' Theorem to compute $P(z_1|w_1)$:*

$$P(z_1|w_1) = \frac{P(w_1|z_1)P(z_1)}{P(w_1)} = \frac{(.5)(.652)}{.5348} = .6096.$$

Then to compute $P(y_1|w_1)$, we again apply Bayes' Theorem:

$$P(y_1|w_1) = \frac{P(w_1|y_1)P(y_1)}{P(w_1)}.$$

We cannot yet complete this computation because we do not know $P(w_1|y_1)$. We can obtain this value using downward propagation as follows:

$$P(w_1|y_1) = (P(w_1|z_1)P(z_1|y_1) + P(w_1|z_2)P(z_2|y_1).$$

After doing this computation, also computing $P(w_1|y_2)$ (because X will need this value), and then determining $P(y_1|w_1)$, we pass $P(w_1|y_1)$ and $P(w_1|y_2)$ to X. We then compute $P(w_1|x_1)$ and $P(x_1|w_1)$ in sequence:

$$P(w_1|x_1) = (P(w_1|y_1)P(y_1|x_1) + P(w_1|y_2)P(y_2|x_1)$$

$$P(x_1|w_1) = \frac{P(w_1|x_1)P(x_1)}{P(w_1)}.$$

It is left as an exercise to perform these computations. Clearly, this upward propagation scheme applies to an arbitrarily long linked list.

The next example shows how to turn corners in a tree.

Example 3.12 *Consider the Bayesian network in Figure 3.17. Suppose W is instantiated for w_1. We compute $P(y_1|w_1)$ followed by $P(x_1|w_1)$ using the upward propagation algorithm just described. Then we proceed to compute $P(z_1|w_1)$ followed by $P(t_1|w_1)$ using the downward propagation algorithm. It is left as an exercise to do this.*

3.4.2 Inference Algorithms and Packages

By exploiting local independencies as we did in the previous subsection, Pearl [1986, 1988] developed a message-passing algorithm for inference in Bayesian networks. Based on a method originated in [Lauritzen and Spiegelhalter, 1988], Jensen et al. [1990] developed an inference algorithm that involves the extraction of an undirected triangulated graph from the DAG in a Bayesian network

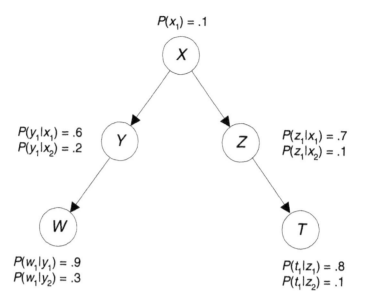

Figure 3.17: A Bayesian network. Each variable only has two possible values; so only the probability of one is shown.

and the creation of a tree whose vertices are the cliques of this triangulated graph. Such a tree is called a junction tree. Conditional probabilities are then computed by passing messages in the junction tree. Li and D'Ambrosio [1994] took a different approach. They developed an algorithm which approximates finding the optimal way to compute marginal distributions of interest from the joint probability distribution. They call this symbolic probabilistic inference (SPI).

All these algorithms are worst-case nonpolynomial time. This is not surprising since the problem of inference in Bayesian networks has been shown to be NP-hard [Cooper, 1990]. In light of this result, approximation algorithms for inference in Bayesian networks have been developed. One such algorithm, likelihood weighting, was developed independently in [Fung and Chang, 1990] and [Shachter and Peot, 1990]. It is proven in [Dagum and Luby, 1993] that the problem of approximate inference in Bayesian networks is also NP-hard. However, there are restricted classes of Bayesian networks which are provably amenable to a polynomial-time solution (see [Dagum and Chavez, 1993]). Indeed, a variant of the likelihood weighting algorithm, which is worst-case polynomial time as long as the network does not contain extreme conditional probabilities, appears in [Pradham and Dagum, 1996].

Practitioners need not concern themselves with all these algorithms as a number of packages for doing inference in Bayesian networks have been developed. A few of them are shown below.

1. Netica (www.norsys.com)

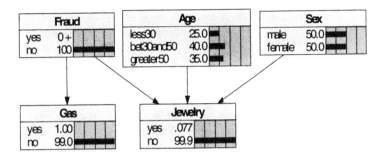

Figure 3.18: The fraud detection Bayesian network in Figure 3.15 implemented using Netica.

2. GeNIe (http://genie.sis.pitt.edu/)

3. HUGIN (http://www.hugin.dk)

4. Elvira (http://www.ia.uned.es/~elvira/index-en.html)

5. BUGS (http://www.mrc-bsu.cam.ac.uk/bugs/)

In this book we use Netica to do inference. You can download a free version that allows up to 13 nodes in a network.

3.4.3 Inference Using Netica

Next, we illustrate inference in a Bayesian network using Netica. Figure 3.18 shows the fraud detection network in Figure 3.15 implemented using Netica. Note that Netica computes and shows the prior probabilities of the variables rather than showing the conditional probability distributions. Probabilities are shown as percentages. For example, the fact that there is a .077 next to *yes* in the *Jewelry* node means

$$P(Jewelry = yes) = .00077.$$

This is the prior probability of a jewelry purchase in the past 24 hours being charged to any particular credit card.

After variables are instantiated, Netica shows the conditional probabilities of the other variables given these instantiations. In Figure 3.19 (a) we instantiated *Age* to *less*30 and *Sex* to *male*. So the fact that there is .010 next to *yes* in the *Jewelry* node means

$$P(Jewelry = yes|Age = less30, Sex = male) = .00010.$$

Notice that the probability of *Fraud* has not changed. This is what we would expect. First, the Markov condition says that *Fraud* should be independent of *Age* and *Sex*. Second, it seems they should be independent. That is, the fact

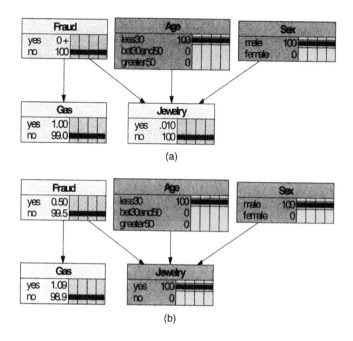

Figure 3.19: In (a) *Age* has been instantiated to *less*30 and *Sex* has been instantiated to *male*. In (b) *Age* has been instantiated to *less*30, *Sex* has been instantiated to *male*, and *Jewelry* has been instantiated to *yes*.

that the card holder is a young man should not make it more or less likely that the card is being used fraudulently. Figure 3.19 (b) has the same instantiations as Figure 3.19 (a) except that we have also instantiated *Jewelry* to *yes*. Notice that the probability of *Fraud* has now changed. First, the jewelry purchase makes *Fraud* more likely to be *yes*. Second, the fact that the card holder is a young man means it is less likely the card holder would make the purchase, thereby making *Fraud* even more likely to be *yes*.

In Figures 3.20 (a) and (b) *Gas* and *Jewelry* have both been instantiated to *yes*. However, in Figure 3.20 (a) the card holder is a young man, while in Figure 3.20 (b) it is an older woman. This illustrates discounting of the jewelry purchase. When the card holder is a young man the probability of *Fraud* being *yes* is high (.0909). However, when it is an older woman, it is still low (.0099) because the fact that the card holder is an older woman explains away the jewelry purchase.

3.5   How Do We Obtain the Probabilities?

So far we have simply shown the conditional probability distributions in the Bayesian networks we have presented. We have not been concerned with how

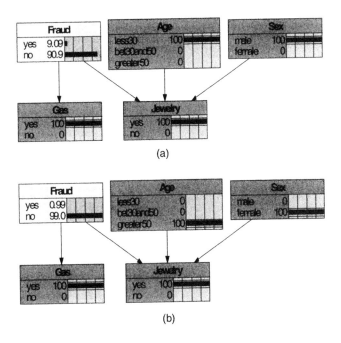

Figure 3.20: *Sex* and *Jewelry* have both been instantiated to *yes* in both (a) and (b). However, in (a) the card holder is a young man, while in (b) it is an older woman.

we obtained them. For example, in the credit card fraud example we simply stated that $P(Age = less30) = .25$. However, how did we obtain this and other probabilities? As mentioned at the beginning of this chapter, they can either be obtained from the subjective judgements of an expert in the area, or they can be learned from data. In Chapter 4 we discuss techniques for learning them from data, while in Parts II and III we show examples of obtaining them from experts and of learning them from data. Here, we show two techniques for simplifying the process of ascertaining them. The first technique concerns the case where a node has multiple parents, while the second technique concerns nodes that represent continuous random variables.

3.5.1 The Noisy OR-Gate Model

After discussing a problem in obtaining the conditional probabilities when a node has multiple parents, we present models that address this problem.

Difficulty Inherent in Multiple Parents

Suppose lung cancer, bronchitis, and tuberculosis all cause fatigue, and we need to model this relationship as part of a system for medical diagnosis. The portion of the DAG concerning only these four variables appears in Figure 3.21.

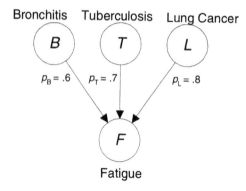

Figure 3.21: We need to assess eight conditional probabilities for node F.

We need to assess eight conditional probabilities for node F, one for each of the eight combinations of that node's parents. That is, we need to assess the following:

$$P(F = yes | B = no, T = no, L = no)$$

$$P(F = yes | B = no, T = no, L = yes)$$

$$\cdots$$

$$P(F = yes | B = yes, T = yes, L = yes).$$

It would be quite difficult to obtain these values either from data or from an expert physician. For example, to obtain the value of $P(F = yes | B = yes, T = yes, L = no)$ directly from data, we would need a sufficiently large population of individuals who are known to have both bronchitis and tuberculosis, but not lung cancer. To obtain this value directly from an expert, the expert would have to be familiar with the likelihood of being fatigued when two diseases are present and the third is not. Next, we show a method for obtaining these conditional probabilities in an indirect way.

The Basic Noisy OR-Gate Model

The noisy OR-gate model concerns the case where the relationships between variables ordinarily represent causal influences, and each variable has only two values. The situation shown in Figure 3.21 is a typical example. Rather than assessing all eight probabilities, we assess the causal strength of each cause for its effect. The **causal strength** is the probability of the cause resulting in the effect whenever the cause is present. In Figure 3.21 we have shown the causal strength p_B of bronchitis for fatigue to be .6. *The assumption is that bronchitis will always result in fatigue unless some unknown mechanism inhibits this from taking place,* and this inhibition takes place 40% of the time. So 60% of the time bronchitis will result in fatigue. Presently, *we assume that all causes of* *the effect are articulated in the DAG*, and the effect cannot occur unless at least one of its causes is present. In this case, mathematically we have

$$p_B = P(F = yes | B = yes, T = no, L = no).$$

The causal strengths of tuberculosis and lung cancer for fatigue are also shown in Figure 3.21. These three causal strengths should not be as difficult to ascertain as all eight conditional probabilities. For example, to obtain p_B from data we only need a population of individuals who have lung bronchitis and do not have the other diseases. To obtain p_B from an expert, the expert need only ascertain the frequency with which bronchitis gives rise to fatigue.

We can obtain the eight conditional probabilities we need from the three causal strengths if we make one additional assumption. *We need to assume that the mechanisms that inhibit the causes act independently from each other.* For example, the mechanism that inhibits bronchitis from resulting in fatigue acts independently from the mechanism that inhibits tuberculosis from resulting in fatigue. Mathematically, this assumption is as follows:

$$\begin{aligned} P(F = no | B = yes, T = yes, L = no) &= (1 - p_B)(1 - p_T) \\ &= (1 - .6)(1 - .7) = .12. \end{aligned}$$

Note that in the previous equality we are conditioning on bronchitis and tuberculosis both being present and lung cancer being absent. In this case, fatigue should occur unless the causal effects of bronchitis and tuberculosis are both inhibited. Since we have assumed these inhibitions act independently, the probability that both effects are inhibited is the product of the probabilities that each is inhibited, which is $(1 - p_B)(1 - p_T)$.

In this same way, if all three causes are present, we have

$$\begin{aligned} P(F = no | B = yes, T = yes, L = yes) &= (1 - p_B)(1 - p_T)(1 - p_L) \\ &= (1 - .6)(1 - .7)(1 - .8) = .024. \end{aligned}$$

Notice that when more causes are present, it is less probable that fatigue will be absent. This is what we would expect. In the following example we compute all eight conditional probabilities needed for node F in Figure 3.21.

Example 3.13 *Suppose we make the assumptions in the noisy OR-gate model, and the causal strengths of bronchitis, tuberculosis, and lung cancer for fatigue are the ones shown in Figure 3.21. Then*

$$P(F = no | B = no, T = no, L = no) = 1$$

$$\begin{aligned} P(F = no | B = no, T = no, L = yes) &= (1 - p_L) \\ &= (1 - .8) = .2 \end{aligned}$$

$$\begin{aligned} P(F = no | B = no, T = yes, L = no) &= (1 - p_T) \\ &= (1 - .7) = .3 \end{aligned}$$

$$P(F = no | B = no, T = yes, L = yes) = (1 - p_T)(1 - p_L)$$
$$= (1 - .7)(1 - .8) = .06$$

$$P(F = no | B = yes, T = no, L = no) = (1 - p_B)$$
$$= (1 - .6) = .4$$

$$P(F = no | B = yes, T = no, L = yes) = (1 - p_B)(1 - p_L)$$
$$= (1 - .6)(1 - .8) = .08$$

$$P(F = no | B = yes, T = yes, L = no) = (1 - p_B)(1 - p_T)$$
$$= (1 - .6)(1 - .7) = .12$$

$$P(F = no | B = yes, T = yes, L = yes) = (1 - p_B)(1 - p_T)(1 - p_L)$$
$$= (1 - .6)(1 - .7)(1 - .8) = .024.$$

Note that since the variables are binary, these are the only values we need to ascertain. The remaining probabilities are uniquely determined by these. For example,

$$P(F = yes | B = yes, T = yes, L = yes) = 1 - .024 = .976.$$

Although we illustrated the model for three causes, it clearly extends to an arbitrary number of causes. We showed the assumptions in the model in italics when we introduced them. Next, we summarize them and show the general formula.

The **noisy OR-gate model** makes the following three assumptions:

1. **Causal inhibition:** This assumption entails that there is some mechanism which inhibits a cause from bringing about its effect, and the presence of the cause results in the presence of the effect if and only if this mechanism is disabled (turned off).

2. **Exception independence:** This assumption entails that the mechanism that inhibits one cause is independent of the mechanism that inhibits other causes.

3. **Accountability:** This assumption entails that an effect can happen only if at least one of its causes is present and is not being inhibited.

The **general formula for the noisy OR-gate model** is as follows: Suppose Y has n causes $X_1, X_2, \ldots, X_n$, all variables are binary, and we assume the noisy OR-gate model. Let p_i be the causal strength of X_i for Y. That is,

$$p_i = P(Y = yes | X_1 = no, X_2 = no, \ldots X_i = yes, \ldots X_n = no).$$

Then if X is a set of nodes that are instantiated to yes,

$$P(Y = no | \mathsf{X}) = \prod_{i \text{ such that } X_i \in \mathsf{X}} (1 - p_i).$$

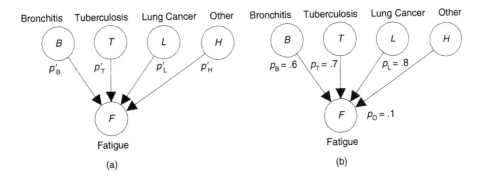

Figure 3.22: The probabilities in (a) are the causal strengths in the noisy OR-gate model. The probabilities in (b) are the ones we ascertain.

The Leaky Noisy OR-Gate Model

Of the three assumptions in the noisy OR-gate model, the assumption of accountability seems to be justified least often. For example, in the case of fatigue there are certainly other causes of fatigue such as listening to a lecture by Professor Neapolitan. So the model in Figure 3.21 does not contain all causes of fatigue, and the assumption of accountability is not justified. It seems in many, if not most, situations we would not be certain that we have elaborated all known causes of an effect. Next, we show a version of the model that does not assume accountability. The derivation of the formula for this model is not simple and intuitive like the one for the basic noisy OR-gate model. So we first present the model without deriving it and then show the derivation.

The Leaky Noisy OR-Gate Formula The leaky noisy OR-gate model assumes that all causes that have not been articulated can be grouped into one other cause H and that the articulated causes, along with H, satisfy the three assumptions in the noisy OR-gate model. This is illustrated for the fatigue example in Figure 3.22 (a). The probabilities in that figure are the causal strengths in the noisy OR-gate model. For example,

$$p'_B = P(F = yes | B = yes, T = no, L = no, H = no).$$

We could not ascertain these values because we do know whether or not H is present. The probabilities in Figure 3.22 (b) are the ones we actually ascertain. For each of the three articulated causes, the probability shown is the probability the effect is present given the remaining two articulated causes are not present. For example,

$$p_B = P(F = yes | B = yes, T = no, L = no).$$

Note the difference in the probabilities p'_B and p_B. The latter one does not condition on a value of H, while the former one does. The probability p_0 is different from the other probabilities. It is the probability that the effect will be present given none of the articulated causes are present. That is,

$$p_0 = P(F = yes | B = no, T = no, L = no).$$

Note again that we are not conditioning on a value of H.

The **general formula for the leaky noisy OR-gate model** is as follows (a derivation appears in the next subsection): Suppose Y has n causes $X_1, X_2, \ldots, X_n$, all variables are binary, and we assume the leaky noisy OR-gate model. Let

$$p_i = P(Y = yes | X_1 = no, X_2 = no, \ldots X_i = yes, \ldots X_n = no) \qquad (3.1)$$

$$p_0 = P(Y = yes | X_1 = no, X_2 = no, \ldots X_n = no). \qquad (3.2)$$

Then if X is a set of nodes that are instantiated to yes,

$$P(Y = no | \mathsf{X}) = (1 - p_0) \prod_{i \text{ such that } X_i \in \mathsf{X}} \frac{1 - p_i}{1 - p_0}.$$

Example 3.14 *Let's compute the conditional probabilities for the network in Figure 3.22 (b). We have*

$$
\begin{aligned}
P(F = no | B = no, T = no, L = no) &= 1 - p_0 \\
&= 1 - .1 = .9
\end{aligned}
$$

$$
\begin{aligned}
P(F = no | B = no, T = no, L = yes) &= (1 - p_0) \frac{1 - p_L}{1 - p_0} \\
&= 1 - .8 = .2
\end{aligned}
$$

$$
\begin{aligned}
P(F = no | B = no, T = yes, L = no) &= (1 - p_0) \frac{1 - p_T}{1 - p_0} \\
&= 1 - .7 = .3
\end{aligned}
$$

$$
\begin{aligned}
P(F = no | B = no, T = yes, L = yes) &= (1 - p_0) \frac{1 - p_T}{1 - p_0} \frac{1 - p_L}{1 - p_0} \\
&= \frac{(1 - .7)(1 - .8)}{1 - .1} = .067
\end{aligned}
$$

$$
\begin{aligned}
P(F = no | B = yes, T = no, L = no) &= (1 - p_0) \frac{1 - p_B}{1 - p_0} \\
&= 1 - .6 = .4
\end{aligned}
$$

$$
\begin{aligned}
P(F = no | B = yes, T = no, L = yes) &= (1 - p_0) \frac{1 - p_B}{1 - p_0} \frac{1 - p_L}{1 - p_0} \\
&= \frac{(1 - .6)(1 - .8)}{1 - .1} = .089
\end{aligned}
$$

$$P(F = no|B = yes, T = yes, L = no) \quad = \quad (1 - p_0)\frac{1 - p_B}{1 - p_0}\frac{1 - p_T}{1 - p_0}$$

$$= \quad \frac{(1 - .6)(1 - .7)}{1 - .1} = .133$$

$$P(F = no|B = yes, T = yes, L = yes) \quad = \quad (1 - p_0)\frac{1 - p_B}{1 - p_0}\frac{1 - p_T}{1 - p_0}\frac{1 - p_L}{1 - p_0}$$

$$= \quad \frac{(1 - .6)(1 - 7)(1 - .8)}{(1 - .1)(1 - .1)} = .030.$$

A Derivation of the Formula★   The following lemmas and theorem derive the formula in the leaky noisy OR-gate model.

Lemma 3.1 *Given the assumptions and notation shown above for the leaky noisy OR-gate model,*

$$p_0 = p_H' \times P(H = yes).$$

Proof. *Owing to Equality 3.2, we have*

$$
\begin{aligned}
p_0 \quad &= \quad P(Y = yes|X_1 = no, X_2 = no, \ldots X_n = no) \\
&= \quad P(Y = yes|X_1 = no, X_2 = no, \ldots X_n = no, H = yes)P(H = yes) + \\
&\quad\quad P(Y = yes|X_1 = no, X_2 = no, \ldots X_n = no, H = no)P(H = no) \\
&= \quad p_H' \times P(H = yes) + 0 \times P(H = no).
\end{aligned}
$$

This completes the proof. ■

Lemma 3.2 *Given the assumptions and notation shown above for the leaky noisy OR-gate model,*

$$1 - p_i' = \frac{1 - p_i}{1 - p_0}.$$

Proof. *Owing to Equality 3.1, we have*

$1 - p_i$

$$
\begin{aligned}
&= \quad P(Y = no|X_1 = no, \ldots X_i = yes, \ldots X_n = no) \\
&= \quad P(Y = no|X_1 = no, \ldots X_i = yes, \ldots X_n = no, H = yes)P(H = yes) + \\
&\quad\quad P(Y = no|X_1 = no, \ldots X_i = yes, \ldots X_n = no, H = no)P(H = no) \\
&= \quad (1 - p_i')(1 - p_H')P(H = yes) + (1 - p_i')P(H = no) \\
&= \quad (1 - p_i')(1 - p_H' \times P(H)) \\
&= \quad (1 - p_i')(1 - p_0).
\end{aligned}
$$

The last equality is due to Lemma 3.1. This completes the proof. ■

Theorem 3.2 *Given the assumptions and notation shown above for the leaky noisy OR-gate model,*

$$P(Y = no|\mathsf{X}) = (1 - p_0) \prod_{i \text{ such that } X_i \in \mathsf{X}} \frac{1 - p_i}{1 - p_0}.$$

Proof. We have

$$
\begin{aligned}
P(Y = no|\mathsf{X}) &= P(Y = no|\mathsf{X}, H = yes)P(H = yes) + \\
&\qquad P(Y = no|\mathsf{X}, H = yes)P(H = yes) \\
&= P(H = yes)(1 - p'_H) \prod_{i \text{ such that } X_i \in \mathsf{X}} (1 - p'_i) + \\
&\qquad P(H = no) \prod_{i \text{ such that } X_i \in \mathsf{X}} (1 - p'_i) \\
&= (1 - p_0) \prod_{i \text{ such that } X_i \in \mathsf{X}} (1 - p'_i) \\
&= (1 - p_0) \prod_{i \text{ such that } X_i \in \mathsf{X}} \frac{1 - p_i}{1 - p_0}.
\end{aligned}
$$

The second to the last equality is due to Lemma 3.1, and the last is due to Lemma 3.2. ∎

Further Models

A generalization of the noisy OR-gate model to the case of more than two values appears in [Srinivas, 1993]. Diez and Druzdzel [2006] propose a general framework for canonical models, classifying them into three categories: deterministic, noisy, and leaky. They then analyze the most common families of canonical models, namely the noisy OR/MAX, the noisy AND/MIN, and the noisy XOR. Other models for succinctly representing the conditional distributions use the **sigmoid** function [Neal, 1992] and the **logit** function [McLachlan and Krishnan, 1997]. Another approach to reducing the number of parameter estimates is the use of **embedded Bayesian networks**, which is discussed in [Heckerman and Meek, 1997].

3.5.2   Methods for Discretizing Continuous Variables★

Often, a Bayesian network contains both discrete and continuous random variables. For example, the Bayesian network in Figure 3.2 contains four random variables that are discrete and one, namely *Age*, that is continuous.[5] However, notice that a continuous probability density function for node *Age* does not appear in the network. Rather, the possible values of the node are three ranges for ages, and the probability of each of these ranges is specified in the network. This is called **discretizing** the continuous variables. Although many Bayesian

---

[5]Technically, if we count age only by years it is discrete. However, even in this case, it is usually represented by a continuous distribution because there are so many values.

network inference packages allow the user to specify both continuous variables and discrete variables in the same network, we can sometimes obtain simpler and better inference results by representing the variables as discrete. One reason for this is that, if we discretize the variables, we do not need to assume any particular continuous probability density function. Examples of this appear in Chapter 4, Section 4.7.1, and Chapter 10. Next, we present two of the most popular methods for discretizing continuous random variables.

Bracket Medians Method

In the **Bracket Medians Method** the mass in a continuous probability distribution function $F(x) = P(X \leq x)$ is divided into n equally spaced intervals. The method proceeds as follows ($n = 5$ in this explanation):

1. Determine n equally spaced intervals in the interval $[0, 1]$. If $n = 5$, the intervals are $[0, .2]$, $[.2, .4]$, $[.4, .6]$, $[.6, .8]$, and $[.8, 1.0]$.

2. Determine points x_1, x_2, x_3, x_4, x_5, and x_6 such that

$$
\begin{aligned}
P(X &\leq x_1) = .0 \\
P(X &\leq x_2) = .2 \\
P(X &\leq x_3) = .4 \\
P(X &\leq x_4) = .6 \\
P(X &\leq x_5) = .8 \\
P(X &\leq x_6) = 1.0,
\end{aligned}
$$

   where the values on the right in these equalities are the endpoints of the five intervals.

3. For each interval $[x_i, x_{i+1}]$ compute the bracket median d_i, which is the value such that

$$
P(x_i \leq X \leq d_i) = P(d_i \leq X \leq x_{i+1}).
$$

4. Define the discrete variable D with the following probabilities:

$$
\begin{aligned}
P(D &= d_1) = .2 \\
P(D &= d_2) = .2 \\
P(D &= d_3) = .2 \\
P(D &= d_4) = .2 \\
P(D &= d_5) = .2.
\end{aligned}
$$

Example 3.15 *Recall that the normal density function is given by*

$$
f(x) = \frac{1}{\sqrt{2\pi}\sigma} e^{-\frac{(x-\mu)^2}{2\sigma^2}} \qquad -\infty < x < \infty,
$$

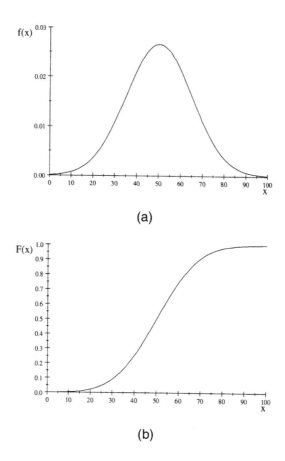

(a)

(b)

Figure 3.23: The normal density function with $\mu = 50$ and $\sigma = 15$ appears in (a), while the corresponding normal cumulative distribution function appears in (b).

where

$$E(X) = \mu \qquad \text{and} \qquad Var(X) = \sigma^2,$$

and the cumulative distribution function for this density function is given by

$$F(x) = P(X \le x) = \frac{1}{\sqrt{2\pi}\sigma} \int_{-\infty}^{x} e^{-\frac{(t-\mu)^2}{2\sigma^2}} dt \qquad -\infty < x < \infty.$$

These functions for $\mu = 50$ and $\sigma = 15$ are shown in Figure 3.23. This might be the distribution of age for some particular population. Next we use the Bracket Medians Method to discretize it into three ranges. Then $n = 3$ and our four steps are as follows:

1. Since there is essentially no mass < 0 or > 100, our three intervals are

$[0, .333]$, $[.333, .666]$, and $[.666, 1]$.

2. We need to find points x_1, x_2, x_3, and x_4 such that

$$
\begin{aligned}
P(X \leq x_1) &= .0 \\
P(X \leq x_2) &= .333 \\
P(X \leq x_3) &= .666 \\
P(X \leq x_4) &= 1.
\end{aligned}
$$

Clearly, $x_1 = 0$ and $x_4 = 100$. To determine x_2 we need to determine

$$x_2 = F^{-1}(.333).$$

Using the mathematics package Maple, we have

$$x_2 = \text{NormalInv}(.333; 50, 15) = 43.5.$$

Similarly,
$$x_3 = \text{NormalInv}(.666; 50, 15) = 56.4.$$

In summary, we have

$$x_1 = 0 \qquad x_2 = 43.5 \qquad x_3 = 56.4 \qquad x_4 = 1.$$

3. Compute the bracket medians. We compute them using Maple by solving the following equations:

$$\text{NormalDist}(d_1; 50, 15)$$

$$= \text{NormalDist}(43.5; 50, 15) - \text{NormalDist}(d_1; 50, 15)$$

Solution is $d_1 = 35.5$.

$$\text{NormalDist}(d_2; 50, 15) - \text{NormalDist}(43.5; 50, 15)$$

$$= \text{NormalDist}(56.4; 50, 15) - \text{NormalDist}(d_2; 50, 15)$$

Solution is $d_2 = 50.0$.

$$\text{NormalDist}(d_3; 50, 15) - \text{NormalDist}(56.4; 50, 15)$$

$$= 1 - \text{NormalDist}(d_3; 50, 15)$$

Solution is $d_3 = 64.5$.

4. Finally, we set

$$
\begin{aligned}
P(D = 35.5) &= .333 \\
P(D = 50.0) &= .333 \\
P(D = 64.5) &= .333.
\end{aligned}
$$

If, for example, a data item's continuous value is between 0 and 43.5, we assign the data item a discrete value of 35.5.

The variable D requires a numeric value if we need to perform computations using it. An example would be if the variable were used in a decision analysis application (Chapter 5). However, if the variable does not require a numeric value for computational purposes, we need not perform Step 3 in the Bracket Medians Method. Rather, we just show ranges as the values of D. In the previous example, we would set

$$P(D = \ < 43.5) = .333$$

$$P(D = 43.5 \text{ to } 56.4) = .333$$

$$P(D = \ > 56.4) = .333.$$

Recall that this is what we did for *Age* in the Bayesian network in Figure 3.2. In this case, if a data item's continuous value is between 0 and 43.5, we simply assign the data item that range.

Pearson-Tukey Method

In some applications we want to give special attention to the case when a data item falls in the tail of a density function. For example, if we are trying to predict whether a company will go bankrupt, then unusually low cash flow is indicative they will, while unusually high cash flow is indicative they will not [McKee and Lensberg, 2002]. Values in the middle are not indicative one way or the other. In such cases we want to group the values in each tail together. The Bracket Medians Method does not do this. However, the Pearson-Tukey Method [Keefer, 1983], which we describe next, does. In Chapter 10 we will discuss the bankruptcy prediction application in detail.

In the **Pearson-Tukey Method** the mass in a continuous probability distribution function $F(x) = P(X \leq x)$ is divided into three intervals. The method proceeds as follows:

1. Determine points x_1, x_2, and x_3 such that

$$P(X \ \leq \ x_1) = .05$$
$$P(X \ \leq \ x_2) = .50$$
$$P(X \ \leq \ x_3) = .95.$$

2. Define the discrete variable D with the following probabilities:

$$P(D \ = \ x_1) = .185$$
$$P(D \ = \ x_2) = .63$$
$$P(D \ = \ x_3) = .185.$$

Example 3.16 *Suppose we have the normal distribution discussed in Example 3.15. Next, we apply the Pearson-Tukey Method to that distribution.*

1. *Using Maple, we have*

$$x_1 = \text{NormalInv}(.05; 50, 15) = 25.3$$

$$x_2 = \text{NormalInv}(.50; 50, 15) = 50$$

$$x_3 = \text{NormalInv}(.95; 50, 15) = 74.7.$$

2. *We set*

$$P(D = 25.3) = .185$$
$$P(D = 50.0) = .63$$
$$P(D = 74.7) = .185.$$

To assign data items discrete values, we need to determine the range of values corresponding to each of the cutoff points. That is, we compute the following:

$$\text{NormalInv}(.185; 50, 15) = 36.6$$

$$\text{NormalInv}(1 - .185; 50, 15) = 63.4.$$

If a data item's continuous value is < 36.6, we assign the data item the value 25.3; if the value is in [36.6, 63.4], we assign the value 50; and if the value is > 63.4, we assign the value 74.7.

Notice that when we used the Pearson-Tukey Method, the middle discrete value represented numbers in the interval [36.6, 63.4], while when we used the Bracket's Median Method, the middle discrete value represented numbers in the interval [43.5, 56.4]. The interval for the Pearson-Tukey Method is larger, meaning more numbers in the middle are treated as the same discrete value, and the other two discrete values represent values only in the tails.

If the variable does not require a numeric value for computational purposes, we need not perform Steps 1 and 2, but rather just determine the range of values corresponding to each of the cutoff points and just show ranges as the values of D. In the previous example, we would set

$$P(D = \ <36.6) = .185$$

$$P(D = 36.6 \text{ to } 63.4) = .63$$

$$P(D = \ >63.4) = .185.$$

In this case if a data item's continuous value is between 0 and 36.6, we simply assign the data item that range.

3.6   Entailed Conditional Independencies★

If $(\mathbb{G}, P)$ satisfies the Markov condition, then each node in $\mathbb{G}$ is conditionally independent of the set of all its nondescendents given its parents. Do these conditional independencies entail any other conditional independencies? That is, if $(\mathbb{G}, P)$ satisfies the Markov condition, are there any other conditional independencies which P must satisfy other than the one based on a node's parents? The answer is yes. Such conditional independencies are called entailed conditional independencies. Specifically, we say a DAG **entails a conditional independency** if every probability distribution, which satisfies the Markov condition with the DAG, must have the conditional independency. Before explicitly showing all entailed conditional independencies, we illustrate that one would expect them.

3.6.1   Examples of Entailed Conditional Independencies

Suppose some distribution P satisfies the Markov condition with the DAG in Figure 3.24 (a). Then we know $I_P(C, \{F, G\}|, B)$ because B is the parent of C, and F and G are nondescendents of C. Furthermore, we know $I_P(B, G|F)$ because F is the parent of B, and G is a nondescendent of B. These are the only conditional independencies according to the statement of the Markov condition. However, can any other conditional independencies be deduced from them? For example, can we conclude $I_P(C, G|F)$? Let's first give the variables meaning and the DAG a causal interpretation to see if we would expect this conditional independency.

Suppose we are investigating how professors obtain citations, and the variables represent the following:

G: Graduate Program Quality
F: First Job Quality
B: Number of Publications
C: Number of Citations.

Further suppose the DAG in Figure 3.24 (a) represents the causal relationships among these variables, and there are no hidden common causes.[6] Then it is reasonable to make the causal Markov assumption, and we would feel that the probability distribution of the variables satisfies the Markov condition with the DAG. Suppose we learn that Professor La Budde attended a graduate program (G) of high quality. We would now expect his first job (F) may well be of high quality, which means that he should have a large number of publications (B), which in turn implies he should have a large number of citations (C). Therefore, we would not expect $I_P(C, G)$.

Suppose we next learn that Professor Pellegrini's first job (F) was of high quality. That is, we instantiate F to "high quality." The cross through the node F in Figure 3.24 (b) indicates it is instantiated. We would now expect his

---

[6]We make no claim that this model accurately represents the causal relationships among the variables. See [Spirtes et al., 1993, 2000] for a detailed discussion of this problem.

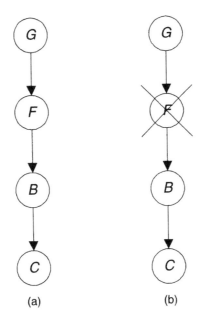

Figure 3.24: A causal DAG is shown in (a). The variable F is instantiated in (b).

number of publications (B) to be large and, in turn, his number of citations (C) to be large. If Professor Pellegrini then tells us he attended a graduate program (G) of high quality, would we expect the number of citations to be even higher than we previously thought? It seems not. The graduate program's high quality implies the number of citations is probably large because it implies the first job is probably of high quality. Once we already know the first job is of high quality, the information concerning the graduate program should be irrelevant to our beliefs concerning the number of citations. Therefore, we would expect C to not only be conditionally independent of G given its parent B, but also its grandparent F. Either one seems to block the dependency between G and C that exists through the chain $[G, F, B, C]$. So we would expect $I_P(C, G|F)$.

It is straightforward to show that the Markov condition does indeed entail $I_P(C, G|F)$ for the DAG $\mathbb{G}$ in Figure 3.24. We show this next. If $(\mathbb{G}, P)$ satisfies the Markov condition,

$$
\begin{aligned}
P(c|g, f) &= \sum_b P(c|b, g, f)P(b|g, f) \\
&= \sum_b P(c|b, f)P(b|f) \\
&= P(c|f).
\end{aligned}
$$

The first equality is due to the law of total probability (in the background space that we know the values of g and f), the second equality is due to the Markov

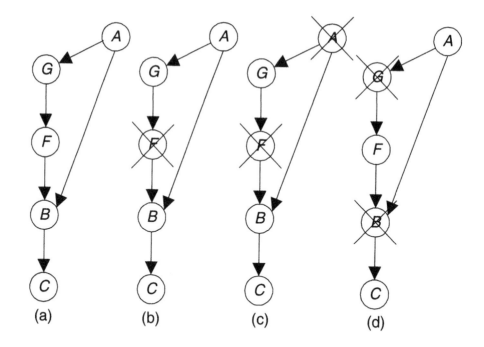

Figure 3.25: A causal DAG is shown in (a). The variable F is instantiated in (b). The variables A and F are instantiated in (c). The variables B and G are instantiated in (d).

condition, and the last equality is again due to the law of total probability.

If we have an arbitrarily long directed linked list of variables, and P satisfies the Markov condition with that list, in the same way as above, we can show that for any variable in the list, the set of variables above it is conditionally independent of the set of variables below it given that variable. That is, the variable blocks the dependency transmitted through the chain.

Suppose now that P does not satisfy the Markov condition with the DAG in Figure 3.24 (a) because there is a common cause A of G and B. For the sake of illustration, let's say A represents the following in the current example:

$$A: \qquad \text{Ability.}$$

Further, suppose there are no other hidden common causes, so we would now expect P to satisfy the Markov condition with the DAG in Figure 3.25 (a). Would we still expect $I_P(C, G|F)$? It seems not. For example, as before, suppose that we initially learn Professor Pellegrini's first job (F) was of high quality. This instantiation is shown in Figure 3.25 (b). We learn next that his graduate program (G) was of high quality. Given the current model, the fact that G is of high quality is indicative of his having high ability (A), which can directly affect his publication rate (B) and therefore his citation rate (C). So we now would feel his citation rate (C) may be even higher than what we thought when we only knew his first job (F) was of high quality. This means we would not feel $I_P(C, G|F)$ as we did with the previous model. Suppose next that we know Professor Pellegrini's first job (F) was of high quality and that he has high ability (A). These instantiations are shown in Figure 3.25 (c). In this case, his attendance at a high-quality graduate program (G) can no longer be indicative of his ability (A), and therefore, it cannot affect our belief concerning his citation rate (C) through the chain $[G, A, B, C]$. That is, this chain is blocked at A. So we would expect $I_P(C, G|\{A, F\})$. Indeed, it is possible to prove that the Markov condition does entail $I_P(C, G|\{A, F\})$.

Finally, consider the conditional independency $I_P(F, A|G)$. This independency is obtained directly by applying the Markov condition to the DAG in Figure 3.25 (a). So we will not offer an intuitive explanation for it. Rather, we discuss whether we would expect the conditional independency to still exist if we also knew the state of B. Suppose we first learn that Professor Georgakis has a high publication rate (B) and attended a high-quality graduate program (G). These instantiations are shown in Figure 3.25 (d). We later learn she also has high ability (A). In this case, her high ability (A) could explain away her high publication rate (B), thereby making it less probable she had a high-quality first job (F). As mentioned in Section 3.1, psychologists call this discounting. So the chain $[A, B, F]$ is opened by instantiating B, and we would not expect $I_P(F, A|\{B, G\})$. Indeed, the Markov condition does not entail $I_P(F, A|\{B, G\})$. Note that the instantiation of C should also open the chain $[A, B, F]$. That is, if we know the citation rate (C) is high, then it is probable the publication rate (B) is high, and each of the causes of B can explain away this high probability. Indeed, the Markov condition does not entail $I_P(F, A|\{C, G\})$ either.

3.6.2 d-Separation

Figure 3.26 shows the chains that can transmit a dependency between variables X and Y in a Bayesian network. To discuss these dependencies intuitively, we give the edges in that figure causal interpretations as follows:

1. The chain $[X, B, C, D, Y]$ is a causal path from X to Y. In general, there is a dependency between X and Y on this chain, and the instantiation of any intermediate cause on the chain blocks the dependency.

2. The chain $[X, F, H, I, Y]$ is a chain in which H is a common cause of X and Y. In general, there is a dependency between any X and Y on this chain, and the instantiation of the common cause H or either of the intermediate causes F and I blocks the dependency.

3. The chain $[X, J, K, L, Y]$ is a chain in which X and Y both cause K. There is no dependency between X and Y on this chain. However, if we instantiate K or M, in general a dependency would be created. We would then need to also instantiate J or L.

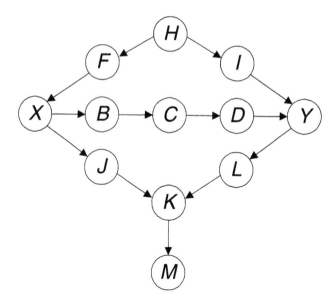

Figure 3.26: This DAG illustrates the chains that can transmit a dependency between X and Y.

To render X and Y conditionally independent we need to instantiate at least one variable on all the chains that transmit a dependency between X and Y. So we would need to instantiate at least one variable on the chain $[X, B, C, D, Y]$, at least one variable on the chain $[X, F, H, I, Y]$, and, if K or M are instantiated, at least one other variable on the chain $[X, J, K, L, Y]$.

Now that we've discussed intuitively how dependencies can be transmitted and blocked in a DAG, we show precisely what conditional independencies are entailed by the Markov condition. To do this, we need the notion of d-separation, which we define shortly. First, we present some preliminary concepts. We say there is a **head-to-head meeting** at X on a chain if the edges incident to X both have their arrows into X. For example, the chain $Y \leftarrow W \rightarrow X \leftarrow V$ has a head-to-head meeting at X. We say there is **head-to-tail meeting** at X on a chain if precisely one of the edges incident to X has its arrows into X. For example, the chain $Y \leftarrow W \leftarrow X \leftarrow V$ has a head-to-tail meeting at X. We say there is **tail-to-tail meeting** at X on a chain if neither of the edges incident to X has its arrows into X. For example, the chain $Y \leftarrow W \leftarrow X \rightarrow V$ has a tail-to-tail meeting at X. We now have the following definition:

Definition 3.2 *Suppose we have a DAG* $\mathbb{G} = (\mathsf{V}, \mathsf{E})$, *a chain* ρ *in the DAG connecting two nodes* X *and* Y, *and a subset of nodes* $\mathsf{W} \subseteq \mathsf{V}$. *We say that the chain* ρ *is **blocked** by* W *if at least one of the following is true:*

1. *There is a node* $Z \in \mathsf{W}$ *that has a head-to-tail meeting on* ρ.

2. There is a node $Z \in W$ that has a tail-to-tail meeting on ρ.

3. There is a node Z, such that Z and all Z's descendents are not in W, that has a head-to-head meeting on ρ.

Example 3.17 *For the DAG in Figure 3.26, the following are some examples of chains that are blocked and that are not blocked:*

1. *The chain* $[X, B, C, D, Y]$ *is blocked by* $W = \{C\}$ *because there is a head-to-tail meeting at* C.

2. *The chain* $[X, B, C, D, Y]$ *is blocked by* $W = \{C, H\}$ *because there is a head-to-tail meeting at* C.

3. *The chain* $[X, F, H, I, Y]$ *is blocked by* $W = \{C, H\}$ *because there is a tail-to-tail meeting at* H.

4. *The chain* $[X, J, K, L, Y]$ *is blocked by* $W = \{C, H\}$ *because there is a head-to-head meeting at* K, *and* K *and* M *are both not in* W.

5. *The chain* $[X, J, K, L, Y]$ *is not blocked by* $W = \{C, H, K\}$ *because there is a head-to-head meeting at* K, *and* K *is not in* W.

6. *The chain* $[X, J, K, L, Y]$ *is blocked by* $W = \{C, H, K, L\}$ *because there is a head-to-tail meeting at* L.

We can now define d-separation.

Definition 3.3 *Suppose we have a DAG* $\mathbb{G} = (V, E)$ *and a subset of nodes* $W \subseteq V$. *Then* X *and* Y *are* **d-separated** *by* W *if every chain between* X *and* Y *is blocked by* W.

Definition 3.4 *Suppose we have a DAG* $\mathbb{G} = (V, E)$ *and three subsets of nodes* X, $Y \subseteq V$, *and* W. *We say* X *and* Y *are d-separated by* W *if for every* $X \in X$ *and* $Y \in Y$, X *and* Y *are d-separated by* W.

As you may have already suspected, d-separation recognizes all the conditional independencies entailed by the Markov condition. Specifically, we have the following theorem:

Theorem 3.3 *Suppose we have a DAG* $\mathbb{G} = (V, E)$ *and three subsets of nodes* X, Y, *and* $W \subseteq V$. *Then* $\mathbb{G}$ *entails the conditional independency* $I_P(X, Y|W)$ *if and only if* X *and* Y *are d-separated by* W.
Proof. *The proof can be found in [Neapolitan, 1990].* ∎

We stated the theorem for sets of variables, but it also applies to single variables. That is, if X contains a single variable X and Y contains a single variable Y, then $I_P(X, Y|W)$ is the same as $I_P(X, Y|W)$. We show examples of this simpler case next and investigate more complex sets in the exercises.

Example 3.18 *Owing to Theorem 3.3, the following are some conditional in-dependencies the Markov condition entails for the DAG in Figure 3.26:*

Conditional Independency	Reason Conditional Independency Is Entailed
$I_P(X,Y\|\{H,C\})$	$[X,F,H,I,Y]$ is blocked at H.
	$[X,B,C,D,Y]$ is blocked at C.
	$[X,J,K,L,Y]$ is blocked at K.
$I_P(X,Y\|\{F,D\})$	$[X,F,H,I,Y]$ is blocked at F.
	$[X,B,C,D,Y]$ is blocked at D.
	$[X,J,K,L,Y]$ is blocked at K.
$I_P(X,Y\|\{H,C,K,L\})$	$[X,F,H,I,Y]$ is blocked at H.
	$[X,B,C,D,Y]$ is blocked at C.
	$[X,J,K,L,Y]$ is blocked at L.
$I_P(X,Y\|\{H,C,M,L\})$	$[X,F,H,I,Y]$ is blocked at H.
	$[X,B,C,D,Y]$ is blocked at C.
	$[X,J,K,L,Y]$ is blocked at L.

In the third row it is necessary to include L to obtain the independency because there is a head-to-head meeting at K on the chain $[X,J,K,L,Y]$, and $K \in \{H,C,K,L\}$. Similarly, in the fourth row, it is necessary to include L to obtain the independency because there is a head-to-head meeting at K on the chain $[X,J,K,L,Y]$, M is a descendent of K, and $M \in \{H,C,M,L\}$.

Example 3.19 *Owing to Theorem 3.3, the following are some conditional in-dependencies the Markov condition does not entail for the DAG in Figure 3.26:*

Conditional Independency	Reason Conditional Independency Is Not Entailed
$I_P(X,Y\|H)$	$[X,B,C,D,Y]$ is not blocked.
$I_P(X,Y\|,D)$	$[X,F,H,I,Y]$ is not blocked.
$I_P(X,Y\|\{H,C,K\})$	$[X,J,K,L,Y]$ is not blocked.
$I_P(X,Y\|\{H,C,M\})$	$[X,J,K,L,Y]$ is not blocked.

Example 3.20 *Owing to Theorem 3.3, the Markov condition entails the following conditional independency for the DAG in Figure 3.27:*

Conditional Independency	Reason Conditional Independency Is Entailed
$I_P(W,X)$	$[W,Y,X]$ is blocked at Y.
	$[W,Y,R,Z,X]$ is blocked at R.
	$[W,Y,R,Z,S,X]$ is blocked at S.

Note that $I_P(W,X)$ is the same as $I_P(W,X|\varnothing)$, where $\varnothing$ is the empty set, and Y, R, S, and T are all not in $\varnothing$.

It is left as an exercise to show that the Markov condition also entails the following conditional independencies for the DAG in Figure 3.27:

1. $I_P(X,R|\{Y,Z\})$

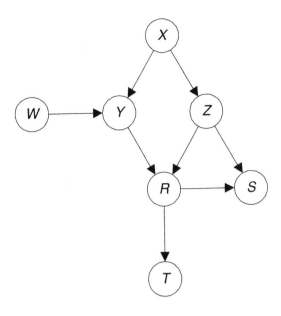

Figure 3.27: The Markov condition entails $I_P(W, X)$ for this DAG.

2. $I_P(X, T|\{Y, Z\})$

3. $I_P(W, T|R)$

4. $I_P(Y, Z|X)$

5. $I_P(W, S|\{R, Z\})$

6. $I_P(W, S|\{Y, Z\})$

7. $I_P(W, S|\{Y, X\})$.

It is also left as an exercise to show that the Markov condition does not entail the following conditional independencies for the DAG in Figure 3.27:

1. $I_P(W, X|Y)$

2. $I_P(W, T|Y)$.

3.6.3 Faithful and Unfaithful Probability Distributions

Recall that a DAG entails a conditional independency if every probability distribution, which satisfies the Markov condition with the DAG, must have the conditional independency. Theorem 3.3 states that all and only d-separations are entailed conditional independencies. Do not misinterpret this result. It does not say that if some particular probability distribution P satisfies the Markov condition with a DAG $\mathbb{G}$, then P cannot have conditional independencies that

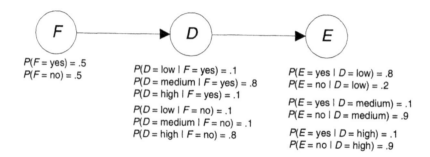

Figure 3.28: For the distribution P in this Bayesian network we have $I_P(E, F)$, but the Markov condition does not entail this conditional independency.

$\mathbb{G}$ does not entail. Rather, it only says that P must have all the conditional independencies that are entailed. We illustrate the difference next.

An Unfaithful Distribution

The following example shows that, even when we obtain a distribution by assigning conditional probabilities in a DAG, we can end up with a distribution that has a conditional independency that is not entailed by the Markov condition.

Example 3.21 *Consider the Bayesian network in Figure 3.28. The only conditional independency entailed by the Markov condition for the DAG $\mathbb{G}$ in that figure is $I_P(E, F|D)$. So Theorem 3.3 says all probability distributions that satisfy the Markov condition with $\mathbb{G}$ must have $I_P(E, F|D)$, which means the probability distribution P in the Bayesian network in Figure 3.28 must have $I_P(E, F|D)$. However, the theorem does not say that P cannot have other independencies. Indeed, it is left as an exercise to show that we have $I_P(E, F)$ for the distribution P in that Bayesian network.*

We purposefully assigned values to the conditional distributions in the network in Figure 3.28 to achieve $I_P(E, F)$. If we randomly assigned values, we would be almost certain to obtain a probability distribution that does not have this independency. That is, Meek [1995] proved that almost all assignments of values to the conditional distributions in a Bayesian network will result in a probability distribution that only has conditional independencies entailed by the Markov condition.

Could actual phenomena in nature result in a distribution like that in Figure 3.28? Although we made up the numbers in the network in that figure, we patterned them after something that actually occurred in nature. Let the variables in the figure represent the following:

Variable	What the Variable Represents
F	Whether the subject takes finasteride
D	Subject's dihydro-testosterone level
E	Whether the subject suffers erectile dysfunction

As shown in Example 3.5, dihydro-testosterone seems to be the hormone necessary for erectile function. Recall from Section 3.3.1 that Merck performed a study indicating that finasteride has a positive causal effect on hair growth. Finasteride accomplishes this by inhibiting the conversion of testosterone to dihydro-testosterone, and dihydro-testosterone is the hormone responsible for hair loss. Given this, Merck feared that dihydro-testosterone would cause erectile dysfunction. That is, ordinarily if X has a causal influence on Y and Y has a causal influence on Z, then X has a causal influence on Z through Y. However, in a manipulation study Merck found that F does not appear to have a causal influence on E. That is, they learned $I_P(E, F)$. The explanation for this is that finasteride does not lower dihydro-testosterone levels beneath some threshold level, and that threshold level is all that is necessary for erectile function. The numbers we assigned in the Bayesian network in Figure 3.28 reflect this. The value of F has no effect on whether D is *low*, and D must be *low* to make the probability that E is *yes* high.

Faithfulness

The probability distribution in the Bayesian network in Figure 3.28 is said to be unfaithful to the DAG in that figure because it contains a conditional independency that is not entailed by the Markov condition. We have the following definition:

Definition 3.5 *Suppose we have a joint probability distribution P of the random variables in some set* V *and a DAG* $\mathbb{G} = (V, E)$*. We say that* $(\mathbb{G}, P)$ *satisfies the* **faithfulness condition** *if all and only the conditional independencies in P are entailed by* $\mathbb{G}$*. Furthermore, we say that P and $\mathbb{G}$ are faithful to each other.*

Notice that the faithfulness condition includes the Markov condition because *only* conditional independencies in P can be entailed by $\mathbb{G}$. However, it requires more; that is, it requires that *all* conditional independencies in P must be entailed by $\mathbb{G}$. As noted previously, almost all assignments of values to the conditional distributions will result in a faithful distribution. For example, it is left as an exercise to show that the probability distribution P in the Bayesian network in Figure 3.29 is faithful to the DAG in that figure. We arbitrarily assigned values to the conditional distributions in the figure. However, owing to the result in [Meek, 1995] that almost all assignments will lead to a faithful distribution, we were willing to bet the farm that this assignment would.

Is there some DAG faithful to the probability distribution in the Bayesian network in Figure 3.28? The answer is "no," but it is beyond the scope of this book to show this. See [Neapolitan, 2004] for a proof of this fact. Intuitively,

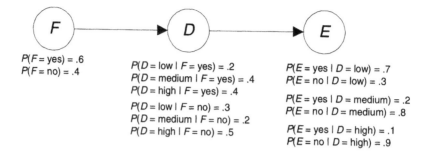

Figure 3.29: The probability distribution in this Bayesian network is faithful to the DAG in the network.

the DAG in Figure 3.28 represents the causal relationships among the variables, which means we should not be able to find a DAG which better represents the probability distribution.

3.6.4 Markov Blankets and Boundaries

A Bayesian network can have a large number of nodes, and the probability of a given node can be affected by instantiating a distant node. However, it turns out that the instantiation of a set of close nodes can shield a node from the effect of all other nodes. The next definition and theorem show this.

Definition 3.6 *Let* V *be a set of random variables,* P *be their joint probability distribution, and* $X \in V$. *Then a* **Markov blanket** M *of* X *is any set of variables such that* X *is conditionally independent of all the other variables given* M. *That is,*

$$I_P(X, V - (M \cup \{X\})|M).$$

Theorem 3.4 *Suppose* (G, P) *satisfies the Markov condition. Then for each variable* X, *the set of all parents of* X, *children of* X, *and parents of children of* X *is a Markov blanket of* X.

Proof. *It is straightforward that this set d-separates* X *from the set of all other nodes in* V. *The proof therefore follows from Theorem 3.3.* ∎

Example 3.22 *Suppose* (G, P) *satisfies the Markov condition where* G *is the DAG in Figure 3.30. Then due to Theorem 3.4,* $\{T, Y, Z\}$ *is a Markov blanket of* X. *So we have*

$$I_P(X, \{S, W\}|\{T, Y, Z\}).$$

Example 3.23 *Suppose* (G, P) *satisfies the Markov condition where* G *is the DAG in Figure 3.30, and* P *has the following conditional independency:*

$$I_P(X, \{S, T, W\}|\{Y, Z\}).$$

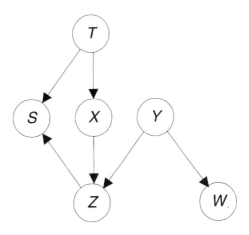

Figure 3.30: If P satisfies the Markov condition with this DAG, then $\{T, Y, Z\}$ is a Markov blanket of X.

Then the Markov blanket $\{T, Y, Z\}$ is not minimal in the sense that its subset $\{Y, Z\}$ is also a Markov blanket of X.

The last example motivates the following definition:

Definition 3.7 *Let* V *be a set of random variables,* P *be their joint probability distribution, and* $X \in \mathsf{V}$. *Then a* **Markov boundary** *of* X *is any Markov blanket such that none of its proper subsets is a Markov blanket of* X.

We have the following theorem:

Theorem 3.5 *Suppose* $(\mathbb{G}, P)$ *satisfies the faithfulness condition. Then for each variable* X, *the set of all parents of* X, *children of* X, *and parents of children of* X *is the unique Markov boundary of* X.

Proof. *The proof can be found in [Neapolitan, 2004].* ∎

Example 3.24 *Suppose* $(\mathbb{G}, P)$ *satisfies the faithfulness condition where* $\mathbb{G}$ *is the DAG in Figure 3.30. Then due to Theorem 3.5,* $\{T, Y, Z\}$ *is the unique Markov boundary of* X.

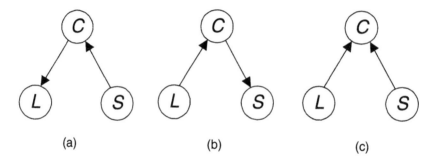

Figure 3.31: The probability distribution discussed in Example 3.1 satisfies the Markov condition with the DAGs in (a) and (b), but not with the DAG in (c).

EXERCISES

Section 3.1

Exercise 3.1 In Example 3.3 it was left as an exercise to show for all value of s, l, and c that

$$P(s,l,c) = P(s|c)P(l|c)P(c).$$

Show this.

Exercise 3.2 Consider the joint probability distribution P in Example 3.1.

1. Show that P satisfies the Markov condition with the DAG in Figure 3.31 (a) and that P is equal to the product of its conditional distributions in that DAG.

2. Show that P satisfies the Markov condition with the DAG in Figure 3.31 (b) and that P is equal to the product of its conditional distributions in that DAG.

3. Show that P does not satisfy the Markov condition with the DAG in Figure 3.31 (c) and that P is not equal to the product of its conditional distributions in that DAG.

Exercise 3.3 Create an arrangement of objects similar to the one in Figure 3.4, but with a different distribution of letters, shapes, and colors, so that, if random variables L, S, and C are defined as in Example 3.1, then the only independency or conditional independency among the variables is $I_P(L, S)$. Does this distribution satisfy the Markov condition with any of the DAGs in Figure 3.31? If so, which one(s)?

Exercise 3.4 Consider the joint probability distribution of the random variables defined in Example 3.1 relative to the objects in Figure 3.4. Suppose we compute that distribution's conditional distributions for the DAG in Figure 3.31 (c), and we take their product. Theorem 3.1 says this product is a joint probability distribution that constitutes a Bayesian network with that DAG. Is this the actual joint probability distribution of the random variables?

Section 3.2

Exercise 3.5 Professor Morris investigated gender bias in hiring in the following way. He gave hiring personnel equal numbers of male and female resumes to review, and then he investigated whether their evaluations were correlated with gender. When he submitted a paper summarizing his results to a psychology journal, the reviewers rejected the paper because they said this was an example of fat hand manipulation. Investigate the concept of fat hand manipulation, and explain why they might have thought this.

Exercise 3.6 Consider the following piece of medical knowledge: tuberculosis and lung cancer can each cause shortness of breath (dyspnea) and a positive chest X-ray. Bronchitis is another cause of dyspnea. A recent visit to Asia could increase the probability of tuberculosis. Smoking can cause both lung cancer and bronchitis. Create a DAG representing the causal relationships among these variables. Complete the construction of a Bayesian network by determining values for the conditional probability distributions in this DAG either based on your own subjective judgement or from data.

Exercise 3.7 Explain why, if we reverse the edges in the DAG in Figure 3.12 to obtain the DAG $E \rightarrow DHT \rightarrow T$, the new DAG also satisfies the Markov condition with the probability distribution of the variables.

Section 3.3

Exercise 3.8 Compute $P(x_1|w_1)$ assuming the Bayesian network in Figure 3.16.

Exercise 3.9 Compute $P(t_1|w_1)$ assuming the Bayesian network in Figure 3.17.

Exercise 3.10 Compute $P(x_1|t_2, w_1)$ assuming the Bayesian network in Figure 3.17.

Exercise 3.11 Using Netica develop the Bayesian network in Figure 3.2, and use that network to determine the following conditional probabilities:

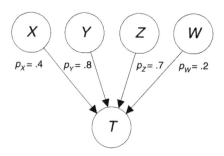

Figure 3.32: The noisy OR-gate model is assumed.

1. $P(F = yes|Sex = male)$. Is this conditional probability different from $P(F = yes)$? Explain why it is or is not.

2. $P(F = yes|J = yes)$. Is this conditional probability different from $P(F = yes)$? Explain why it is or is not.

3. $P(F = yes|Sex = male, J = yes)$. Is this conditional probability different from $P(F = yes|J = yes)$? Explain why it is or is not.

4. $P(G = yes|F = yes)$. Is this conditional probability different from $P(G = yes)$? Explain why it is or is not.

5. $P(G = yes|J = yes)$. Is this conditional probability different from $P(G = yes)$? Explain why it is or is not.

6. $P(G = yes|J = yes, F = yes)$. Is this conditional probability different from $P(G = yes|F = yes)$? Explain why it is or is not.

7. $P(G = yes|A = < 30)$. Is this conditional probability different from $P(G = yes)$? Explain why it is or is not.

8. $P(G = yes|A = < 30, J = yes)$. Is this conditional probability different from $P(G = yes|J = yes)$? Explain why it is or is not.

Section 3.4

Exercise 3.12 Assume the noisy OR-gate model and the causal strengths are those shown in Figure 3.32. Compute the probability $T = yes$ for all combinations of values of the parents.

Exercise 3.13 Assume the leaky noisy OR-gate model and the relevant probabilities are those shown in Figure 3.33. Compute the probability $T = yes$ for all combinations of values of the parents. Compare the results to those obtained in Exercise 3.12.

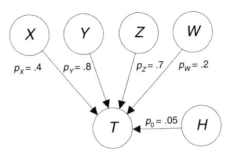

Figure 3.33: The leaky noisy OR-gate model is assumed.

Exercise 3.14 Suppose we have the normal density function with $\mu = 100$ and $\sigma = 20$.

1. Discretize this function into four ranges using the Brackets Median Method.

2. Discretize this function using the Pearson-Tukey Method.

Section 3.5

Exercise 3.15 Consider the DAG $\mathbb{G}$ in Figure 3.25 (a). Prove that the Markov condition entails $I_P(C, G|\{A, F\})$ for $\mathbb{G}$.

Exercise 3.16 Suppose we add another variable R, an edge from F to R, and an edge from R to C to the DAG $\mathbb{G}$ in Figure 3.25 (a). The variable R might represent the professor's initial reputation. State which of the following conditional independencies you would feel are entailed by the Markov condition for $\mathbb{G}$. For each that you feel is entailed, try to prove it actually is.

1. $I_P(R, A)$

2. $I_P(R, A|F)$

3. $I_P(R, A|\{F, C\})$.

Exercise 3.17 Show that the Markov condition entails the following conditional independencies for the DAG in Figure 3.27:

1. $I_P(X, R|\{Y, Z\})$

2. $I_P(X, T|\{Y, Z\})$

3. $I_P(W, T|R)$

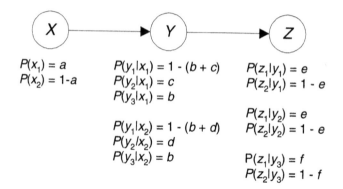

Figure 3.34: Any probability distribution P obtained by assigning values to the parameters in this network is not faithful to the DAG in the network because we have $I_P(X, Z)$.

4. $I_P(Y, Z|X)$

5. $I_P(W, S|\{R, Z\})$

6. $I_P(W, S|\{Y, Z\})$

7. $I_P(W, S|\{Y, X\})$.

Exercise 3.18 Show that the Markov condition does not entail the following conditional independencies for the DAG in Figure 3.27:

1. $I_P(W, X|Y)$

2. $I_P(W, T|Y)$.

Exercise 3.19 State which of the following conditional independencies are entailed by the Markov condition for the DAG in Figure 3.27:

1. $I_P(W, S|\{R, X\})$

2. $I_P(\{W, X\}, \{S, T\}|\{R, Z\})$

3. $I_P(\{Y, Z\}, T|\{R, S\})$

4. $I_P(\{X, S\}, \{W, T\}|\{R, Z\})$

5. $I_P(\{X, S, Z\}, \{W, T\}|R)$

6. $I_P(\{X, Z\}, W)$

7. $I_P(\{X, S, Z\}, W)$.

Does the Markov condition entail $I_P(\{X, S, Z\}, W|\mathsf{U})$ for any subset of variables U in that DAG?

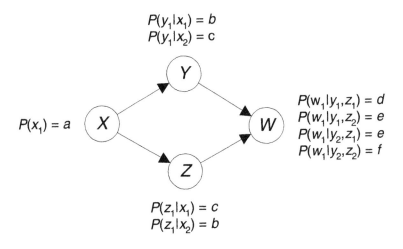

Figure 3.35: Any probability distribution P obtained by assigning values to the parameters in this network is not faithful to the DAG in the network because we have $I_P(X, W)$.

Exercise 3.20 Show $I_P(F, E)$ for the distribution P in the Bayesian network in Figure 3.28.

Exercise 3.21 Consider the Bayesian network in Figure 3.34. Show that for all assignments of values to a, b, c, d, e, and f that yield a probability distribution P, we have $I_P(X, Z)$. Such probability distributions are not faithful to the DAG in that figure because X and Z are not d-separated by the empty set. Note that the probability distribution in Figure 3.28 is a member of this family of distributions.

Exercise 3.22 Assign arbitrary values to the conditional distributions for the DAG in Figure 3.34, and see if the resultant distribution is faithful to the DAG. Try to find an unfaithful distribution besides ones in the family shown in that figure.

Exercise 3.23 Consider the Bayesian network in Figure 3.35. Show that for all assignments of values to a, b, c, d, e, and f that yield a probability distribution P, we have $I_P(X, W)$. Such probability distributions are not faithful to the DAG in that figure because X and W are not d-separated by the empty set.

If the edges in the DAG in Figure 3.35 represent causal influences, X and W would be independent if the causal effect of X on W through Y negated the causal effect of X on W through Z. If X represents an individual's age, W represents the individual's typing ability, Y represent's the individual's experience, and Z represents the individual's manual dexterity, do you feel X and W might be independent for this reason?

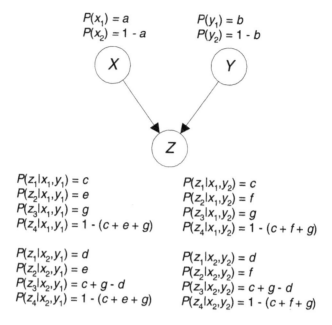

$P(x_1) = a$ $P(y_1) = b$
$P(x_2) = 1 - a$ $P(y_2) = 1 - b$

$P(z_1|x_1,y_1) = c$ $P(z_1|x_1,y_2) = c$
$P(z_2|x_1,y_1) = e$ $P(z_2|x_1,y_2) = f$
$P(z_3|x_1,y_1) = g$ $P(z_3|x_1,y_2) = g$
$P(z_4|x_1,y_1) = 1 - (c + e + g)$ $P(z_4|x_1,y_2) = 1 - (c + f + g)$

$P(z_1|x_2,y_1) = d$ $P(z_1|x_2,y_2) = d$
$P(z_2|x_2,y_1) = e$ $P(z_2|x_2,y_2) = f$
$P(z_3|x_2,y_1) = c + g - d$ $P(z_3|x_2,y_2) = c + g - d$
$P(z_4|x_2,y_1) = 1 - (c + e + g)$ $P(z_4|x_2,y_2) = 1 - (c + f + g)$

Figure 3.36: Any probability distribution P obtained by assigning values to the parameters in this network is not faithful to the DAG in the network because we have $I_P(X, Y|Z)$.

Exercise 3.24 Consider the Bayesian network in Figure 3.36. Show that for all assignments of values to a, b, c, d, e, f, and g that yield a probability distribution P, we have $I_P(X, Y|Z)$. Such probability distributions are not faithful to the DAG in that figure because X and Y are not d-separated by Z.

   If the edges in the DAG in Figure 3.36 represent causal influences, X and Y would be independent given Z if no discounting occurred. Try to find some causal influences that might behave like this.

Exercise 3.25 Apply Theorem 3.4 to find a Markov blanket for each node in the DAG in Figure 3.30.

Exercise 3.26 Apply Theorem 3.4 to find a Markov blanket for each node in the DAG in Figure 3.27.

Chapter 4

Learning Bayesian Networks

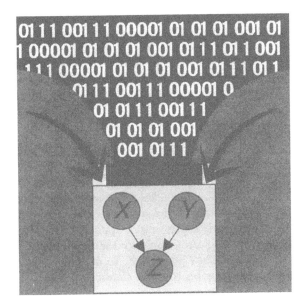

Up until the early 1990s the DAG in a Bayesian network was ordinarily hand-constructed by a domain expert. Then the conditional probabilities were assessed by the expert, learned from data, or obtained using a combination of both techniques. Eliciting Bayesian networks from experts can be a laborious and difficult process in the case of large networks. As a result, researchers developed methods that could learn the DAG from data. Furthermore, they formalized methods for learning the conditional probabilities from data. We discuss these methods next. In a Bayesian network the DAG is called the **structure** and the conditional probability distributions are called the **parameters**. In Section 4.1 we address the problem of learning the parameters from data. Then after introducing the problem of learning structure in Section 4.2, we discuss score-based structure learning in Section 4.3 and constraint-based structure learning in Section 4.4. In Section 4.5 we apply structure learning to inferring causal influences from data. Section 4.6 presents learning packages that implement the methods discussed in the first three sections. Finally, in Section 4.7 we show examples of learning Bayesian networks and of causal learning.

4.1   Parameter Learning

We can only learn parameters from data when the probabilities are relative frequencies, which were discussed in Chapter 2, Section 2.3.1. So this discussion pertains only to such probabilities. Although the method is based on rigorous mathematical results obtained by modeling an individual's subjective belief concerning a relative frequency, the method itself is quite simple. Here, we merely present the method. See [Neapolitan, 2004] for the mathematical development. First, we discuss learning a single parameter, and then we show how to learn all the parameters in a Bayesian network.

4.1.1   Learning a Single Parameter

After presenting a method for learning the probability of a binomial variable, we extend the method to multinomial variables. Finally, we provide guidelines for articulating our prior beliefs concerning probabilities.

Binomial Variables

We illustrate learning with a sequence of examples.

Example 4.1 *Recall the discussion concerning a thumbtack at the beginning of Chapter 2, Section 2.3.1. We noted that a thumbtack could land on its flat end, which we call "heads," or it could land with the edge of the flat end and the point touching the ground, which we call "tails." Because the thumbtack is not symmetrical, we have no reason to apply the Principle of Indifference and assign probabilities of .5 to both outcomes. So we need data to estimate the probability of heads. Suppose we toss the thumbtack 100 times, and it lands heads 65 of those times. Then we can estimate*

$$P(heads) \approx \frac{65}{100} = .65.$$

This estimate, obtained by dividing the number of heads by the number of trials, is called the **maximum likelihood estimate (MLE)** of the probability. In general, if there are s heads in n trials, the MLE of the probability is

$$P(heads) \approx \frac{s}{n}.$$

Using the MLE seems reasonable when we have no prior belief concerning the probability. However, it is not so reasonable when we do have prior belief. Consider the next example.

Example 4.2 *Suppose you take a coin from your pocket, toss it 10 times, and it lands heads all of those times. Then using the MLE we estimate*

$$P(heads) \approx \frac{10}{10} = 1.$$

After the coin landed heads 10 times, we would not bet as if we were certain that the outcome of the 11th toss will be heads. So our belief concerning the $P(heads)$ is not the MLE value of 1. Assuming we believe the coins in our pockets are fair, should we instead maintain $P(heads) = .5$ after all 10 tosses landed heads? This may seem reasonable for 10 tosses, but it does not seem so reasonable if 1000 straight tosses landed heads. At some point we would start suspecting the coin was weighted to land heads. We need a method that incorporates one's prior belief with the data. The standard way to do this is for the probability assessor to ascertain integers a and b such that the assessor's experience is equivalent to having seen the first outcome (heads in this case) occur a times and the second outcome occur b times in $m = a + b$ trials. Then the assessor's prior probabilities are

$$P(heads) = \frac{a}{m} \qquad P(tails) = \frac{b}{m}.$$

After observing s heads and t tails in $n = s + t$ trials, the assessor's posterior probabilities are

$$P(heads|s,t) = \frac{a+s}{m+n} \qquad P(tails|s,t) = \frac{a+t}{m+n}. \qquad (4.1)$$

This probability is called the **maximum a posterior probability (MAP)**. Note that we have used the symbol "=" rather than "≈," and we have written the probability as a conditional probability rather than as an estimate. The reason is that this is a Bayesian technique, and Bayesians say that the value is their probability (belief) based on the data rather than saying it is an estimate of a probability (relative frequency). It is the assumption of exchangeability that enables us to update our beliefs using Equality 4.1. The assumption of **exchangeability**, which was first developed by de Finetti in 1937, is that an individual assigns the same probability to all sequences of the same length containing the same number of each outcome. For example, the individual assigns the same probability to these two sequences of heads (H) and tails (T):

$$H, T, H, T, H, T, H, T, T, T \qquad \text{and} \qquad H, T, T, T, T, H, H, T, H, T.$$

Furthermore, the individual assigns the same probability to any other sequence of 10 tosses that has 4 heads and 6 tails. Neapolitan [2004] discusses exchangeability in detail and derives Equality 4.1. Here, we accept that equality on intuitive grounds.

Next, we show more examples. In these examples we only compute the probability of the first outcome because the probability of the second outcome is uniquely determined by it.

Example 4.3 *Suppose you are going to repeatedly toss a coin from your pocket. Since you would feel it highly probable that the relative frequency is around .5, you might feel your prior experience is equivalent to having seen 50 heads in 100 tosses. Therefore, you could represent your belief with $a = 50$ and $b = 50$. Then $m = 50 + 50 = 100$, and your prior probability of heads is*

$$P(heads) = \frac{a}{m} = \frac{50}{100} = .5.$$

After seeing 48 heads in 100 tosses, your posterior probability is

$$P(heads|48, 52) = \frac{a+s}{m+n} = \frac{50+48}{100+100} = .49.$$

The notation $48, 52$ *on the right of the conditioning bar in* $P(heads|48, 52)$ *represents the event that 48 heads and 52 tails have occurred.*

Example 4.4 *Suppose you are going to repeatedly toss a thumbtack. Based on its structure, you might feel it should land heads about half the time, but you are not nearly so confident as you were with the coin from your pocket. So you might feel your prior experience is equivalent to having seen 3 heads in 6 tosses. Then your prior probability of heads is*

$$P(heads) = \frac{a}{m} = \frac{3}{6} = .5.$$

After seeing 65 heads in 100 tosses, your posterior probability is

$$P(heads|65,35) = \frac{a+s}{m+n} = \frac{3+65}{6+100} = .64.$$

Example 4.5 *Suppose you are going to sample individuals in the United States and determine whether they brush their teeth. In this case, you might feel your prior experience is equivalent to having seen 18 individuals brush their teeth out of 20 sampled. Then your prior probability of brushing is*

$$P(brushes) = \frac{a}{m} = \frac{18}{20} = .9.$$

After sampling 100 individuals and learning 80 brush their teeth, your posterior probability is

$$P(brushes|80, 20) = \frac{a+s}{m+n} = \frac{18+80}{20+100} = .82.$$

You may feel that if we have complete prior ignorance as to the probability we should take $a = b = 0$. However, consider the next example.

Example 4.6 *Suppose we are going to sample dogs and determine whether or not they eat the potato chips which we offer them. Since we have no idea whether a particular dog would eat potato chips, we assign $a = b = 0$, which* means $m = 0 + 0 = 0$. *Since we cannot divide a by m, we have no prior probability. Suppose next that we sample one dog, and that dog eats the potato chips. Our probability of the next dog eating potato chips is now*

$$P(eats|1,0) = \frac{a+s}{m+n} = \frac{0+1}{0+1} = 1.$$

This belief is not very reasonable as we are now certain that all dogs eat potato chips. Owing to difficulties such as this and more rigorous mathematical results, prior ignorance to a probability is usually modeled by taking $a = b = 1$, which means $m = 1 + 1 = 2$. If we use these values instead, our posterior probability when the first dog sampled was found to eat potato chips is given by

$$P(eats|1,0) = \frac{a+s}{m+n} = \frac{1+1}{2+1} = \frac{2}{3}.$$

Sometimes we want fractional values for a and b. Consider this example.

Example 4.7 *This example is taken from [Berry, 1996]. Glass panels in high-rise buildings sometimes break and fall to the ground. A particular case involved 39 broken panels. In their quest for determining why the panels broke, the owners wanted to analyze the broken panels, but they could only recover three of them. These three were found to contain Nickel Sulfide (NiS), a manufacturing flaw which can cause panels to break. In order to determine whether they should hold the manufacturer responsible, the owners then wanted to determine how probable it was that all 39 panels contained NiS. So they contacted a glass expert.*

The glass expert testified that among glass panels that break, only 5% contain NiS. However, NiS tends to be pervasive in production lots. So given that the first panel sampled, from a particular production lot of broken panels, contains NiS, the expert felt the probability was .95 that the second panel sampled also contains NiS. It was known that all 39 panels came from the same production lot. So, if we model the expert's prior belief using values of a, b, and $m = a + b$ as discussed above, we must have that the prior probability is given by

$$P(NiS) = \frac{a}{m} = .05.$$

Furthermore, the expert's posterior probability after finding the first panel contains NiS must be given by

$$P(NiS|1,0) = \frac{a+1}{m+1} = .95.$$

Solving these last two equations for a and m yields

$$a = \frac{1}{360} \qquad m = \frac{20}{360}.$$

This is an alternative technique for assessing a and b. Namely, we assess the probability for the first trial. Then we assess the conditional probability for the second trial given the first one is a "success." Once we have values of a and b, we can determine how likely it is that any one of the other 36 panels (the next one sampled) contains NiS after the first three sampled were found to contain it. We have that this probability is given by

$$P(NiS|3,0) = \frac{a+s}{m+n} = \frac{1/360+3}{20/360+3} = .983.$$

Notice how the expert's probability of NiS quickly changed from being very small to be being very large. This is because the values of a and m are so small. We are really most interested in whether all 36 remaining panels contain NiS. It is left as an exercise to show that this probability is given by

$$\prod_{i=0}^{35} \frac{1/360+3+i}{20/360+3+i} = .866.$$

Multinomial Variables

The method just discussed readily extends to multinomial variables. Suppose k is the number of values the variable can assume, and $x_1, x_2, \ldots, x_k$ are the k outcomes. We then ascertain numbers $a_1, a_2, \ldots, a_k$ such that our experience is equivalent to having seen the first outcome occur a_1 times, the second outcome occur a_2 times, $\ldots$, and the last outcome occur a_k times. Our prior probabilities before any trials are then

$$P(x_1) = \frac{a_1}{m} \qquad P(x_2) = \frac{a_2}{m} \qquad \cdots \qquad P(x_k) = \frac{a_k}{m},$$

where $m = a_1 + a_2 + \cdots + a_k$. After seeing x_1 occur s_1 times, x_2 occur s_2 times, $\ldots$, and x_n occur s_n times in $n = s_1 + s_2 + \cdots + s_k$ trials, our posterior probabilities are as follows:

$$P(x_1|s_1, s_2, \ldots, s_k) = \frac{a_1 + s_1}{m+n}$$
$$P(x_2|s_1, s_2, \ldots, s_k) = \frac{a_2 + s_2}{m+n}$$
$$\vdots$$
$$P(x_k|s_1, s_2, \ldots, s_k) = \frac{a_k + s_k}{m+n}.$$

Example 4.8 *Suppose we have an asymmetrical-sided six-sided die, and we have little idea of the probability of each side coming up. However, it seems that all sides are equally likely. So we assign*

$$a_1 = a_2 = \cdots = a_6 = 3.$$

Then our prior probabilities are as follows:

$$P(1) = P(2) = \cdots = P(6) = \frac{a_i}{n} = \frac{3}{18} = .16667.$$

Suppose next we throw the die 100 times with the following result:

Outcome	Number of Occurrences
1	10
2	15
3	5
4	30
5	13
6	27

We then have

$$P(1|10, 15, 5, 30, 13, 27) = \frac{a_1 + s_1}{m + n} = \frac{3 + 10}{18 + 100} = .110$$

$$P(2|10, 15, 5, 30, 13, 27) = \frac{a_2 + s_2}{m + n} = \frac{3 + 15}{18 + 100} = .153$$

$$P(3|10, 15, 5, 30, 13, 27) = \frac{a_3 + s_3}{m + n} = \frac{3 + 5}{18 + 100} = .067$$

$$P(4|10, 15, 5, 30, 13, 27) = \frac{a_4 + s_4}{m + n} = \frac{3 + 30}{18 + 100} = .280$$

$$P(5|10, 15, 5, 30, 13, 27) = \frac{a_5 + s_5}{m + n} = \frac{3 + 13}{18 + 100} = .136$$

$$P(6|10, 15, 5, 30, 13, 27) = \frac{a_6 + s_6}{m + n} = \frac{3 + 27}{18 + 100} = .254.$$

Guidelines for Articulating Prior Belief

Next, we give some guidelines for choosing the values that represent our prior beliefs.

Binary Variables   The guidelines for binary variables are as follows.

1. $a = b = 1$: We use these values when we feel we have no knowledge at all concerning the value of the probability. We might also use these values to try to achieve objectivity in the sense that we impose none of our beliefs concerning the probability on the learning algorithm. We only impose the fact that we know, at most, two things can happen. An example might be when we are learning the probability of lung cancer given smoking from data, and we want to communicate our result to the scientific community. The scientific community would not be interested in our prior belief; rather it would be interested only in what the data had to say. Essentially, when we use these values, the posterior probability represents the belief of an individual who has no prior belief concerning the probability.

2. $a, b > 1$: These values mean we feel it is likely that the probability of the first outcome is a/m. The larger the values of a and b are, the more we believe this. We would use such values when we want to impose our beliefs concerning the relative frequency on the learning algorithm. For

example, if we were going to toss a coin taken from the pocket, we might take $a = b = 50$.

3. $a, b < 1$: These values mean we feel it is likely that the probability of one of the outcomes is high, although we are not committed to which one it is. If we take $a = b \approx 0$ (almost 0), then we are almost certain the probability of one of the outcomes is very close to 1. We would also use values like these when we want to impose our beliefs concerning the probability on the learning algorithm. Example 4.7 shows a case in which we would choose values less than 1. Notice that such prior beliefs are quickly overwhelmed by data. For example, if $a = b = .1$, and we saw the first outcome x_1 occur in a single trial, we have

$$P(x_1|1,0) = \frac{.1+1}{.2+1} = .917. \qquad (4.2)$$

Intuitively, we thought *a priori* that the probability of one of the outcomes was high. The fact that it took the value x_1 once makes us believe it is probably that outcome.

Multinomial Variables   The guidelines are essentially the same as those for binomial variables, but we restate them for the sake of clarity.

1. $a_1 = a_2 = \cdots = a_k = 1$: We use these values when we feel we have no knowledge at all concerning the probabilities. We might also use these values to try to achieve objectivity in the sense that we impose none of our beliefs concerning the probability on the learning algorithm. We only impose the fact that we know, at most, k things can happen. An example might be learning the probability of low, medium, and high blood pressure from data, which we want to communicate to the scientific community.

2. $a_1 = a_2 = \cdots = a_k > 1$: These values mean we feel it more likely that the probability of the kth value is around a_k/m. The larger the values of a_k are, the more we believe this. We would use such values when we want to impose our beliefs concerning the probability on the learning algorithm. For example, if we were going to toss an ordinary die, we might take $a_1 = a_2 = \cdots = a_6 = 50$.

3. $a_1 = a_2 = \cdots = a_k < 1$: These values mean we feel that it is likely that only a few outcomes are probable. We would use such values when we want to impose our beliefs concerning the probabilities on the learning algorithm. For example, suppose we know there are 1,000,000 different species in the world, and we are about to land on an uncharted island. We might feel it probable that not very many of the species are present. So if we considered the probabilities with which we encountered different species, we would not consider probabilities, which resulted in a lot of different species, likely. Therefore, we might take $a_i = 1/1,000,000$ for all i.

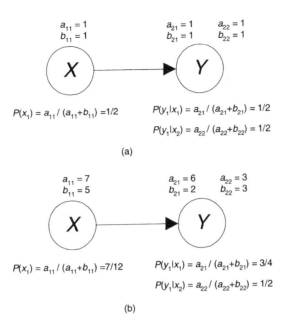

Figure 4.1: A Bayesian network initialized for parameter learning appears in (a), while the updated network based on the data in Figure 4.2 appears in (b).

4.1.2 Learning All Parameters in a Bayesian Network

The method for learning all parameters in a Bayesian network follows readily from the method for learning a single parameter. We illustrate the method with binomial variables. It extends readily to the case of multinomial variables (see [Neapolitan, 2004]). After showing the method, we discuss equivalent sample sizes.

Procedure for Learning Parameters

Consider the two-node Bayesian network in Figure 4.1 (a). It has been initialized for parameter learning. For each probability in the network there is a pair (a_{ij}, b_{ij}). The i indexes the variable, while the j indexes the value of the parent(s) of the variable. For example, the pair (a_{11}, b_{11}) is for the 1st variable (X), and the 1st value of its parent (in this case there is a default of one parent value since X has no parent). The pair (a_{21}, b_{21}) is for the 2nd variable (Y), and the 1st value of its parent, namely x_1. The pair (a_{22}, b_{22}) is for the 2nd variable (Y), and the 2nd value of its parent, namely x_2. We have attempted to represent prior ignorance as to the value of all probabilities by taking $a_{ij} = b_{ij} = 1$. We compute the prior probabilities using these pairs just as we did when we were considering a single parameter. We have

$$P(x_1) = \frac{a_{11}}{a_{11} + b_{11}} = \frac{1}{1+1} = \frac{1}{2}$$

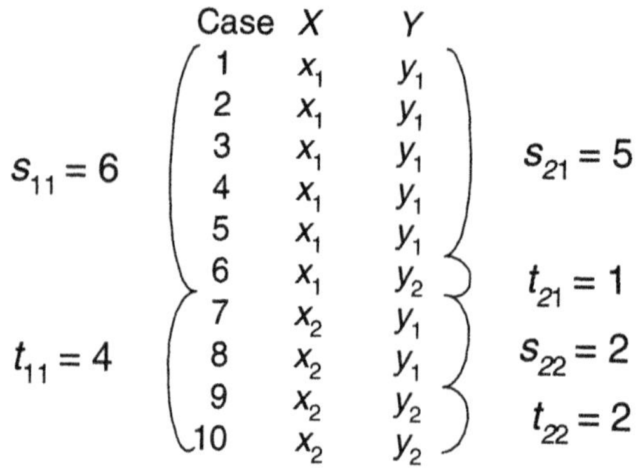

$$\begin{array}{ccc}
\text{Case} & X & Y \\
1 & x_1 & y_1 \\
2 & x_1 & y_1 \\
3 & x_1 & y_1 \\
4 & x_1 & y_1 \\
5 & x_1 & y_1 \\
6 & x_1 & y_2 \\
7 & x_2 & y_1 \\
8 & x_2 & y_1 \\
9 & x_2 & y_2 \\
10 & x_2 & y_2
\end{array}$$

$s_{11} = 6$ $s_{21} = 5$

$t_{11} = 4$ $t_{21} = 1$

$s_{22} = 2$

$t_{22} = 2$

Figure 4.2: Data on 10 cases.

$$P(y_1|x_1) = \frac{a_{21}}{a_{21} + b_{21}} = \frac{1}{1+1} = \frac{1}{2}$$

$$P(y_1|x_2) = \frac{a_{22}}{a_{22} + b_{22}} = \frac{1}{1+1} = \frac{1}{2}.$$

When we obtain data, we use an (s_{ij}, t_{ij}) pair to represent the counts for the ith variable when the variable's parents have their jth value. Suppose we obtain the data in Figure 4.2. The values of the (s_{ij}, t_{ij}) pairs are shown in that figure. We have that $s_{11} = 6$ because x_1 occurs 6 times, and $t_{11} = 4$ because x_2 occurs 4 times. Of the 6 times that x_1 occurs, y_1 occurs 5 times and y_2 occurs 1 time. So $s_{21} = 5$ and $t_{21} = 12$. Of the 4 times that x_2 occurs, y_1 occurs 2 times and y_2 occurs 2 times. So $s_{22} = 2$ and $t_{22} = 2$. To determine the posterior probability distribution based on the data we update each conditional probability with the counts relative to that conditional probability. Since we want an updated Bayesian network, we recompute the values of the (a_{ij}, b_{ij}) pairs. We therefore have

$$\begin{aligned}
a_{11} &= a_{11} + s_{11} = 1 + 6 = 7 \\
b_{11} &= b_{11} + t_{11} = 1 + 4 = 5
\end{aligned}$$

$$\begin{aligned}
a_{21} &= a_{21} + s_{21} = 1 + 5 = 6 \\
b_{21} &= b_{21} + t_{21} = 1 + 1 = 2
\end{aligned}$$

$$\begin{aligned}
a_{22} &= a_{22} + s_{22} = 1 + 2 = 3 \\
b_{22} &= b_{22} + t_{22} = 1 + 2 = 3.
\end{aligned}$$

We then compute the new values of the parameters:

$$P(x_1) = \frac{a_{11}}{a_{11} + b_{11}} = \frac{7}{7+5} = \frac{7}{12}$$

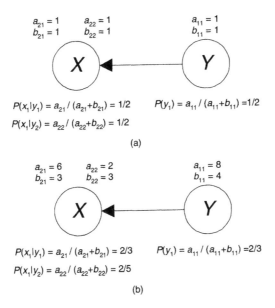

Figure 4.3: A Bayesian network initialized for parameter learning appears in (a), while the updated network based on the data in Figure 4.2 appears in (b).

$$P(y_1|x_1) = \frac{a_{21}}{a_{21} + b_{21}} = \frac{6}{6+2} = \frac{3}{4}$$

$$P(y_1|x_2) = \frac{a_{22}}{a_{22} + b_{22}} = \frac{3}{3+3} = \frac{1}{2}.$$

The updated network is shown in Figure 4.1 (b).

Equivalent Sample Size

There is a problem with the way we represented prior ignorance in the preceding subsection. Although it seems natural to set $a_{ij} = b_{ij} = 1$ to represent prior ignorance of all the conditional probabilities, such assignments are not consistent with the metaphor we used for articulating these values. Recall that we said the probability assessor is to choose values of a and b such that the assessor's experience is equivalent to having seen the first outcome occur a times in $a + b$ trials. Therefore, if we set $a_{11} = b_{11} = 1$, the assessor's experience is equivalent to having seen x_1 occur 1 time in 2 trials. However, if we set $a_{21} = b_{21} = 1$, the assessor's experience is equivalent to having seen y_1 occur 1 time out of the 2 times x_1 occurred. This is not consistent. First, we are saying x_1 occurred once; then we are saying it occurred twice. Aside from this inconsistency, we obtain odd results if we use these priors. Consider the Bayesian network for parameter learning in Figure 4.3 (a). If we update that network with the data in Figure 4.2, we obtain the network in Figure 4.3 (b). The DAG in Figure 4.3 (a) is equivalent to the one in Figure 4.1 (a) in a certain sense. That is, we say two DAGs are **Markov equivalent** if they entail the same conditional independencies. The DAGs in Figures 4.1 (a) and 4.3 (a) are Markov equivalent because both DAGs entail no conditional independencies. It seems that if we represent the same prior beliefs with equivalent DAGs, then the posterior distributions based on data should be the same. In this case we have attempted to represent prior ignorance as to all probabilities with the networks in Figure 4.1 (a) and Figure 4.3 (a). So the posterior distributions based on the data in Figure 4.2 should be the same. However, from the Bayesian network in Figure 4.1 (b) we have

$$P(x_1) = \frac{7}{12} = .583,$$

while from the Bayesian network in Figure 4.3 (b) we have

$$
\begin{aligned}
P(x_1) &= P(x_1|y_1)P(y_1) + P(x_1|y_2)P(y_2) \\
&= \frac{2}{3} \times \frac{2}{3} + \frac{2}{5} \times \frac{1}{3} = .578.
\end{aligned}
$$

We see that we obtain different posterior probabilities. Such results are not only odd, but unacceptable since we have attempted to model the same prior belief with the Bayesian networks in Figures 4.1 (a) and 4.3 (a), but end up with different posterior beliefs.

We can eliminate this difficulty by using a prior equivalent sample size. That is, we specify values of a_{ij} and b_{ij} that could actually occur in a prior sample that exhibit the conditional independencies entailed by the DAG. For example, given the network $X \to Y$, if we specify that $a_{21} = b_{21} = 1$, this means our prior sample must have x_1 occurring 2 times. So we need to specify $a_{11} = 2$. Similarly, if we specify that $a_{22} = b_{22} = 1$, this means our prior sample must have x_2 occurring 2 times. So we need to specify $b_{11} = 2$. Note that we are not saying we actually have a prior sample. We are saying the probability assessor's beliefs are represented by a prior sample. Figure 4.4 shows prior Bayesian networks using equivalent sample sizes. Notice that the values of a_{ij} and b_{ij} in these networks represent the following prior sample:

Case	X	Y
1	x_1	y_1
2	x_1	y_2
3	x_2	y_1
4	x_2	y_2

It is left as an exercise to show that if we update both the Bayesian networks in Figure 4.4 using the data in Figure 4.2, we obtain the same posterior probability distribution. This result is true, in general. We state it as a theorem, but first give a formal definition of a prior equivalent sample size.

Definition 4.1 *Suppose we specify a Bayesian network for the purpose of learning parameters in the case of binomial variables. If there is a number*

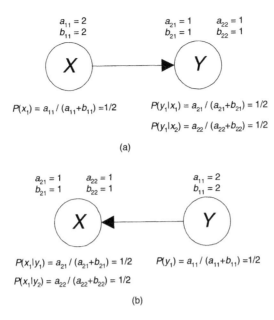

Figure 4.4: Bayesian networks for parameter learning containing prior equivalent sample sizes.

N such that for all i and j

$$a_{ij} + b_{ij} = P(\mathsf{pa}_{ij}) \times N,$$

where pa_{ij} denotes the jth instantiation of the parents of the ith variable, then we say the network has prior **equivalent sample size** *N.*

This definition is a bit hard to grasp by itself. The following theorem, whose proof can be found in [Neapolitan, 2004], yields a way to represent uniform prior distributions, which is often what we want to do.

Theorem 4.1 *Suppose we specify a Bayesian network for the purpose of learning parameters in the case of binomial variables and assign for all i and j*

$$a_{ij} = b_{ij} = \frac{N}{2q_i}.$$

where N is a positive integer and q_i is the number of instantiations of the parents of the ith variable. Then the resultant Bayesian network has equivalent sample size N, and the joint probability distribution in the Bayesian network is uniform.

We can represent prior ignorance by applying the preceding theorem with $N = 2$. The next example illustrates the technique.

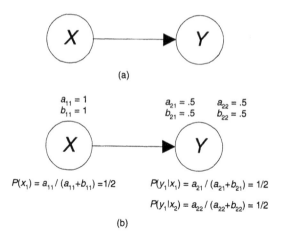

Figure 4.5: Given the DAG in (a) and that X and Y are binomial variables, the Bayesian network for parameter learning in (b) represents prior ignorance.

Example 4.9 *Suppose we start with the DAG in Figure 4.5 (a) and X and Y are binary variables. Set $N = 2$. Then using the method in Theorem 4.1 we have*

$$a_{11} = b_{11} = \frac{N}{2q_1} = \frac{2}{2 \times 1} = 1$$

$$a_{21} = b_{21} = a_{22} = b_{22} = \frac{N}{2q_2} = \frac{2}{2 \times 2} = .5.$$

We obtain the Bayesian network for parameter learning in Figure 4.5 (b).

Note that we obtained fractional values for a_{21}, b_{21}, a_{22}, and b_{22} in the preceding example, which may seem odd. However, the sum of these values is

$$a_{21} + b_{21} + a_{22} + b_{22} = .5 + .5 + .5 + .5 = 2.$$

So these fractional values are consistent with the metaphor that says we represent prior ignorance of the $P(Y)$ by assuming the assessor's experience is equivalent to having seen two trials (see Section 4.1.1). The following is an intuitive justification for why these values should be fractional. Recall from Section 4.1.1 that we said we use fractional values when we feel it is likely that the probability of one of the outcomes is high, although we are not committed to which one it is. The smaller the values are, the more likely we feel this is the case. Now the more parents a variable has the smaller are the values of a_{ij} and b_{ij} when we set $N = 2$. Intuitively, this seems reasonable because when a variable has many parents and we know the values of the parents, we know a lot about the state of the variable, and therefore, it is more likely the probability of one of the outcomes is high.

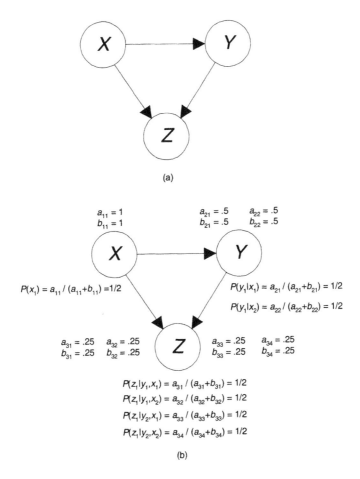

Figure 4.6: Given the DAG in (a) and that X, Y, and Z are binomial variables, the Bayesian network for parameter learning in (b) represents prior ignorance.

Example 4.10 *Suppose we start with the DAG in Figure 4.6 (a), and X and Y are binary variables. If we set $N = 2$ and use the method in Theorem 4.1, then*

$$a_{11} = b_{11} = \frac{N}{2q_1} = \frac{2}{2 \times 1} = 1$$

$$a_{21} = b_{21} = a_{22} = b_{22} = \frac{N}{2q_2} = \frac{2}{2 \times 2} = .5$$

$$a_{31} = b_{31} = a_{32} = b_{32} = a_{33} = b_{33} = a_{34} = b_{34} = \frac{N}{2q_3} = \frac{2}{2 \times 4} = .25.$$

We obtain the Bayesian network for parameter learning in Figure 4.6 (b).

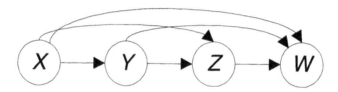

Figure 4.7: Every probability distribution of X, Y, Z, and W satisfies the Markov condition with this complete DAG.

Case	Sex	Height (inches)	Wage ($)
1	female	64	30,000
2	female	64	30,000
3	female	64	40,000
4	female	64	40,000
5	female	68	30,000
6	female	68	40,000
7	male	64	40,000
8	male	64	50,000
9	male	68	40,000
10	male	68	50,000
11	male	70	40,000
12	male	70	50,000

Table 4.1: Data on 12 workers.

4.2   Learning Structure (Model Selection)

Structure learning consists of learning the DAG in a Bayesian network from data. We want to learn a DAG that satisfies the Markov condition with the probability distribution P that is generating the data. Note that we do not know P; all we know are the data. The process of learning such a DAG is called **model selection**.

Example 4.11 *Suppose we want to model the probability distribution P of sex, height, and wage for American workers. We may obtain the data on 12 workers appearing in Table 4.1. We don't know the probability distribution P. However, from these data we want to learn a DAG that is likely to satisfy the Markov condition with P.*

If our only goal was simply learning some DAG that satisfied the Markov condition with P, our task would be trivial because a probability distribution P satisfies the Markov condition with every complete DAG containing the variables over which P is defined. We illustrate this with a DAG containing four variables. First, recall that a complete DAG is a DAG that has an edge between every pair of variables. Next consider the complete DAG in Figure 4.7. Each variable in the DAG has no nondescendents. (Recall that in this book we do not consider parents nondescendents.) So each variable is trivially independent of its nondescendents given its parents, and the Markov condition is satisfied. Another way to look at this is to notice that the chain rule (see Chapter 2, Section 2.2.1) says that for all values x, y, z, and w of X, Y, Z, and W

$$P(x, y, z, w) = P(w|z, y, x)P(z|y, x)P(x).$$

So P is equal to the product of its conditional distributions in the DAG in Figure 4.7, which means, owing to Chapter 3, Theorem 3.1, P satisfies the Markov condition with that DAG.

Recall that our goal with a Bayesian network is to represent a probability distribution succinctly. A complete DAG does not accomplish this because, if there are n binomial variables, the last variable in a complete DAG would require 2^{n-1} conditional distributions. To represent a distribution P succinctly, we need to find a sparse DAG (one containing few edges) that satisfies the Markov condition with P. The next two sections present two methods for doing this.

4.3 Score-Based Structure Learning★

After illustrating score-based structure learning using the Bayesian score, we discuss model averaging.

4.3.1 Learning Structure Using the Bayesian Score

We say a DAG **includes** a probability distribution P if the DAG does not entail any conditional independencies that are not in P. In **score-based structure learning**, we assign a score to each DAG based on the data such that in the limit (of the size of the data set) we have the following: (1) DAGs that include P score higher than DAGs that do not include P, and (2) smaller DAGs that include P score higher than larger DAGs that include P. After scoring the DAGs, we use the score, possibly along with prior probabilities, to learn a DAG.

The most straightforward score, called the **Bayesian score**, is the probability of the data D given the DAG. That is,

$$score_{Bayesian}(\mathbb{G}) = P(\mathsf{D}|\mathbb{G}).$$

We present this score shortly. However, first we need to discuss the probability of data.

Probability of Data

Suppose we are going to toss the same coin two times in a row. Let X_1 be a random variable whose value is the result of the first toss, and let X_2 be a random variable whose value is the result of the second toss. If we know that the probability of heads for this coin is .5 and make the usual assumption that the outcomes of the two tosses are independent, we have

$$P(X_1 = heads, X_2 = heads) = \frac{1}{2} \times \frac{1}{2} = \frac{1}{4}.$$

This is a standard result. Suppose now that we are going to toss a thumbtack two times in a row, and we represent prior ignorance of the probability of heads by taking $a = b = 1$. If the outcome of the first toss is heads, using the notation developed in Section 4.1, our updated probability of heads is

$$P(heads|1,0) = \frac{a+1}{a+b+1} = \frac{1+1}{1+1+1} = \frac{2}{3}.$$

Heads is more probable for the second toss because our belief has changed owing to heads occurring on the first toss. So using our current notation in which we have articulated two random variables, we have that

$$\begin{aligned}
P(X_1 = heads, X_2 = heads) &= P(X_2 = heads|X_1 = heads)P(X_1 = heads) \\
&= \frac{2}{3} \times \frac{1}{2} = \frac{1}{3}.
\end{aligned}$$

In the same way

$$\begin{aligned}
P(X_1 = heads, X_2 = tails) &= P(X_2 = tails|X_1 = heads)P(X_1 = heads) \\
&= \frac{1}{3} \times \frac{1}{2} = \frac{1}{6}
\end{aligned}$$

$$\begin{aligned}
P(X_1 = tails, X_2 = heads) &= P(X_2 = heads|X_1 = tails)P(X_1 = tails) \\
&= \frac{1}{3} \times \frac{1}{2} = \frac{1}{6}
\end{aligned}$$

$$\begin{aligned}
P(X_1 = tails, X_2 = tails) &= P(X_2 = tails|X_1 = tails)P(X_1 = tails) \\
&= \frac{2}{3} \times \frac{1}{2} = \frac{1}{3}.
\end{aligned}$$

It may seem odd to you that the four outcomes do not have the same probability. However, recall that we do not know the probability of heads. Therefore, we learn something about the probability of heads from the result of the first toss. If heads occurs on the first toss, that probability goes up, while if tails occurs, it goes down. So two consecutive heads or two consecutive tails are more probable *a priori* than a head followed by a tail or a tail followed by a head.

The above result extends readily to a sequence of tosses. For example, suppose we toss the thumbtack three times. Then owing to the chain rule,

$$\begin{aligned}
&P(X_1 = heads, X_2 = tails, X_3 = tails) \\
&= P(X_3 = tails|X_2 = tails, X_1 = heads)P(X_2 = tails|X_1 = heads) \\
&\quad P(X_1 = heads) \\
&= \frac{b+1}{a+b+2} \times \frac{b}{a+b+1} \times \frac{a}{a+b} \\
&= \frac{1+1}{1+1+2} \times \frac{1}{1+1+1} \times \frac{1}{1+1} = .0833.
\end{aligned}$$

We get the same probability regardless of the order of the outcomes as long as the number of heads and tails is the same. For example,

$$P(X_1 = tails, X_2 = tails, X_3 = heads)$$

$$= P(X_3 = heads|X_2 = tails, X_1 = tails)P(X_2 = tails|X_1 = tails)$$
$$P(X_1 = tails)$$

$$= \frac{a}{a+b+2} \times \frac{b+1}{a+b+1} \times \frac{b}{a+b}$$

$$= \frac{1}{1+1+2} \times \frac{2}{1+1+1} \times \frac{1}{1+1} = .0833.$$

This result is actually the assumption of exchangeability, which was discussed in Section 4.1.1.

We now have the following theorem:

Theorem 4.2 *Suppose we are about to repeatedly toss a thumbtack (or perform any repeatable experiment with two outcomes), and our prior experience is equivalent to having seen a heads and b tails in m trials, where a and b are positive integers. Let D be data that consist of s heads and t tails in n trials.*

$$P(D) = \frac{(m-1)!}{(m+n-1)!} \times \frac{(a+s-1)!(b+t-1)!}{(a-1)!(b-1)!}.$$

Proof. *Since the probability is the same regardless of the order in which the heads and tails occur, we can assume all the heads occur first. We therefore have (as before the notation s, t on the right side of the conditioning bar means we have seen s heads and t tails)*

$P(D)$

$$= P(X_{s+t} = tails|s, t-1) \cdots P(X_{s+2} = tails|s, 1)P(X_{s+1} = tails|s, 0)$$
$$P(X_s = heads|s-1, 0) \cdots P(X_2 = heads|1, 0)P(X_1 = heads)$$

$$= \frac{b+t-1}{a+b+s+t-1} \times \cdots \frac{b+1}{a+b+s+1} \times \frac{b}{a+b+s} \times$$
$$\frac{a+s-1}{a+b+s-1} \times \cdots \frac{a+1}{a+b+1} \times \frac{a}{a+b}$$

$$= \frac{(a+b-1)!}{(a+b+s+t-1)!} \times \frac{(a+s-1)!}{(a-1)!} \times \frac{(b+t-1)!}{(b-1)!}$$

$$= \frac{(m-1)!}{(m+n-1)!} \times \frac{(a+s-1)!(b+t-1)!}{(a-1)!(b-1)!}.$$

This completes the proof. ∎

Example 4.12 *Suppose, before tossing a thumbtack, we assign a = 3 and b = 5 to model the slight belief that tails is more probable than heads. We then* *toss the coin* 10 *times and obtain* 4 *heads and* 6 *tails. Owing to the preceding theorem, the probability of obtaining these data D is given by*

$$
\begin{aligned}
P(D) &= \frac{(m-1)!}{(m+n-1)!} \times \frac{(a+s-1)!(b+t-1)!}{(a-1)!(b-1)!} \\
&= \frac{(8-1)!}{(8+10-1)!} \times \frac{(3+4-1)!(5+6-1)!}{(3-1)!(5-1)!} = .00077.
\end{aligned}
$$

Note that the probability of these data is very small. This is because there are many possible outcomes (namely 2^{10}) of tossing a thumbtack 10 *times.*

Theorem 4.2 *holds even if a and b are not integers. The proof can be found in* [Neapolitan, 2004]. *We merely state the result here.*

Theorem 4.3 *Suppose we are about to repeatedly toss a thumbtack (or perform any repeatable experiment with two outcomes), and we represent our prior experience concerning the probabilities of heads and tails using positive real numbers a and b, where $m = a + b$. Let D be data that consist of s heads and t tails in n trials. Then*

$$
P(D) = \frac{\Gamma(m)}{\Gamma(m+n)} \times \frac{\Gamma(a+s)\Gamma(b+t)}{\Gamma(a)\Gamma(b)}. \tag{4.3}
$$

In the preceding theorem Γ denotes the gamma function. When n is an integer ≥ 1, we have that

$$
\Gamma(n) = (n-1)!
$$

So the preceding theorem generalizes Theorem 4.2.

Example 4.13 *Recall that in Example* 4.7 *we set $a = 1/360$ and $b = 19/360$. Then after seeing 3 windows containing NiS, our updated values of a and b became as follows:*

$$
\begin{aligned}
a &= \frac{1}{360} + 3 = \frac{1081}{360} \\
b &= \frac{19}{360} + 0 = \frac{19}{360}.
\end{aligned}
$$

We then wanted the probability that the next 36 windows would contain NiS. This is the probability of obtaining data with $s = 36$ and $t = 0$. Owing to the previous theorem, the probability of these data D is given by

$$
\begin{aligned}
P(D) &= \frac{\Gamma(m)}{\Gamma(m+n)} \times \frac{\Gamma(a+s)\Gamma(b+t)}{\Gamma(a)\Gamma(b)} \\
&= \frac{\Gamma(\frac{1100}{360})}{\Gamma(\frac{1100}{360}+36)} \times \frac{\Gamma(\frac{1081}{360}+36)\Gamma(\frac{19}{360}+0)}{\Gamma(\frac{1081}{360})\Gamma(\frac{19}{360})} \\
&= .866.
\end{aligned}
$$

Recall that we obtained this same result by direct computation at the end of Example 4.7.

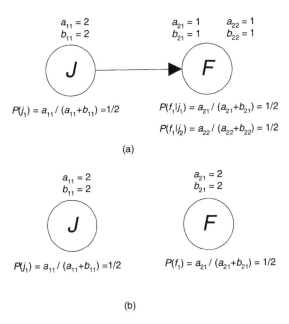

$a_{11} = 2$
$b_{11} = 2$

$a_{21} = 1$ $a_{22} = 1$
$b_{21} = 1$ $b_{22} = 1$

$P(j_1) = a_{11} / (a_{11}+b_{11}) = 1/2$

$P(f_1|j_1) = a_{21} / (a_{21}+b_{21}) = 1/2$

$P(f_1|j_2) = a_{22} / (a_{22}+b_{22}) = 1/2$

(a)

$a_{11} = 2$
$b_{11} = 2$

$a_{21} = 2$
$b_{21} = 2$

$P(j_1) = a_{11} / (a_{11}+b_{11}) = 1/2$

$P(f_1) = a_{21} / (a_{21}+b_{21}) = 1/2$

(b)

Figure 4.8: The network in (a) models that J has a causal effect on F, while the network in (b) models that neither variable causes the other.

We developed the method for computing the probability of data for the case of binomial variables. It extends readily to multinomial variables. See [Neapolitan, 2004] for that extension.

Probability of DAG Models and DAG Learning

Next, we show how we score a DAG model by computing the probability of the data given the model and then how we use that score to learn a DAG.

Models with Two Nodes   First, we show how to learn two-node networks from data. Then we discuss using the learned network to do inference.

Learning Two-Node Networks   Suppose we have a Bayesian network for learning as discussed in Section 4.1.2. For example, we may have the network in Figure 4.8 (a). Here, we call such a network a **DAG model**. We can score the DAG model $\mathbb{G}$ based on data D by determining how probable the data are given the DAG model. That is, we compute $P(D|\mathbb{G})$. The formula for this probability is the same as that developed in the previous subsection, except there is a term of the form in Equality 4.3 for each probability in the network. So the probability of data D given the DAG model $\mathbb{G}_1$ in Figure 4.8 (a) is given by

$$P(\mathsf{D}|\mathbb{G}_1) = \frac{\Gamma(m_{11})}{\Gamma(m_{11}+n_{11})} \times \frac{\Gamma(a_{11}+s_{11})\Gamma(b_{11}+t_{11})}{\Gamma(a_{11})\Gamma(b_{11})} \times \qquad (4.4)$$

$$\frac{\Gamma(m_{21})}{\Gamma(m_{21}+n_{21})} \times \frac{\Gamma(a_{21}+s_{21})\Gamma(b_{21}+t_{21})}{\Gamma(a_{21})\Gamma(b_{21})} \times$$

$$\frac{\Gamma(m_{22})}{\Gamma(m_{22}+n_{22})} \times \frac{\Gamma(a_{22}+s_{22})\Gamma(b_{22}+t_{22})}{\Gamma(a_{22})\Gamma(b_{22})}.$$

The data used in each term include only the data relevant to the conditional probability the term represents. This is exactly the same scheme that was used to learn parameters in Section 4.1.2. For example, the value of s_{21} is the number of cases that have J equal to j_2 and F equal to f_1.

Similarly, the probability of data D given the DAG model $\mathbb{G}_2$ in Figure 4.8 (b) is given by

$$P(\mathsf{D}|\mathbb{G}_2) = \frac{\Gamma(m_{11})}{\Gamma(m_{11}+n_{11})} \times \frac{\Gamma(a_{11}+s_{11})\Gamma(b_{11}+t_{11})}{\Gamma(a_{11})\Gamma(b_{11})} \times \qquad (4.5)$$

$$\frac{\Gamma(m_{21})}{\Gamma(m_{21}+n_{21})} \times \frac{\Gamma(a_{21}+s_{21})\Gamma(b_{21}+t_{21})}{\Gamma(a_{21})\Gamma(b_{21})}.$$

Note that the values of a_{11}, s_{11}, etc. in Equality 4.5 are the ones relevant to $\mathbb{G}_2$ and are not the same values as those in Equality 4.4. We have not explicitly shown their dependence on the DAG model for the sake of notational simplicity.

Example 4.14 *Suppose we wish to determine whether job status (J) has a causal effect on whether someone defaults on a loan (F). Furthermore, we articulate just two values for each variable as follows:*

Variable	Value	When the Variable Takes This Value
J	j_1	Individual is a white collar worker.
	j_2	Individual is a blue collar worker.
F	f_1	Individual has defaulted on a loan at least once.
	f_2	Individual has never defaulted on a loan.

We represent the assumption that J has a causal effect on F with the DAG model $\mathbb{G}_1$ in Figure 4.8 (a) and the assumption that neither variable has a causal effect on the other with the DAG model $\mathbb{G}_2$ in Figure 4.8 (b). We assume that F does not have a causal effect on J; so we do not model this situation. Note that in both models we used a prior equivalent sample size of 4 and we represented the prior belief that all probabilities are .5. In general, we can use whatever prior sample size and prior belief that best model our prior knowledge. The only requirement is that both DAG models have the same prior equivalent sample size.

Suppose that next we obtain the data D in the following table:

Case	J	F
1	j_1	f_1
2	j_1	f_1
3	j_1	f_1
4	j_1	f_1
5	j_1	f_2
6	j_2	f_1
7	j_2	f_2
8	j_2	f_2

Then, owing to Equality 4.4,

$$
\begin{aligned}
P(\mathsf{D}|\mathbb{G}_1) &= \frac{\Gamma(m_{11})}{\Gamma(m_{11}+n_{11})} \times \frac{\Gamma(a_{11}+s_{11})\Gamma(b_{11}+t_{11})}{\Gamma(a_{11})\Gamma(b_{11})} \times \\
& \quad \frac{\Gamma(m_{21})}{\Gamma(m_{21}+n_{21})} \times \frac{\Gamma(a_{21}+s_{21})\Gamma(b_{21}+t_{21})}{\Gamma(a_{21})\Gamma(b_{21})} \times \\
& \quad \frac{\Gamma(m_{22})}{\Gamma(m_{22}+n_{22})} \times \frac{\Gamma(a_{22}+s_{22})\Gamma(b_{22}+t_{22})}{\Gamma(a_{22})\Gamma(b_{22})} \\
&= \frac{\Gamma(4)}{\Gamma(4+8)} \times \frac{\Gamma(2+5)\,\Gamma(2+3)}{\Gamma(2)\Gamma(2)} \times \\
& \quad \frac{\Gamma(2)}{\Gamma(2+5)} \times \frac{\Gamma(1+4)\,\Gamma(1+1)}{\Gamma(1)\Gamma(1)} \times \\
& \quad \frac{\Gamma(2)}{\Gamma(2+3)} \times \frac{\Gamma(1+1)\,\Gamma(1+2)}{\Gamma(1)\Gamma(1)} \\
&= 7.2150 \times 10^{-6}.
\end{aligned}
$$

Furthermore,

$$
\begin{aligned}
P(\mathsf{D}|\mathbb{G}_2) &= \frac{\Gamma(m_{11})}{\Gamma(m_{11}+n_{11})} \times \frac{\Gamma(a_{11}+s_{11})\Gamma(b_{11}+t_{11})}{\Gamma(a_{11})\Gamma(b_{11})} \times \\
& \quad \frac{\Gamma(m_{21})}{\Gamma(m_{21}+n_{21})} \times \frac{\Gamma(a_{21}+s_{21})\Gamma(b_{21}+t_{21})}{\Gamma(a_{21})\Gamma(b_{21})} \times \\
&= \frac{\Gamma(4)}{\Gamma(4+8)} \times \frac{\Gamma(2+5)\,\Gamma(2+3)}{\Gamma(2)\Gamma(2)} \times \\
& \quad \frac{\Gamma(4)}{\Gamma(4+8)} \times \frac{\Gamma(2+5)\,\Gamma(2+3)}{\Gamma(2)\Gamma(2)} \\
&= 6.7465 \times 10^{-6}.
\end{aligned}
$$

If our prior belief is that neither model is more probable than the other, we assign

$$
P(\mathbb{G}_1) = P(\mathbb{G}_2) = .5.
$$

Then, owing to Bayes' Theorem,

$$
\begin{aligned}
P(\mathbb{G}_1|\mathsf{D}) &= \frac{P(\mathsf{D}|\mathbb{G}_1)P(\mathbb{G}_1)}{P(\mathsf{D}|\mathbb{G}_1)P(\mathbb{G}_1) + P(\mathsf{D}|\mathbb{G}_2)P(\mathbb{G}_2)} \\
&= \frac{7.2150 \times 10^{-6} \times .5}{7.2150 \times 10^{-6} \times .5 + 6.7465 \times 10^{-6} \times .5} \\
&= .517
\end{aligned}
$$

and

$$
\begin{aligned}
P(\mathbb{G}_2|\mathsf{D}) &= \frac{P(\mathsf{D}|\mathbb{G}_2)P(\mathbb{G}_2)}{P(\mathsf{D})} \\
&= \frac{6.7465 \times 10^{-6}(.5)}{7.2150 \times 10^{-6} \times .5 + 6.7465 \times 10^{-6} \times .5} \\
&= .483.
\end{aligned}
$$

We select (learn) DAG $\mathbb{G}_1$ and conclude that it is more probable that job status does have a causal effect on whether someone defaults on a loan.

Example 4.15 *Suppose we are doing the same study as in Example 4.14, and we obtain the data in the following table:*

Case	J	F
1	j_1	f_1
2	j_1	f_1
3	j_1	f_1
4	j_1	f_1
5	j_2	f_2
6	j_2	f_2
7	j_2	f_2
8	j_2	f_2

Then it is left as an exercise to show

$$P(\mathsf{D}|\mathbb{G}_1) = 8.6580 \times 10^{-5}$$

$$P(\mathsf{D}|\mathbb{G}_2) = 4.6851 \times 10^{-6},$$

and if we assign the same prior probability to both DAG models,

$$
\begin{aligned}
P(\mathbb{G}_1|\mathsf{D}) &= .949 \\
P(\mathbb{G}_2|\mathsf{D}) &= .051.
\end{aligned}
$$

We select (learn) DAG $\mathbb{G}_1$. Notice that the causal model is substantially more probable. This makes sense because even though there is not much data, it exhibits complete dependence.

Example 4.16 *Suppose we are doing the same study as in Example 4.14, and we obtain the data D in the following table:*

Case	J	F
1	j_1	f_1
2	j_1	f_1
3	j_1	f_2
4	j_1	f_2
5	j_2	f_1
6	j_2	f_1
7	j_2	f_2
8	j_2	f_2

Then it is left as an exercise to show

$$P(\mathsf{D}|\mathbb{G}_1) = 2.4050 \times 10^{-6}$$

$$P(\mathsf{D}|\mathbb{G}_2) = 4.6851 \times 10^{-6},$$

and if we assign the same prior probability to both DAG models,

$$P(\mathbb{G}_1|\mathsf{D}) = .339$$
$$P(\mathbb{G}_2|\mathsf{D}) = .661.$$

We select (learn) DAG $\mathbb{G}_2$. Notice that it is somewhat more probable that the two variables are independent. This makes sense since the data exhibit complete independence.

Using the Learned Network to Do Inference Once we learn a DAG from data, we can then learn the parameters. The result will be a Bayesian network which we can use to do inference for the next case. The next example illustrates the technique.

Example 4.17 *Suppose we have the situation and data in Example 4.14. That is, we have the data D in the following table:*

Case	J	F
1	j_1	f_1
2	j_1	f_1
3	j_1	f_1
4	j_1	f_1
5	j_1	f_2
6	j_2	f_1
7	j_2	f_2
8	j_2	f_2

Then, as shown in Example 4.14, we would learn the DAG in Figure 4.8 (a). Next, we can update the conditional probabilities in the Bayesian network for learning in Figure 4.8 (a) using the above data and the parameter learning technique discussed in Section 4.1.2. The result is the Bayesian network in Figure 4.9. Suppose now that we find out that Sam has $F = f_2$. That is, Sam

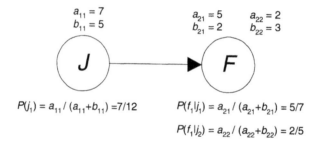

Figure 4.9: An updated Bayesian network for learning based on the data in Example 4.17.

has never defaulted on a loan. We can use the Bayesian network to compute the probability that Sam is a white collar worker. For this simple network we can just use Bayes' Theorem as follows

$$P(j_1|f_1) \;=\; \frac{P(f_1|j_1)P(j_1)}{P(f_1|j_1)P(j_1) + Pf_1|j_2)P(j_2)}$$

$$=\; \frac{(5/7)\,(7/12)}{(5/7)\,(7/12) + (2/5)\,(5/12)} = .714.$$

The probabilities in the previous calculation are all conditional on the data **D** and the DAG that we select. However, once we select a DAG and learn the parameters, we do not bother to show this dependence.

Bigger DAG Models We illustrated scoring using only two variables. Ordinarily, we want to learn structure when there are many more variables. When there are only a few variables, we can exhaustively score all possible DAGs as was done in the previous examples. We then select the DAG(s) with the highest score. However, when the number of variables is not small, to find the maximizing DAGs by exhaustively considering all DAG patterns is computationally unfeasible. That is, Robinson [1977] showed that the number of DAGs containing n nodes is given by the following recurrence:

$$f(n) \;=\; \sum_{i=1}^{n}(-1)^{i+1}\binom{n}{i}2^{i(n-i)}f(n-i) \qquad n > 2$$
$$f(0) \;=\; 1$$
$$f(1) \;=\; 1.$$

It is left as an exercise to show $f(2) = 3$, $f(3) = 25$, $f(5) = 29{,}000$, and $f(10) = 4.2 \times 10^{18}$. Furthermore, Chickering [1996] proved that for certain classes of prior distributions, the problem of finding a highest scoring DAG is NP-hard. So researchers developed heuristic DAG search algorithms. It is beyond the scope of this book to discuss these algorithms. See [Neapolitan, 2004] for a detailed discussion of them.

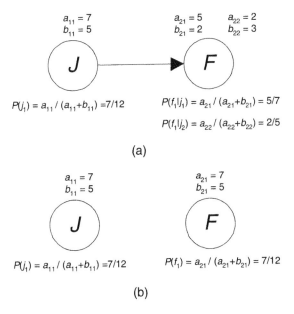

$$P(j_1) = a_{11} / (a_{11}+b_{11}) = 7/12 \qquad P(f_1|j_1) = a_{21} / (a_{21}+b_{21}) = 5/7$$

$$P(f_1|j_2) = a_{22} / (a_{22}+b_{22}) = 2/5$$

(a)

$$P(j_1) = a_{11} / (a_{11}+b_{11}) = 7/12 \qquad P(f_1) = a_{21} / (a_{21}+b_{21}) = 7/12$$

(b)

Figure 4.10: Updated Bayesian network for learning based on the data in Examples 4.14 and 4.17.

4.3.2 Model Averaging

Heckerman et al. [1999] illustrate that when the number of variables is small and the amount of data is large, one structure can be orders of magnitude more likely than any other. In such cases, model selection yields good results. However, recall that in Example 4.14 we had little data, we obtained $P(\mathbb{G}_1|\mathbf{D}) = .517$ and $P(\mathbb{G}_2|\mathbf{D}) = .483$, and we chose (learned) DAG $\mathbb{G}_1$ because it was most probable. Then in Example 4.17 we used a Bayesian network containing DAG $\mathbb{G}_1$ to do inference for Sam. Since the probabilities of the two models are so close, it seems somewhat arbitrary to choose $\mathbb{G}_1$. So model selection does not seem appropriate. Next, we describe another approach.

Instead of choosing a single DAG and then using it to do inference, we could use the law of total probability to do the inference as follows: we perform the inference using each DAG and multiply the result (a probability value) by the posterior probably of the DAG. This is called **model averaging**.

Example 4.18 *Recall that based on the data in Example 4.14 we learned that*

$$P(\mathbb{G}_1|\mathbf{D}) = .517$$

and

$$P(\mathbb{G}_2|\mathbf{D}) = .483.$$

In Example 4.17 we updated a Bayesian network containing $\mathbb{G}_1$ based on the data to obtain the Bayesian network in Figure 4.10 (a). If in the same way we update a Bayesian network containing $\mathbb{G}_2$, we obtain the Bayesian network in Figure 4.10 (b). Given that Sam has never defaulted on a loan ($F = f_2$), we can then use model averaging to compute the probability that Sam is a white collar worker as follows:[1]

$$
\begin{aligned}
P(j_1|f_1, \mathsf{D}) &= P(j_1|f_1, \mathbb{G}_1)P(\mathbb{G}_1|\mathsf{D}) + P(j_1|f_1, \mathbb{G}_2)P(\mathbb{G}_2|\mathsf{D}) \\
&= (.714)(.517) + (7/12)(.483) = .651.
\end{aligned}
$$

The result that $P(j_1|f_1, \mathbb{G}_1) = .714$ was obtained in Example 4.17, although in that example we did not show the dependence on $\mathbb{G}_1$ because that DAG was the only DAG considered. The result that $P(j_1|f_1, \mathbb{G}_2) = 7/12$ is obtained directly from the Bayesian network in Figure 4.10 (b) because J and F are independent in that network.

As is the case for model selection, when the number of possible DAGs is large, we cannot average over all DAGs. In these situations we heuristically search for high probability DAGs, and then we average over them. Such techniques are discussed in [Neapolitan, 2004].

4.4   Constraint-Based Structure Learning

Next, we discuss a quite different structure learning technique called constraint-based learning. In the **constraint-based approach**, we try to find a DAG for which the Markov condition entails all and only those conditional independencies that are in the probability distribution P of the variables of interest. First, we illustrate the constraint-based approach by showing how to learn a DAG faithful to a probability distribution. This is followed by a discussion of embedded faithfulness. Finally, we present causal learning.

4.4.1   Learning a DAG Faithful to P

Recall that $(\mathbb{G}, P)$ satisfies the faithfulness condition if all and only the conditional independencies in P are entailed by $\mathbb{G}$. After discussing why we would want to learn a DAG faithful to a probability distribution P, we illustrate learning such a DAG.

Why We Want a Faithful DAG

Consider again the objects in Chapter 2, Figure 2.1, which appear also in Chapter 3, Figure 3.4. In Chapter 2, Example 2.21, we let P assign 1/13 to each object in the figure, and we defined these random variables on the set containing the objects:

---

[1]Note that we substitute $P(\mathbb{G}_1|\mathsf{D})$ for $P(\mathbb{G}_1|f_1, \mathsf{D})$. They are not exactly equal, but we are assuming that the data set is sufficiently large so that the dependence on F can be ignored.

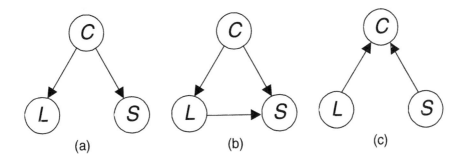

Figure 4.11: If the only conditional independency in P is $I_P(L, S|C)$, then P sastifies the Markov condition with the DAGs in (a) and (b), and P sastifies the faithfulness condition only with the DAG in (a).

Variable	Value	Outcomes Mapped to This Value
L	l_1	All objects containing an "A"
	l_2	All objects containing a "B"
S	s_1	All square objects
	s_2	All diamond-shaped objects
C	c_1	All black objects
	c_2	All white objects

We then showed that L and S are conditionally independent given C. That is, using the notation established in Chapter 2, Section 2.2.2, we showed that

$$I_P(L, S|C).$$

In Chapter 3, Example 3.1, we showed that this implies P satisfies the Markov condition with the DAG in Figure 4.11 (a). However, P also satisfies the Markov condition with the complete DAG in Figure 4.11 (b). P does not satisfy the Markov condition with the DAG in Figure 4.11 (c) because that DAG entails $I_P(L, S)$ and this independency is not in P. The DAG in Figure 4.11 (b) does not represent P very well because it does not entail a conditional independency that is in P, namely $I_P(L, S|C)$. This is a violation of the faithfulness condition. Only the DAG in Figure 4.11 (a) is faithful to P. If we can find a DAG that is faithful to P, we have achieved our goal of representing P succinctly. As we shall see, not every P has a DAG that is faithful to it. However, if there are DAGs (there could be more than one) faithful to P, it is relatively easy to discover them. We discuss learning a faithful DAG next.

Learning a Faithful DAG

We assume that we have a sample of entities from the population over which the random variables are defined, and we know the values of the variables of interest for the entities in the sample. The sample could be a random sample, or it could be obtained from passive data. From this sample, we have deduced

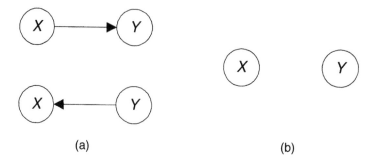

(a)   (b)

Figure 4.12: If the set of conditional independencies is $\{I(X,Y)\}$, we must have the DAG in (b), whereas if it is $\varnothing$, we must have one of the DAGs in (a).

the conditional independencies among the variables. A method for deducing conditional independencies and obtaining a measure of our confidence in them is described in [Spirtes et al., 1993, 2000] and [Neapolitan, 2004]. Our confidence in the DAG we learn is no greater than our confidence in these conditional independencies.

Example 4.19 *It is left as an exercise to show that the data shown in Example 4.11 exhibit this conditional independency:*

$$I_P(Height, Wage | Sex).$$

Therefore, from these data we can conclude, with a certain measure of confidence, that this conditional independency exists in the population at large.

Next, we give a sequence of examples in which we learn a DAG faithful to the probability distribution of interest. These examples illustrate how a faithful DAG can be learned from the conditional independencies if one exists. We stress again that the DAG is faithful to the conditional independencies we have learned from the data. We are not certain that these are the conditional independencies in the probability distribution for the entire population.

Example 4.20 *Suppose* V *is our set of observed variables,* $\mathsf{V} = \{X, Y\}$, *and the set of conditional independencies in P is*

$$\{I_P(X,Y)\}.$$

We want to find a DAG faithful to P. We cannot have either of the DAGs in Figure 4.12 (a). The reason is that these DAGs do not entail that X and Y are independent, which means the faithfulness condition is not satisfied. So we must have the DAG in Figure 4.12 (b). We conclude P is faithful to the DAG in Figure 4.12 (b).

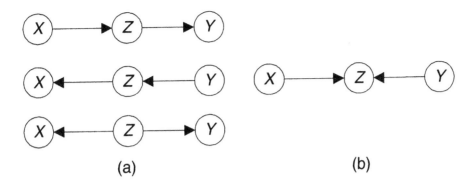

(a)   (b)

Figure 4.13: If the set of conditional independencies is $\{I_P(X,Y)\}$, we must have the DAG in (b); if it is $\{I_P(X,Y|Z)\}$, we must have one of the DAGs in (a).

Example 4.21 *Suppose* $\mathsf{V} = \{X, Y\}$, *and the set of conditional independencies in* P *is the empty set*

$$\varnothing.$$

That is, there are no independencies. We want to find a DAG faithful to P. *We cannot have the DAG in Figure 4.12 (b). The reason is that this DAG entails that* X *and* Y *are independent, which means the Markov condition is not satisfied. So we must have one of the DAGs in Figure 4.12 (a). We conclude* P *is faithful to both the DAGs in Figure 4.12 (a).*

The previous example shows that there is not necessarily a unique DAG with which a probability distribution is faithful.

Example 4.22 *Suppose* $\mathsf{V} = \{X, Y, Z\}$, *and the set of conditional independencies in* P *is*

$$\{I_P(X,Y)\}.$$

We want to find a DAG faithful to P. *There can be no edge between* X *and* Y *in the DAG owing to the reason given in Example 4.20. Furthermore, there must be edges between* X *and* Z *and between* Y *and* Z *owing to the reason given in Example 4.21. We cannot have any of the DAGs in Figure 4.13 (a). The reason is that these DAGs entail* $I(X,Y|Z)$, *and this conditional independency is not present. So the Markov condition is not satisfied. Furthermore, the DAGs do not entail* $I(X,Y)$. *So the DAG must be the one in Figure 4.13 (b). We conclude* P *is faithful to the DAG in Figure 4.13 (b).*

Example 4.23 *Suppose* $\mathsf{V} = \{X, Y, Z\}$, *and the set of conditional independencies in* P *is*

$$\{I_P(X,Y|Z)\}.$$

We want to find a DAG faithful to P. *Owing to reasons similar to those given before, the only edges in the DAG must be between* X *and* Z *and between* Y

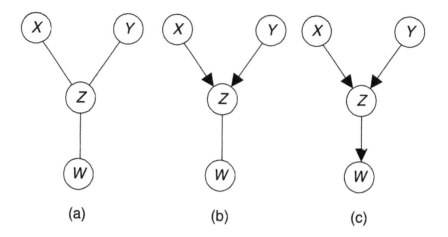

Figure 4.14: If the set of conditional independencies is $\{I_p(X, \{Y, W\}),$ $I_P(Y, \{X, Z\})\}$, we must have the DAG in (c).

and Z. We cannot have the DAG in Figure 4.13 (b). The reason is that this DAG entails $I(X, Y)$, and this conditional independency is not present. So the Markov condition is not satisfied. So we must have one of the DAGs in Figure 4.13 (a). We conclude P is faithful to all the DAGs in Figure 4.13 (a).

We now state a theorem whose proof can be found in [Neapolitan, 2004]. At this point your intuition should suspect that it is true.

Theorem 4.4 If $(\mathbb{G}, P)$ satisfies the faithfulness condition, then there is an edge between X and Y if and only if X and Y are not conditionally independent given any set of variables.

Example 4.24 Suppose $V = \{X, Y, Z, W\}$, and the set of conditional independencies in P is

$$\{I_p(X, Y), \quad I_P(W, \{X, Y\}|Z)\}.$$

We want to find a DAG faithful to P. Owing to Theorem 4.4, the links (edges without regard for direction) must be as shown in Figure 4.14 (a). We must have the directed edges shown in Figure 4.14 (b) because we have $I_p(X, Y)$. Therefore, we must also have the directed edge shown in Figure 4.14 (c) because we do not have $I_P(W, X)$. We conclude P is faithful to the DAG in Figure 4.14 (c).

Example 4.25 Suppose $V = \{X, Y, Z, W\}$, and the set of conditional independencies in P is

$$\{I_P(X, \{Y, W\}), \quad I_P(Y, \{X, Z\})\}.$$

We want to find a DAG faithful to P. Owing to Theorem 4.4, we must have the links shown in Figure 4.15 (a). Now, if we have the chain $X \rightarrow Z \rightarrow W$,

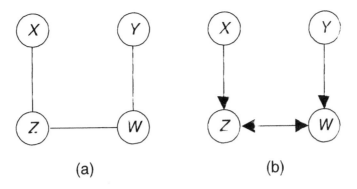

Figure 4.15: If the set of conditional independencies is $\{I_P(X,\{Y,W\}),$ $I_P(Y,\{X,Z\})\}$ and we try to find a DAG faithful to P, we obtain the graph in (b) which is not a DAG.

$X \leftarrow Z \leftarrow W$, or $X \leftarrow Z \rightarrow W$, then we do not have $I_P(X,W)$. So we must have the chain $X \rightarrow Z \leftarrow W$. Similarly, we must have the chain $Y \rightarrow W \leftarrow Z$. So our graph must be the one in Figure 4.15 (b). However, this graph is not a DAG. The problem here is that there is no DAG faithful to P.

Example 4.26 *Again suppose* $V = \{X,Y,Z,W\}$*; and the set of conditional independencies in P is*

$$\{I_P(X,\{Y,W\}), \quad I_P(Y,\{X,Z\})\}.$$

As shown in the previous example, there is no DAG faithful to P. However, this does not mean we cannot find a more succinct way to represent P than using a complete DAG. P satisfies the Markov condition with each of the DAGs in Figure 4.16. That is, the DAG in Figure 4.16 (a) entails

$$\{I_P(X,Y), \quad I_P(Y,Z)\},$$

and these conditional independencies are both in P, while the DAG in Figure 4.16 (b) entails
$$\{I_P(X,Y), \quad I_P(X,W)\},$$

and these conditional independencies are both in P. However, P does not satisfy the faithfulness condition with either of these DAGs because the DAG in Figure 4.16 (a) does not entail $I_P(X,W)$, while the DAG in Figure 4.16 (b) does not entail $I_P(Y,Z)$. Each of these DAGs is as succinctly as we can represent the probability distribution. So when there is no DAG faithful to a probability distribution P, we can still represent P much more succinctly than we would by using the complete DAG. A structure learning algorithm tries to find the most succinct representation. Depending on the number of alternatives of each variable, one of the DAGs in Figure 4.16 may actually be a more succinct representation than the other because it contains fewer parameters. A

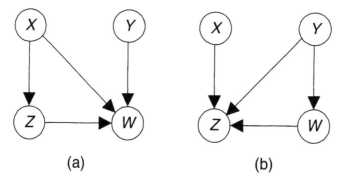

Figure 4.16: If the set of conditional independencies is $\{I_P(X, \{Y, W\}),$ $I_P(Y, \{X, Z\})\}$, P sastisfies the Markov condition with both these DAGs.

constraint-based learning algorithm could not distinguish between the two, but a score-based one could. See [Neapolitan, 2004] for a complete discussion of this matter.

4.4.2   Learning a DAG in Which P Is Embedded Faithfully$\star$

In Example 4.26 in a sense we compromised because the DAG we learned did not entail all the conditional independencies in P. This is fine if our goal is to learn a Bayesian network which will later be used to do inference. However, another application of structure learning is causal learning, which is discussed in the next subsection. When learning causes it would be better to find a DAG in which P is embedded faithfully. We discuss embedded faithfulness next.

Definition 4.2 *Suppose we have a joint probability distribution P of the random variables in some set* V *and a DAG* $\mathbb{G} = (W, E)$ *such that* V $\subseteq$ W. *We say that* $(\mathbb{G}, P)$ *satisfy the **embedded faithfulness condition** if all and only the conditional independencies in P are entailed by* $\mathbb{G}$, *restricted to variables in* V. *Furthermore, we say that P is embedded faithfully in* $\mathbb{G}$.

Example 4.27 *Again suppose* V $= \{X, Y, Z, W\}$, *and the set of conditional independencies in P is*

$$\{I_P(X, \{Y, W\}), \quad I_P(Y, \{X, Z\})\}.$$

Then P is embedded faithfully in the DAG in Figure 4.17. It is left as an exercise to show this. By including the variable H in the DAG, we are able to entail all and only the conditional independencies in P restricted to variables in V.

Variables such as H are called hidden variables because they are not among the observed variables. By including them in the DAG, we can achieve faithfulness.

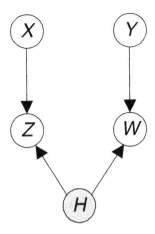

Figure 4.17: If the set of conditional independencies in P is $\{I_P(X, \{Y, W\}),$ $I_P(Y, \{X, Z\})\}$, then P is embedded faithfully in this DAG.

4.5 Causal Learning

In many, if not most, applications the variables of interest have causal relationships to each other. For example, the variables in Example 4.11 are causally related in that sex has a causal effect on height and may have a causal effect on wage. If the variables are causally related, we can learn something about their causal relationships when we learn the structure of the DAG from data. However, we must make certain assumptions to do this. We discuss these assumptions and causal learning next.

4.5.1 Causal Faithfulness Assumption

Recall from Chapter 3, Section 3.3.2, that if we assume the observed probability distribution P of a set of random variables V satisfies the Markov condition with the causal DAG $\mathbb{G}$ containing the variables, we say we are making the **causal Markov assumption**, and we call $(\mathbb{G}, P)$ a **causal network**. Furthermore, we concluded that the causal Markov assumption is justified for a causal graph if the following conditions are satisfied:

1. There are no hidden common causes. That is, all common causes are represented in the graph.

2. There are no causal feedback loops. That is, our graph is a DAG.

3. Selection bias is not present.

Recall the discussion concerning credit card fraud in Chapter 3, Section 3.1. Suppose that both fraud and sex do indeed have a causal effect on whether jewelry is purchased, and there are no other causal relationships among the

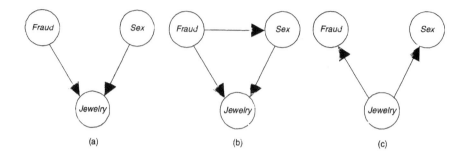

Figure 4.18: If the only causal relationships are that *Fraud* and *Sex* have causal influences on *Jewelry*, then the causal DAG is the one in (a). If we make the causal Markov assumption, only the DAG in (c) is ruled out if we observe $I_P(Fraud, Sex)$.

variables. Then the causal DAG containing these three variables is the one in Figure 4.18 (a). If we make the causal Markov assumption, we must have $I_P(Fraud, Sex)$.

Suppose now that we do not know the causal relationships among the variables, but we make the causal Markov assumption, and we learn only the conditional independency $I_P(Fraud, Sex)$ from data. Can we conclude that the causal DAG must be the one in Figure 4.18 (a)? No, we cannot because P also satisfies the Markov condition with the DAG in Figure 4.18 (b). This is a bit tricky to understand. However, recall that we are assuming that we do not know the causal relationships among the variables. As far as we know, they could be the ones in Figure 4.18 (b). If the DAG in Figure 4.18 (b) were the causal DAG, the causal Markov assumption would still be satisfied when the only conditional independency in P is $I_P(Fraud, Sex)$ because that DAG satisfies the Markov condition with P. So by making only the causal Markov assumption, we cannot distinguish the causal DAGs in Figures 4.18 (a) and (b) based on the conditional independency $I_P(Fraud, Sex)$. The causal Markov assumption only enables us to rule out causal DAGs that contain conditional independencies that are not in P. One such DAG is the one in Figure 4.18 (c). We need to make the causal faithfulness assumption to conclude the causal DAG is the one in Figure 4.18 (a). That assumption is as follows: If we assume the observed probability distribution P of a set of random variables V satisfies the faithfulness condition with the causal $\mathbb{G}$ containing the variables, then we say we are making the **causal faithfulness assumption**. If we make the causal faithfulness assumption, then if we find a unique DAG that is faithful to P, the edges in that DAG must represent causal influences. This is illustrated by the following examples.

Example 4.23 *Recall that in Example 4.22 we showed that if* V = $\{X, Y, Z\}$

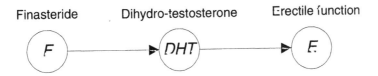

Figure 4.19: Finasteride and erectile function are independent.

and the set of conditional independencies in P is

$$\{I_P(X, Y)\},$$

then the only DAG faithful to P is the one in Figure 4.13 (b). If we make the causal faithfulness assumption, then this DAG must be the causal DAG, which means we can conclude X and Y each cause Z. This is the exact same situation as that illustrated above concerning fraud, sex, and jewelry. Therefore, if we make the causal faithfulness assumption, we can conclude that the causal DAG is the one in Figure 4.18 (a) based on the conditional independency $I_P(Fraud, Sex)$.

Example 4.29 In Example 4.23 we showed that if $V = \{X, Y, Z\}$ and the set of conditional independencies in P is

$$\{I_P(X, Y|Z)\},$$

then all the DAGs in Figure 4.13 (a) are faithful to P. So if we make the causal faithfulness assumption, we can conclude that one of these DAGs is the causal DAG, but we do not know which one.

Example 4.30 In Example 4.24 we showed that if $V = \{X, Y, Z, W\}$ and the set of conditional independencies in P is

$$\{I_p(X, Y), \quad I_P(W, \{X, Y\}|Z)\},$$

the only DAG faithful to P is the one in Figure 4.14 (c). So we can conclude that X and Y each cause Z and Z causes W.

When is the causal faithfulness assumption justified? It requires the three conditions mentioned previously for the causal Markov assumption plus one more, which we discuss next. Recall from Chapter 3, Section 3.3.1, that the pharmaceutical company Merck developed the drug finasteride, which was found to be effective in the treatment of baldness. Recall from Chapter 3, Section 3.6.3, that finasteride accomplishes this by lowering dehydro-testosterone (DHT) levels in men, and DHT is necessary to erectile function. So the causal relationships among finasteride, DHT, and erectile dysfunction are those shown in Figure 4.19. Recall further from Chapter 3, Section 3.6.3, that a large-scale manipulation study indicated that finasteride had a causal effect on DHT level but had no causal effect on erectile dysfunction. How could this be when the causal relationships among finasteride (F), DHT (DHT), and erectile dysfunction (E) have clearly been found to be those depicted in Figure 4.19? We would expect a causal mediary to transmit an effect from its antecedent to its consequence, but in this case it does not. The explanation is that finasteride cannot lower DHT levels beyond a certain threshold level, and that level is all that is needed for erectile function. So we have $I(F, E)$.

The Markov condition does not entail $I(F, E)$ for the causal DAG in Figure 4.19. It only entails $I(F, E|DHT)$. So the causal faithfulness assumption is not justified. If we learned the conditional independencies in the probability distribution of these variables from data, we would learn the following set of independencies:

$$\{I_P(F, E), \quad I_P(F, E|DHT)\}.$$

There is no DAG that entails both these conditional independencies. So no DAG could be learned from such data.

The causal faithfulness assumption is usually justified when the three conditions listed previously for the causal Markov assumption are satisfied and when we do not have unusual causal relationships as in the finasteride example. So the causal faithfulness assumption is ordinarily justified for a causal graph if the following conditions are satisfied:

1. There are no hidden common causes. That is, all common causes are represented in the graph.

2. There are no causal feedback loops. That is, our graph is a DAG.

3. Selection bias is not present.

4. All intermediate causes transmit influences from their antecedents to their consequences.

4.5.2   Causal Embedded Faithfulness Assumption*

It seems that the main exception to the causal faithfulness assumption (and the causal Markov assumption) is the presence of hidden common causes. Even in the example concerning sex, height, and wage (Example 4.11), perhaps there is a genetic trait which makes people grow taller and also gives them some personality trait that helps them compete better in the job market. Our next assumption eliminates the requirement that there are no hidden common causes. If we assume that the observed probability distribution P of a set of random variables V is embedded faithfully in a causal DAG containing the variables, we say we are making the **causal embedded faithfulness assumption**. The causal embedded faithfulness assumption is usually justified when the conditions for the causal faithfulness assumption are satisfied, except that hidden common causes may be present.

Next we illustrate the causal embedded faithfulness assumption. Suppose the causal DAG in Figure 4.20 (a) satisfies the causal faithfulness assumption. However, we only observe V $= \{N, F, C, T\}$. Then the causal DAG containing

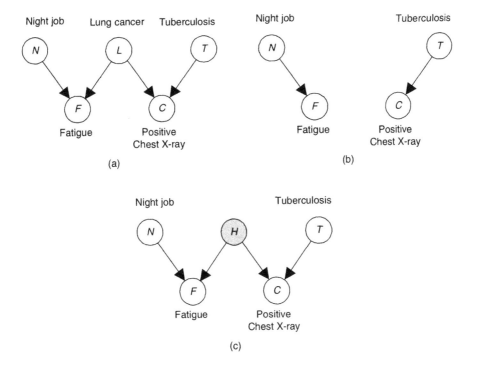

Figure 4.20: If the causal relationships are those shown in (a), P is not faithful to the DAG in (b), but P is embedded faithfully in the DAG in (c).

the observed variables is the one in Figure 4.20 (b). The DAG in Figure 4.20 (b) entails $I(F, C)$, and this conditional independency is not entailed by the DAG in Figure 4.20 (a). Therefore, the observed distribution $P(\mathsf{V})$ does not satisfy the Markov condition with the causal DAG in Figure 4.20 (b), which means the causal faithfulness assumption is not warranted. However, $P(\mathsf{V})$ is embedded faithfully in the DAG in Figure 4.20 (c). So the causal embedded faithfulness assumption is warranted. Note that this example illustrates a situation in which we identify four variables and two of them have a hidden common cause. That is, we have not identified lung cancer as a feature of humans.

Let's see how much we can learn about causal influences when making only the causal embedded faithfulness assumption.

Example 4.31 *Recall that in Example 4.28 $\mathsf{V} = \{X, Y, Z\}$, our set of conditional independencies was*

$$\{I_P(X, Y)\},$$

and we concluded that X and Y each caused Z while making the causal faithfulness assumption. However, the probability distribution is embedded faithfully in all the DAG in Figure 4.21. So if we make only the causal embedded faithfulness assumption, it could be that X causes Z or it could be that X and Z

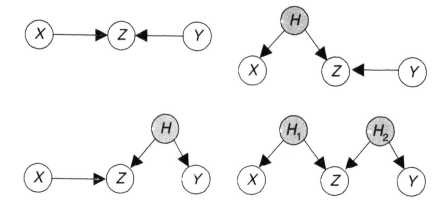

Figure 4.21: If we make the causal embedded faithfulness assumption and our set of conditional independencies is $\{I_P(X,Y)\}$, the causal relationships could be the ones in any of these DAGs.

have a hidden common cause. The same holds for Y and Z.

While making only the more reasonable causal embedded faithfulness assumption, we were not able to learn any causal influences in the previous example. Can we ever learn a causal influence while making only this assumption? The next example shows that we can.

Example 4.32 *Recall that in Example 4.30 $V = \{X, Y, Z, W\}$, and our set of conditional independencies was*

$$\{I_P(X,Y),\quad I_P(W,\{X,Y\}|Z)\}.$$

In this case the probability distribution P is embedded faithfully in the DAGs in Figures 4.22 (a) and (b). However, it is not embedded faithfully in the DAGs in Figure 4.22 (c) or (d). The reason is that these latter DAGs entail $I(X,W)$, and we do not have this conditional independency. That is, the Markov condition says X must be independent of its nondescendents conditional on its parents. Since X has no parents, this means X must simply be independent of its nondescendents, and W is one of its nondescendents. If we make the causal embedded faithfulness assumption, we conclude that Z causes W.

Example 4.33 *Recall that in Example 4.26 $V = \{X, Y, Z, W\}$, our set of conditional independencies was*

$$\{I(X,\{Y,W\}),\quad I(Y,\{X,Z\}),$$

and we obtained the graph in Figure 4.23 (a) when we tried to learn a DAG faithful to P. We concluded that there is no DAG faithful to P. Then in Example 4.27 we showed that P is embedded faithfully in the DAG in Figure

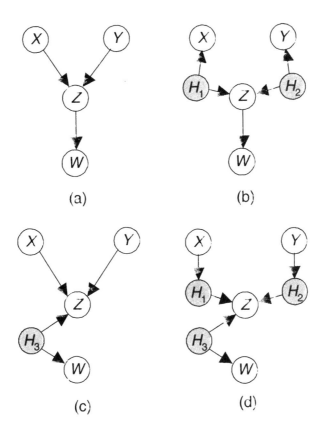

Figure 4.22: If our set of conditional independencies is $\{I_P(X,Y),$ $I_P(W,\{X,Y\}|Z)\}$, then P is embedded faithfully in the DAGs in (a) and (b), but not in the DAGs in (c) and (d).

4.23 (c). P is also embedded faithfully in the DAG in Figure 4.23 (b). If we make the causal embedded faithfulness assumption, we conclude that Z and W have a hidden common cause.

4.6 Software Packages for Learning

Based on considerations such as those illustrated in Section 4.4.1, Spirtes et al. [1993, 2000] developed an algorithm which finds the DAG faithful to P from the conditional independencies in P when there is a DAG faithful to P. Spirtes et al. [1993, 2000] further developed an algorithm which learns a DAG in which P is embedded faithfully from the conditional independencies in P when such a DAG exists. These algorithms have been implemented in the Tetrad software package [Scheines et al., 1994], which can be downloaded for free from the following site:

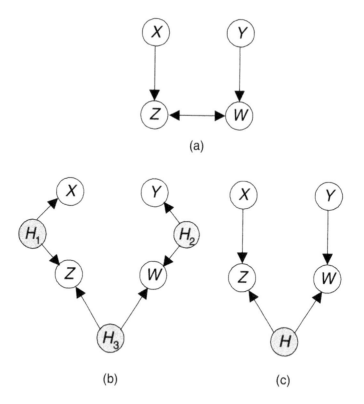

Figure 4.23: If our set of conditional independencies is $\{I_P(X, \{Y, W\}),$ $I_P(Y, \{X, Z\})$, we can conclude that Z and W have a hidden common cause.

http://www.phil.cmu.edu/projects/tetrad/.

In 1997 Meek developed a heuristic search algorithm called **Greedy Equivalent Search (GES)** which has the following property: if there were a DAG faithful to P, the limit, as the size of the data set approaches infinity, of the probability of finding a DAG faithful to P is equal to 1. The Tetrad software package also has a module which uses that algorithm along with the **Bayesian information criterion (BIC)** to learn a Bayesian network from data.

Other Bayesian network learning packages include the following:

1. Belief Network Power Constructor (constraint-based approach),

    http://www.cs.ualberta.ca/~jcheng/bnpc.htm.

2. Bayesware (structure and parameters), http://www.bayesware.com/.

3. Bayes Net Toolbox, http://bnt.sourceforge.net/.

4. Probabilistic Net Library,

    http://www.intel.com/technology/computing/pnl/.

4.7 Examples of Learning

Next we show some examples of learning using Bayesian network learning packages. First we show examples of learning Bayesian networks, and then we illustrate causal learning. You don't need to study all this material in order to understand the chapters that follow. However, it should be worth your while to look at a few examples.

4.7.1 Learning Bayesian Networks

We show several examples in which useful Bayesian networks were learned from data.

Cervical Spinal-Cord Trauma

Physicians face the problem of assessing cervical spinal-cord trauma. To learn a Bayesian network which could assist physicians in this task, Herskovits and Dagher [1997] obtained a data set from the Regional Spinal Cord Injury Center of the Delaware Valley. The data set consisted of 104 records of patients with spine injury, who were evaluated acutely and at one year follow-up. Each record consisted of the following variables:

Variable	What the Variable Represents
UE_F	Upper extremity functional score
LE_F	Lower extremity functional score
$Rostral$	Most superior point of cord edema as demonstrated by MRI
$Length$	Length of cord edema as demonstrated by MRI
$Heme$	Cord hemorrhage as demonstrated by MRI
UE_R	Upper extremity recovery at one year
LE_R	Lower extremity recovery at one year

They discretized the data and used the Bayesian network learning program Cogito^{TM} to learn a Bayesian network. Cogito, which was developed by E. H. Herskovits and A. P. Dagher, performs model selection using the score-based method presented in Section 4.3. The structure that Cogito learned is shown in Figure 4.24.

Herskovits and Dagher [1997] compared the performance of their Bayesian network to that of a regression model that had independently been developed by other researchers from the same data set [Flanders et al., 1996]. The other researchers did not discretize the data, but rather they assumed normal distributions. The comparison consisted of evaluating 40 new cases not present in the original data set. They entered the values of all variables except the outcome variables, which are UE_R (upper extremity recovery at one year) and LE_R (lower extremity recovery at one year), and used the Bayesian network inference program Ergo^{TM} [Beinlich and Herskovits, 1990] to predict the values of the outcome variables. They also used the regression model to predict these values. Finally, they compared the predictions of both models to the actual values for each case. They found that the Bayesian network correctly predicted

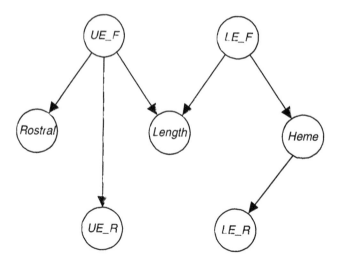

Figure 4.24: The structure learned by Cogito for assessing cervical spinal-cord trauma.

the degree of upper-extremity recovery three times as often as the regression model. They attributed part of this result to the fact that the original data did not follow a normal distribution, which the regression model assumed. An advantage of Bayesian networks is that they need not assume any particular distribution and therefore can accommodate unusual distributions.

Forecasting Sea Breezes

Next, we describe Bayesian networks, developed by Kennett et al. [2001], for forecasting sea breezes. They describe the sea breeze prediction problem as follows:

> Sea breezes occur because of the unequal heating and cooling of neighboring sea and land areas. As warm air rises over the land, cool air is drawn in from the sea. The ascending air returns seaward in the upper current, building a cycle and spreading the effect over a large area. If wind currents are weak, a sea breeze will usually commence soon after the temperature of the land exceeds that of the sea, peaking in mid-afternoon. A moderate to strong prevailing offshore wind will delay or prevent a sea breeze from developing, while a light to moderate prevailing offshore wind at 900 meters (known as the gradient level) will reinforce a developing sea breeze. The sea breeze process is affected by time of day, prevailing weather, seasonal changes, and geography.

Kennett et al. [2001] note that forecasting in the Sydney area was currently being done using a simple rule-based system. The rule is as follows:

If the wind is offshore
and the wind is less than 23 knots
and part of the timeslice falls in the afternoon,
then a sea breeze is likely to occur.

The Australian Bureau of Meteorology (BOM) provides a data set of meteorological information obtained from three different sensor sites in the Sydney area. Kennett et al. [2001] used 30 MB of data obtained from October 1997 to October 1999. Data on ground-level wind speed (ws) and direction (wd) at 30-minute intervals (date and time stamped) were obtained from automatic weather stations (AWS). Olympic sites provided ground-level wind speed (ws), wind direction (wd), gust strength, temperature, dew temperature, and rainfall. Weather balloon data from a Sydney airport, which was collected at 5 AM and 11 PM daily, provided vertical readings for gradient-level wind speed (gws) and direction (gwd), temperature, and rainfall. Predicted variables are wind speed prediction (wsp) and wind direction prediction (wdp). The variables used in the networks are summarized in the following table:

Variable	What the Variable Represents
gwd	Gradient-level wind direction
gws	Gradient-level wind speed
wd	Ground-level wind direction
ws	Ground-level wind speed
$date$	Date
$time$	Time
wdp	Wind direction prediction (predicted variable)
wsp	Wind speed prediction (predicted variable)

From this data set, Kennett et al. [2001] used Tetrad II, both with and without a prior temporal ordering, to learn a Bayesian network. They also learned a Bayesian network by searching the space of causal models and using a data compression scoring criterion called **minimum message length (MML)** [Wallace and Korb, 1999] to score DAGs. They called this method **causal minimum message length (CaMML)**. Furthermore, they constructed a Bayesian network using expert elicitation with meteorologists at the BOM. The links between the variables represent the experts' beliefs concerning the causal relationships among the variables. The networks learned using CaMML, Tetrad II with a prior temporal ordering, and expert elicitation are shown in Figure 4.25.

Next, Kennett et al. [2001] learned the values of the parameters in each Bayesian network by inputting 80% of the data from 1997 and 1998 to the learning package Netica (http://www.norsys.com). Netica uses the techniques discussed in Section 4.1 for learning parameters from data. Finally, they evaluated the predictive accuracy of all four networks and the rule-based system using the remaining 20% of the data. All four Bayesian networks had almost identical predictive accuracies, and all significantly outperformed the rule-based system. Figure 4.26 plots the predictive accuracy of CaMML and the rule-based system.

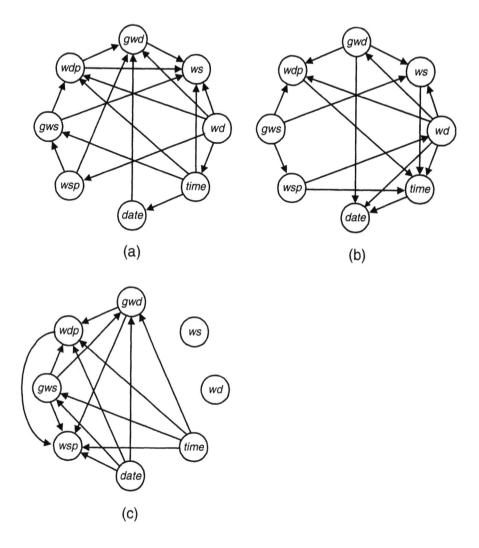

Figure 4.25: The sea breeze forecasting Bayesian networks learned by (a) CaMML, (b) Tetrad II with a prior temporal ordering, and (c) expert elicitation.

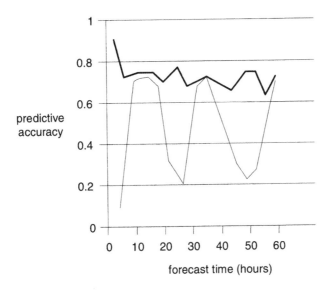

Figure 4.26: The thick curve represents the predictive accuracy of CaMML, and the thin curve represents that of the rule-based system.

Note the periodicity in the prediction rates and the extreme fluctuations for the rule-based system.

MENTOR

Mani et al. [1997] developed MENTOR, a system that predicts the risk of mental retardation (MR) in infants. Specifically, the system can determine the probabilities of the child later obtaining scores in four different ranges on the Raven Progressive Matrices Test, which is a test of cognitive function. The probabilities are conditional on values of variables such as the mother's age at time of birth, whether the mother had recently had an X-ray, whether labor was induced, etc.

Developing the Network The structure of the Bayesian network used in MENTOR was created in the following three steps:

1. Mani et al. [1997] obtained the Child Health and Development Study (CHDS) data set, which is the data set developed in a study concerning pregnant mothers and their children. The children were followed through their teen years and included numerous questionnaires, physical and psychological exams, and special tests. The study was conducted by the University of California at Berkeley and the Kaiser Foundation. It started in 1959 and continued into the 1980s. There are approximately 6000 children and 3000 mothers with IQ scores in the data set. The children were either

Variable	What the Variable Represents
MOM_RACE	Mother's race classified as White (European or White and American Indian or others considered to be of white stock) or non-White (Mexican, Black, Oriental, interracial mixture, South-East Asian).
MOMAGE_BR	Mother's age at time of child's birth classified as 14-19 years, 20-34 years, or $\geq$ 35 years.
MOM_EDU	Mother's education classified as $\leq$ 12 and did not graduate high school, graduated high school, and $>$ high school (attended college or trade school).
DAD_EDU	Father's education classified same as mother's education.
MOM_DIS	Yes if mother had one or more of lung trouble, heart trouble, high blood pressure, kidney trouble, convulsions, diabetes, thyroid trouble, anemia, tumors, bacterial disease, measles, chicken pox, herpes simplex, eclampsia, placenta previa, any type of epilepsy, or malnutrition; no otherwise.
FAM_INC	Family income classified as $<$ \$10,000 or $\geq$ \$10,000.
MOM_SMOK	Yes if mother smoked during pregnancy; no otherwise.
MOM_ALC	Mother's alcoholic drinking level classified as mild (0-6 drinks per week), moderate (7-20 per week), or severe ($>$ 20 per week).
PREV_STILL	Yes if mother previously had a stillbirth; no otherwise.
PN_CARE	Yes if mother had prenatal care; no otherwise.
MOM_XRAY	Yes if mother had been X-rayed in the year prior to or during the pregnancy; no otherwise.
GESTATN	Period of gestation categorized as premature ($\leq$ 258 days), normal (259-294 days), or postmature ($\geq$ 295 days).
FET_DIST	Fetal distress classified as yes if there was prolapse of cord, mother had a history of uterine surgery, there was uterine rupture or fever at or just before delivery, or there was an abnormal fetal heart rate; no otherwise.
INDUCE_LAB	Yes if mother had induced labor; no otherwise.
C_SECTION	Yes if delivery was a caesarean section; no if it was vaginal.
CHLD_GEND	Gender of child (male or female).
BIRTH_WT	Birth weight classified as low ($<$ 2500 g) or normal ($\geq$ 2500 g).
RESUSCITN	Yes if child had resuscitation; no otherwise.
HEAD_CIRC	Normal if head circumference is 20-21cm; abnormal otherwise.
CHLD_ANOM	Child anomaly classified as yes if child has cerebral palsy, hypothyroidism, spina bifida, Down's syndrome, chromosomal abnormality, anencephaly, hydrocephalus, epilepsy, Turner's syndrome, cerebellar ataxia, speech defect, Klinefelter's syndrome, or convulsions; no otherwise.
CHILD_HPRB	Child's health problem categorized as having a physical problem, having a behavior problem, having both a physical and a behavioral problem, or having no problem.
CHLD_RAVN	Child's cognitive level, measured by the Raven test, categorized as mild MR, borderline MR, normal, or superior.
P_MOM	Mother's cognitive level, measured by the Peabody test, categorized as mild MR, borderline MR, normal, or superior.

Table 4.2: The variables used in MENTOR.

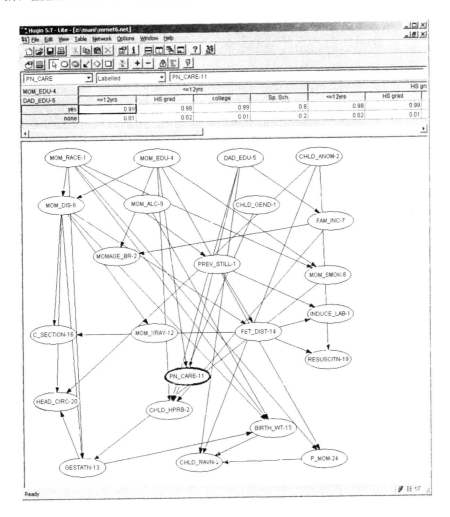

Figure 4.27: The DAG used in MENTOR (displayed using HUGIN).

5 years old or 9 years old when their IQs were tested. The IQ test used for the children was the Raven Progressive Matrices Test. The mothers' IQs were also tested, and the test used was the Peabody Picture Vocabulary Test.

Initially, Mani et al. [1997] identified 50 variables in the data set that were thought to play a role in the causal mechanism of mental retardation. However, they eliminated those with weak associations to the Raven score, and finally used only 22 in their model. The variables used are shown in Table 4.2.

After the variables were identified, they used the CB Algorithm to learn a network structure from the data set. The CB Algorithm, which is discussed in [Singh and Valtorta, 1995], uses the constraint-based method

Cognitive Level	Avg. Probability for Controls ($n = 13{,}019$)	Avg. Probability for Subjects ($n = 3598$)
Mild MR	.06	.09
Borderline MR	.12	.16
Mild or Borderline MR	.18	.25

Table 4.3: Average probabilities, as determined by MENTOR, of having MR for controls (children identified as having normal cognitive function at age 8) and subjects (children identified as having mild or borderline MR at age 8).

to propose a total ordering of the nodes and then uses a heuristic search and a scoring criterion to learn a DAG.

2. Mani et al. [1997] decided that they wanted the network to be a causal network. So next they modified the DAG according to the following three rules:

   (a) Rule of Chronology: An event cannot be the parent of a second event that preceded the first event in time. For example, CHILD_HPRB (child's health problem) cannot be the parent of MOM_DIS (mother's disease).

   (b) Rule of Commonsense: The causal links should not go against common sense. For example, DAD_EDU (father's education) cannot be a cause of MOM_RACE (mother's race).

   (c) Domain Rule: The causal links should not violate established domain rules. For example, PN_CARE (prenatal care) should not cause MOM_SMOK (maternal smoking).

3. Finally, the DAG was refined by an expert. The expert was a clinician who had 20 years of experience with children with mental retardation and other developmental disabilities. When the expert stated there was no relationship between variables with a causal link, the link was removed and new ones were incorporated to capture knowledge of the domain causal mechanisms.

The final DAG specifications were input to HUGIN (see [Olesen et al., 1992]) using the HUGIN graphic interface. The output is the DAG shown in Figure 4.27.

After the DAG was developed, the conditional probability distributions were learned from the CHDS data set using the techniques discussed in Section 4.1. After that, they too were modified by the expert, resulting finally in the Bayesian network in MENTOR.

Validating the Model Mani et al. [1997] tested their model in a number of different ways. We present two of their results.

The National Collaborative Perinatal Project (NCPP), of the National Institute of Neurological and Communicative Disorders and Strokes, developed a

Variable	Case 1 Variable Value	Case 2 Variable Value	Case 3 Variable Value
MOM_RACE	Non-White	White	White
MOMAGE_BR	14-19		≥ 35
MOM_EDU	≤ 12	> High school	≤ 12
DAD_EDU	≤ 12	> High school	High school
MOM_DIS			No
FAM_INC	< $10,000		< $10,000
MOM_SMOK			Yes
MOM_ALC			Moderate
PREV_STILL			
PN_CARE		Yes	
MOM_XRAY			Yes
GESTATN	Normal	Normal	Premature
FET_DIST		No	Yes
INDUCE_LAB			
C_SECTION			
CHLD_GEND			
BIRTH_WT	Low	Normal	Low
RESUSCITN			
HEAD_CIRC			Abnormal
CHLD_ANOM		No	
CHILD_HPRB			Both
CHLD_RAVN			
P_MOM	Normal	Superior	Borderline

Table 4.4: Generated values for three cases.

data set containing information on pregnancies between 1959 and 1974 and 8 years of follow-up for live-born children. For each case in the data set, the values of all 22 variables except CHLD_RAVN (child's cognitive level as measured by the Raven test) were entered, and the conditional probabilities of each of the four values of CHLD_RAVN were computed. Table 4.3 shows the average values of $P(\text{CHLD\_RAVN} = mildMR|\mathsf{d})$ and $P(\text{CHLD\_RAVN} = borderlineMR|\mathsf{d})$, where d is the set of values of the other 22 variables, for both the controls (children in the study with normal cognitive function at age 8) and the subjects (children in the study with mild or borderline MR at age 8).

In actual clinical cases, the diagnosis of mental retardation is rarely made after only a review of history and a physical examination. Therefore, we cannot expect MENTOR to do more than indicate a risk of mental retardation by computing the probability of it: the higher the probability the greater the risk. Table 4.3 shows that on the average children, who were later determined to have mental retardation, were found to be at greater risk than those who were not. MENTOR can confirm a clinician's assessment by reporting the probability of mental retardation.

The expert was in agreement with MENTOR's assessments (conditional

Value of CHLD_RAVN and Prior Probability	Case 1 Posterior Probability	Case 2 Posterior Probability	Case 3 Posterior Probability
Mild MR (.056)	.101	.010	.200
Borderline MR (.124)	.300	.040	.400
Normal (.731)	.559	.690	.380
Superior (.089)	.040	.260	.200

Table 4.5: Posterior probabilities for three cases.

probabilities) in seven of the nine cases. In the two cases where the expert was not in complete agreement, there were problems with the child. In one case the child had a congenital anomaly, while in the other case the child had a health problem. In both of these cases a review of the medical chart would indicate the exact nature of the problem, and this information would then be used by the expert to determine the probabilities. It is possible that MENTOR's conditional probabilities are accurate given the current information, and the domain expert could not accurately determine probabilities without the additional information.

As another test of the model, Mani et al. [1997] developed a strategy for comparing the results of MENTOR with the judgements of an expert. They generated nine cases, each with some set of variables instantiated to certain values, and let MENTOR compute the conditional probability of the values of CHLD_RAVN. The generated values for three of the cases are shown in Table 4.4, while the conditional probabilities of the values of CHLD_RAVN for those cases are shown in Table 4.5.

4.7.2 Causal Learning

When using a software package to learn causal influences, the graph learned has several different types of edges, with each indicating a different conclusion about the causal relationship between the variables connected by the edge. Different packages may denote these edges differently. Below is the notation used in this book.

Edge	Causal Relationship
$X - Y$	X causes Y or (exclusive) Y causes X; or X and Y have a hidden common cause.
$X \rightarrow Y$	X causes Y or X and Y have a hidden common cause; and Y does not cause X.
$X \rightarrowtail Y$	X causes Y.
$X \leftrightarrow Y$	X and Y have a hidden common cause.

In the above, when we do not denote an "or" as exclusive, it is not exclusive. There is one additional restriction. If we have the chain $X - Y - Z$, we cannot have two causal edges with their heads at node Y. For example, we could not

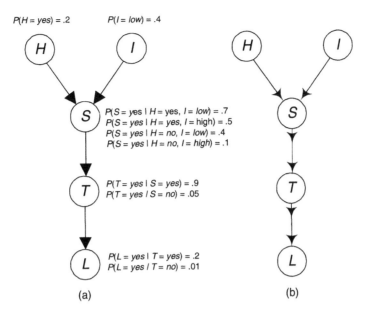

Figure 4.28: If the conditional independencies in P are the ones entailed by the DAG in (a), Tetrad will learn the graph in (b).

have node X causing node Y and nodes Y and Z having a hidden common cause.

Next, we show several examples of causal learning.

Smoking and Lung Cancer

Suppose we have the following variables:

Variable	What the Variable Represents
H	Parents' smoking habits
I	Income
S	Smoking
T	Tar deposits
L	Lung cancer

If we created the Bayesian network shown in Figure 4.28 (a), randomly generated a large amount of data using the probability distribution in that network, and then used Tetrad's constraint-based learning algorithm to learn causal influences from the data, we would obtain the graph in Figure 4.28 (b). We could conclude that either H causes S or they have a hidden common cause, either I causes S or they have a hidden common cause, S causes T, and T causes L. So smoking causes lung cancer indirectly through tar deposits. Note that we are just describing a simulation based on what seems to be reasonable causal relationships. We know of no actual data exhibiting the conditional independencies entailed by the DAG in Figure 4.28 (a).

College Student Retention Rate

Using the data base collected by the *U.S. News and World Record* magazine for the purpose of college ranking, Druzdzel and Glymour [1999] analyzed the influences that affect university student retention rate. By "student retention rate" we mean the percent of entering freshmen who end up graduating from the university at which they initially matriculate. Low student retention rate is a major concern at many American universities, as the mean retention rate over all American universities is only 55%.

The data base provided by the *U.S. News and World Record* magazine contains records for 204 U.S. universities and colleges identified as major research institutions. Each record consists of over 100 variables. The data were collected separately for the years 1992 and 1993. Druzdzel and Glymour [1999] selected the following eight variables as being most relevant to their study:

Variable	What the Variable Represents
grad	Fraction of entering students who graduate from the institution
rejr	Fraction of applicants who are not offered admission
tstsc	Average standardized score of incoming students
*tp*10	Fraction of incoming students in the top 10% in high school
acpt	Fraction of students who accept the institution's admission offer
spnd	Average educational and general expenses per student
sfrat	Student/faculty ratio
salar	Average faculty salary

Druzdzel and Glymour [1999] used Tetrad II [Scheines et al., 1994] to learn causal influences from the data. Tetrad II allows the user to specify a temporal ordering of the variables. If variable Y precedes X in this order, the algorithm assumes there can be no path from X to Y in any DAG in which the probability distribution of the variables is embedded faithfully. It is called a temporal ordering because in applications to causality if Y precedes X in time, we would assume X could not cause Y. Druzdzel and Glymour [1999] specified the following temporal ordering for the variables in this study:

spnd, sfrat, salar
rejr, acpt
tstsc, tp10
grad

Their reasons for this ordering are as follows: They believed the average spending per student (*spnd*), the student/teacher ratio (*sfrat*), and the faculty salary (*salar*) are determined based on budget considerations and are not influenced by any of the other five variables. Furthermore, they placed rejection rate (*rejr*) and the fraction of students who accept the institution's admission offer (*acpt*)

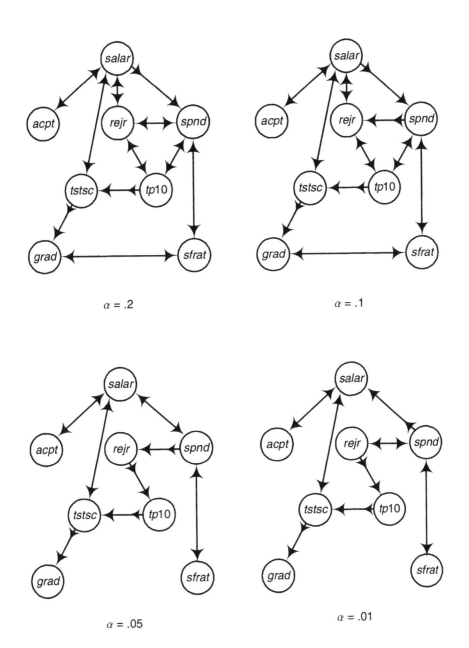

$\alpha = .2$

$\alpha = .1$

$\alpha = .05$

$\alpha = .01$

Figure 4.29: The graphs Tetrad II learned from *U.S. News and World Record*'s 1992 data base.

ahead of average test scores (*tstsc*) and class standing (*tp*10) because the values of these latter two variables are only obtained from matriculating students. Finally, they assumed the graduate rate (*grad*) does not cause any of the other variables.

Tetrad II allows the user to enter a significance level. A significance level of α means the probability of rejecting a conditional independency hypothesis, when it is true, is α. Therefore, the smaller the value of α is, the less likely we are to reject a conditional independency, and therefore, the sparser our resultant graph. Figure 4.29 shows the graphs which Druzdzel and Glymour [1999] learned from the 1992 data base, provided by *U.S. News and World Record*, using significance levels of .2, .1, .05, and .01.

Although different graphs were obtained at different levels of significance, all the graphs in Figure 4.29 show that the average standardized test score (*tstsc*) has a direct causal influence on graduation rate (*grad*), and no other variable has a direct causal influence on *grad*. The results for the 1993 data base were not as overwhelming, but they too indicated *tstsc* to be the only direct causal influence of *grad*.

To test whether the causal structure may be different for top research universities, Druzdzel and Glymour [1999] repeated the study using only the top 50 universities according to the ranking of *U.S. News and World Record*. The results were similar to those for the complete data base.

These results indicate that although factors such as spending per student and faculty salary may have an influence on graduation rates, they do so only indirectly by affecting the standardized test scores of matriculating students. If the results correctly model reality, retention rates can be improved by bringing in students with higher test scores in any way whatsoever. Indeed, in 1994 Carnegie Mellon changed its financial aid policies to assign a portion of its scholarship fund on the basis of academic merit. Druzdzel and Glymour [1999] noted that this resulted in an increase in the average test scores of matriculating freshman classes and an increase in freshman retention.

Before closing, we note that the notion that the average test score has a causal influence on the graduation rate does not fit into common notions of causation such as the one concerning manipulation. For example, if we manipulated a university's average test score by accessing the testing agency's data base and changing the scores of the university's students to much higher values, we would not expect the university's graduation rate to increase. Rather, this study indicates that the test score is a near perfect indicator of some other variable, which we can call "graduation potential."

Harassment in the Military

The next example, taken from [Neapolitan and Morris, 2002], illustrates problems one can encounter when inferring causation from passive data.

Scarville et al. [1996] provide a data base obtained from a survey in 1996 of experiences of racial harassment and discrimination of military personnel in the U.S. Armed Forces. Surveys were distributed to 73,496 members of the

U.S. Army, Navy, Marine Corps, Air Force, and Coast Guard. The survey sample was selected using a nonproportional stratified random sample in order to ensure adequate representation of all subgroups. Usable surveys were received from 39,855 service members (54%). The survey consisted of 81 questions related to experiences of racial harassment and discrimination and job attitudes. Respondents were asked to report incidents that had occurred during the previous 12 months. The questionnaire asked participants to indicate the occurrence of 57 different types of racial/ethnic harassment or discrimination. Incidents ranged from telling offensive jokes to physical violence and included harassment by military personnel as well as the surrounding community. Harassment experienced by family members was also included.

Neapolitan and Morris [2002] used Tetrad III in an attempt to learn causal influences from the data base. For their analysis, 9640 records (13%) were selected which had no missing data on the variables of interest. The analysis was initially based on eight variables. Similar to the situation concerning university retention rates, they found one causal relationship to be present regardless of the significance level. That is, they found that whether or not the individual held the military responsible for the racial incident had a direct causal influence on the individual's race. Since this result made no sense, they investigated which variables were involved in Tetrad III learning this causal influence. The five variables involved are the following:

Variable	What the Variable Represents
race	Respondent's race/ethnicity
yos	Respondent's years of military service
inc	Did the respondent experience a racial incident?
rept	Was the incident reported to military personnel?
resp	Did the respondent hold the military responsible for the incident?

The variable *race* consisted of five categories: White, Black, Hispanic, Asian or Pacific Islander, and Native American or Alaskan Native. Respondents who reported Hispanic ethnicity were classified as Hispanic, regardless of race. Respondents were classified based on self-identification at the time of the survey. Missing data were replaced with data from administrative records. The variable *yos* was classified into four categories: 6 years or less, 7-11 years, 12-19 years, and 20 years or more. The variable *inc* was coded dichotomously to indicate whether any type of harassment was reported on the survey. The variable *rept* indicates responses to a single question concerning whether the incident was reported to military and/or civilian authorities. This variable was coded 1 if an incident had been reported to military officials. It was coded 0 if an individual experienced no incident, did not report the incident, or only reported the incident to civilian officials. The variable *resp* indicates responses to a single question concerning whether the respondent believed the military to be responsible for an incident of harassment. This variable was coded 1 if the respondent indicated that the military was responsible for some or all of a reported incident. If the respondent indicated no incident, unknown responsibility, or that the military was not responsible, the variable was coded 0.

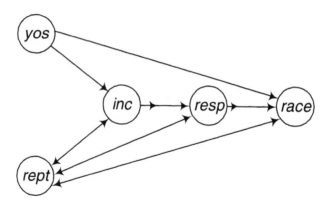

Figure 4.30: The graph Tetrad III learned from the racial harassment survey at the .01 significance level.

They reran the experiment using only these five variables, and again at all levels of significance, they found that *resp* had a direct causal influence on *race*. In all cases, this causal influence was learned because *rept* and *yos* were found to be probabilistically independent, and there was no edge between *race* and *inc*. That is, the causal connection between *race* and *inc* was mediated by other variables. Figure 4.30 shows the graph obtained at the .01 significance level. The edges *yos* → *inc* and *rept* → *inc* are directed toward *inc* because *yos* and *rept* were found to be independent. The edge *yos* → *inc* resulted in the edge *inc* ↦ *resp* being directed the way it was, which in turn resulted in *resp* ↦ *race* being directed the way it was. If there had been an edge between *inc* and *race*, the edge between *resp* and *race* would not have been directed.

It seems suspicious that no direct causal connection between *race* and *inc* was found. Recall, however, that these are the probabilistic relationships among the responses; they are not necessarily the probabilistic relationships among the actual events. There is a problem with using responses on surveys to represent occurrences in nature because subjects may not respond accurately. Let's assume race is recorded accurately. The actual causal relationship between *race*, *inc*, and *says_inc* may be as shown in Figure 4.31. By *inc* we now mean whether there really was an incident, and by *says_inc* we mean the survey response. Perhaps races that experienced higher rates of harassment were less likely to report the incident, and the causal influence of *race* on *says_inc* through *inc* was negated by the direct influence of *race* on *inc*. This would be a case in which faithfulness is violated similar to the situation involving finasteride discussed in Section 4.5.1. The previous conjecture is substantiated by another study. Stangor et al. [2002] found that minority members were more likely to attribute a negative outcome to discrimination when responses were recorded privately, but less likely to report discrimination when they had to express their opinion publicly and there was a member of the non-minority group present. Although the survey of military personnel was intended to be

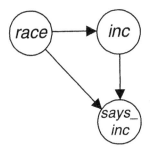

Figure 4.31: Possible causal relationships among race, incidence of harassment, and saying there is an incident of harassment.

confidential, minority members in the military may have had similar feelings about reporting discrimination to the army as the subjects in the study in [Stangor et al., 2002] had about reporting it in the presence of a non-minority individual.

As noted previously, Tetrad II (and III) allows the user to enter a temporal ordering. So one could have put *race* first in such an ordering to avoid it being an effect of another variable. However, one should do this with caution. The fact that the data strongly support that race is an effect indicates there is something wrong with the data, which means we should be dubious of drawing any conclusions from the data. In the present example, Tetrad III actually informed us that we could not draw causal conclusions from the data when we make *race* a root. That is, when Neapolitan and Morris [2002] made *race* a root, Tetrad III concluded that there is no consistent orientation of the edge between *race* and *resp*, which means the probability distribution does not admit an embedded faithful DAG representation unless the edge is directed toward *race*.

EXERCISES

Section 4.1

Exercise 4.1 For some two-outcome experiment, which you can repeat indefinitely (such as the tossing of a thumbtack), determine the number of occurrences a and b of each outcome that you feel your prior experience is equivalent to having seen. Determine the probability of the first outcome.

Exercise 4.2 Assume I feel my prior experience concerning the relative frequency of smokers in a particular bar is equivalent to having seen 14 smokers and 6 nonsmokers.

1. I then decide to poll individuals in the bar and ask them if they smoke. What is my probability of the first individual polled being a smoker?

2. Suppose after polling 10 individuals, I obtain these data (the value 1 means the individual smokes and 2 means the individual does not smoke.):

$$\{1, 2, 2, 2, 2, 1, 2, 2, 2, 1\}.$$

What is my probability that the next individual polled is a smoker?

3. Suppose that after polling 1000 individuals (it is a big bar), I learn that 312 are smokers. What is my probability that the next individual polled is a smoker? How does this probability compare to my prior probability?

Exercise 4.3 In Example 4.7 it was left as an exercise to show that the probability that all 36 remaining panels contain NiS is given by

$$\prod_{i=0}^{35} \frac{1/360 + 3 + i}{20/360 + 3 + i} = .866.$$

Show this.

Exercise 4.4 Suppose I am about to watch Sam and Dave race several times, and Sam looks substantially athletically inferior to Dave. So I give Sam a probability of .1 of winning the first race. However, I feel that if Sam wins once, he should usually win. So given that Sam wins the first race, I give him a .8 probability of winning the next one.

1. Using the method shown in Example 4.7, compute my prior values of a and b.

2. Determine my probability that Sam will win the first 5 races.

3. Suppose next that Sam wins 4 of the first 5 races. Determine my probability that Sam will win the 6th race.

Exercise 4.5 Find a rectangular block (not a cube) and label the sides. Determine values of $a_1, a_2, \ldots, a_6$ that represent your prior probability concerning each side coming up when you throw the block.

1. What is your probability of each side coming up on the first throw?

2. Throw the block 20 times. Compute your probability of each side coming up on the next throw.

Exercise 4.6 Suppose we are going to sample individuals who have smoked two packs of cigarettes or more daily for the past 10 years. We will determine whether each individual's systolic blood pressure is ≤ 100, 101-120, 121-140, 141-160, or ≥ 161. Determine values of $a_1, a_2, \ldots, a_5$ that represent your prior probability of each blood pressure range.

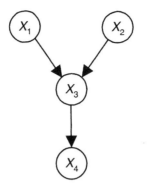

Figure 4.32: A DAG.

1. Next you sample such smokers. What is your probability of each blood pressure range for the first individual sampled?

2. Suppose after sampling 100 individuals, we obtain the following results:

Blood Pressure Range	# of Individuals in This Range
≤ 100	2
101-120	15
121-140	23
141-160	25
≥ 161	35

Compute the probability of each range for the next individual sampled.

Exercise 4.7 Suppose we have the Bayesian network for parameter learning in Figure 4.6 (b), and we have the following data:

Case	X	Y	Z
1	x_1	y_2	z_1
2	x_1	y_1	z_2
3	x_2	y_1	z_1
4	x_2	y_2	z_1
5	x_1	y_2	z_1
6	x_2	y_2	z_2
7	x_1	y_2	z_1
8	x_2	y_1	z_2
9	x_1	y_2	z_1
10	x_1	y_1	z_1

Determine the updated Bayesian network for parameter learning.

Exercise 4.8 Use the method in Theorem 4.1 to develop Bayesian networks for parameter learning with equivalent sample sizes 1, 2, 4, and 8 for the DAG in Figure 4.32.

Section 4.3

Exercise 4.9 Suppose we have the two models in Figure 4.8 and the following data:

Case	J	F
1	j_1	f_1
2	j_1	f_1
3	j_1	f_1
4	j_1	f_1
5	j_1	f_1
6	j_1	f_2
7	j_2	f_2
8	j_2	f_2
9	j_2	f_2
10	j_2	f_1

1. Score each DAG model using the Bayesian scoring criterion, and compute their posterior probabilities assuming the prior probability of each model is .5.

2. Create a data set containing 20 records by duplicating the data in the table above one time, and score the models using this 20-record data set. How have the scores changed?

3. Create a data set containing 100 records by duplicating the data in the table above nine times, and score the models using this 100-record data set. How have the scores changed?

Exercise 4.10 Assume we have the models and data set discussed in Exercise 4.9. Using model averaging, compute the following:

1. $P(j_1|f_1, \mathsf{D})$ when D consists of our original 10 records.

2. $P(j_1|f_1, \mathsf{D})$ when D consists of our original 20 records.

3. $P(j_1|f_1, \mathsf{D})$ when D consists of our original 100 records.

Section 4.4

Exercise 4.11 Suppose $\mathsf{V} = \{X, Y, Z, U, W\}$, and the set of conditional independencies in P is

$$\{I_P(X, Y) \quad I_P(\{W, U\}, \{X, Y\}|Z) \quad I_P(U, \{X, Y, Z\}|W)\}.$$

Find all DAGs faithful to P.

Exercise 4.12 Suppose $V = \{X, Y, Z, U, W\}$, and the set of conditional independencies in P is

$$\{I_P(X, Y) \quad I_P(X, Z) \quad I_P(Y, Z) \quad I_P(U, \{X, Y, Z\}|W)\}.$$

Find all DAGs faithful to P.

Exercise 4.13 Suppose $V = \{X, Y, Z, U, W\}$, and the set of conditional independencies in P is

$$\{I_P(X, Y|U) \quad I_P(U, \{Z, W\}|\{X, Y\}) \quad I_P(\{X, Y, U\}, W|Z)\}.$$

Find all DAGs faithful to P.

Exercise 4.14 Suppose $V = \{X, Y, Z, W, T, V, R\}$, and the set of conditional independencies in P is

$$\{I_P(X, Y|Z) \quad I_P(T, \{X, Y, Z, V\}|W)$$

$$I_P(V, \{X, Z, W, T\}|Y) \quad I_P(R, \{X, Y, Z, W\}|\{T, V\})\}.$$

Find all DAGs faithful to P.

Exercise 4.15 Suppose $V = \{X, Y, Z, W, U\}$, and the set of conditional independencies in P is

$$\{I_P(X, \{Y, W\}|U) \quad I_P(Y, \{X, Z\}|U)\}.$$

1. Is there any DAG faithful to P?

2. Find DAGs in which P is embedded faithfully.

Section 4.5

Exercise 4.16 If we make the causal faithfulness assumption, determine what causal influences we can learn in each of the following cases:

1. Given the conditional independencies in Exercise 4.11

2. Given the conditional independencies in Exercise 4.12

3. Given the conditional independencies in Exercise 4.13

4. Given the conditional independencies in Exercise 4.14

Exercise 4.17 If we make only the causal embedded faithfulness assumption, determine what causal influences we can learn in each of the following cases:

1. Given the conditional independencies in Exercise 4.11

2. Given the conditional independencies in Exercise 4.12

3. Given the conditional independencies in Exercise 4.13

4. Given the conditional independencies in Exercise 4.14

5. Given the conditional independencies in Exercise 4.15

Section 4.6

Exercise 4.18 Using Tetrad (or some other Bayesian network learning algorithm), learn a DAG from the data in Table 4.1. Next, learn the parameters for the DAG. Can you suspect any causal influences from the DAG learned?

Exercise 4.19 Create a data file containing 120 records from the data in Table 4.1 by duplicating the data nine times. Using Tetrad (or some other Bayesian network learning algorithm), learn a DAG from this larger data set. Next, learn the parameters for the DAG. Compare these results to those obtained in Exercise 4.18.

Exercise 4.20 Suppose we have the following variables:

Variable	What the Variable Represents
H	Parents' smoking habits
I	Income
S	Smoking
L	Lung cancer

and the following data:

Case	H	I	S	L
1	Yes	30,000	Yes	Yes
2	Yes	30,000	Yes	Yes
3	Yes	30,000	Yes	No
4	Yes	50,000	Yes	Yes
5	Yes	50,000	Yes	Yes
6	Yes	50,000	Yes	No
7	Yes	50,000	No	No
8	No	30,000	Yes	Yes
9	No	30,000	Yes	Yes
10	No	30,000	Yes	No
11	No	30,000	No	No
12	No	30,000	No	No
14	No	50,000	Yes	Yes
15	No	50,000	Yes	Yes
16	No	50,000	Yes	No
17	No	50,000	No	No
18	No	50,000	No	No
19	No	50,000	No	No

Using Tetrad (or some other Bayesian network learning algorithm) learn a DAG from these data. Next learn the parameters for the DAG. Can you suspect any causal influences from the DAG learned?

Exercise 4.21 Create a data file containing 190 records from the data in Exercise 4.20 by duplicating the data 9 times. Using Tetrad (or some other Bayesian network learning algorithm), learn a DAG from this larger data set. Next, learn the parameters for the DAG. Compare these results to those obtained in Exercise 4.20.

Section 4.7

Exercise 4.22 Obtain some large data set containing records on several or more variables, and use Tetrad (or some other Bayesian network learning algorithm) to learn a DAG from the data set. Next, learn the parameters for the DAG. Determine whether you can suspect any causal influence from the DAG learned.

One possibility would be to try to obtain the large data set that Fidelity (fidelity.com) maintains on mutual funds. For each of thousands of funds they provide information on variables that seem relevant to the fund's future performance. For example, they provide the 5-year return, the 10-year return, the turnover rate, the beta, R^2, the Morningstar rating, and many other seemingly relevant items of information. The investor must analyze all this information and then estimate the fund's performance in the coming year. If you could learn a Markov boundary of the one-year return, and then develop a Bayesian network containing the one-year return and the Markov boundary, that network could be used to compute the probability distribution of the one-year return from the values of the variables in the Markov boundary.

Chapter 5

Decision Analysis Fundamentals

In general, the information obtained by doing inference in a Bayesian network can be used to arrive at a decision, even though the Bayesian network itself does not recommend a decision. In this chapter, we extend the structure of a Bayesian network so that the network actually does recommend a decision. Such a network is called an influence diagram. Section 5.1 introduces decision trees, which are mathematically equivalent to influence diagrams, but which have difficulty representing large instances because their size grows exponentially with the number of variables. In Section 5.2 we discuss influence diagrams, whose size only grows linearly with the number of variables. Finally, in Section

5.3 we introduce dynamic Bayesian networks and influence diagrams, which model relationships among random variables that change over time. Many of the examples in this chapter concern the stock market. If you are not familiar with stocks and investing in stocks, you should first read Chapter 7, Section 7.1.

5.1 Decision Trees

After presenting some simple examples of decision trees, we discuss several issues regarding their use. Then we provide more complex examples.

5.1.1 Simple Examples

We start with the following example:

Example 5.1 *Suppose your favorite stock NASDIP is downgraded by a reputable analyst, and it plummets from $40 to $10 per share. You feel this is a good buy, but there is a lot of uncertainty involved. NASDIP's quarterly earnings are about to be released, and you think they will be good, which should positively influence its market value. However, you also think there is a good chance the whole market will crash, which will negatively influence NASDIP's market value. In an attempt to quantify your uncertainty, you decide there is a .25 probability the market will crash, in which case you feel NASDIP will go to $5 by the end of the month. If the market does not crash, you feel that by the end of the month NASDIP will be either at $10 or at $20 depending on the earnings report. You think it is twice as likely it will be at $20 as at $10. So you assign a .5 probability to NASDIP being at $20 and a .25 probability to it being at $10 at the end of the month. Your decision now is whether to buy 100 shares of NASDIP for $1000 or to leave the $1000 in the bank where it will earn .005 interest in the next month.*

One way to make your decision is to determine the expected value of your investment if you purchase NASDIP and compare that value to the amount of money you would have if you put the money in the bank. Let X be a random variable, whose value is the worth of your $1000 investment in one month if you purchase NASDIP. If NASDIP goes to $5, your investment will be worth $500; if it stays at $10, your investment will be worth $1000; and if it goes to $20, it will be worth $2000. Therefore,

$$E(X) \;=\; .25(\$500) + .25(\$1000) + .5(\$2000)$$
$$\;=\; \$1375.$$

If you leave the money in the bank, your investment will be worth

$$1.005(\$1000) = \$1005.$$

If you are what is called an expected value maximizer, your decision would be to buy NASDIP because that decision has the larger expected value.

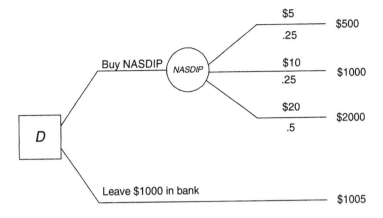

Figure 5.1: A decision tree representing the problem instance in Example 5.1.

The problem instance in the previous example can be represented by a decision tree. That tree is shown in Figure 5.1. A **decision tree** contains two kinds of nodes: **chance (or uncertainty) nodes** representing random variables and **decision nodes** representing decisions to be made. We depict these nodes as follows:

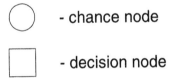

A **decision** represents a set of mutually exclusive and exhaustive actions the decision maker can take. Each action is called an **alternative** in the decision. There is an edge emanating from a decision node for each alternative in the decision. In Figure 5.1, we have the decision node D with two alternatives: "Buy NASDIP" and "Leave $1000 in bank." There is one edge emanating from a chance node for each possible outcome (value) of the random variable. We show the probability of the outcome on the edge and the utility of the outcome to the right of the edge. The **utility** of the outcome is the value of the outcome to the decision maker. When an amount of money is small relative to one's total wealth, we can usually take the utility of an outcome to be the amount of money realized given the outcome. In this chapter, we make this assumption. Handling the case where we do not make this assumption is discussed in Chapter 6, Section 6.1. So, for example, if you buy 100 shares of NASDIP, and NASDIP goes to $20, we assume that the utility of this outcome to you is $2000. In Figure 5.1, we have the chance node $NASDIP$ with three possible outcome utilities, namely $500, $1000, and $2000. The **expected utility** EU of a chance node is defined to be the expected value of the utilities associated with its outcomes. The expected utility of a decision alternative is defined to be the expected utility

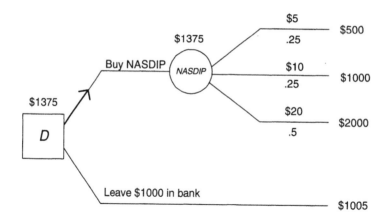

Figure 5.2: The solved decision tree given the decision tree in Figure 5.1.

of the chance node encountered if that decision is made. If there is certainty when the alternative is taken, the expected utility is the value of that certain outcome. So

$$EU(\text{Buy NASDIP}) = EU(NASDIP) \quad = \quad .25(\$500) + .25(\$1000) + .5(\$2000)$$
$$= \quad \$1375$$

$$EU(\text{Leave \$1000 in bank}) = \$1005.$$

Finally, the expected utility of a decision node is defined to be the maximum of the expected utilities of all its alternatives. So

$$EU(D) = \max(\$1375, \ \$1005) = \$1375.$$

The alternative chosen is the one with the largest expected utility. The process of determining these expected utilities is called **solving the decision tree**. After solving it, we show expected utilities above nodes and an arrow to the alternative chosen. The solved decision tree, given the decision tree in Figure 5.1, is shown in Figure 5.2. The entire process of identifying the components of a problem, structuring the problem as a decision tree (or influence diagram), solving the decision tree (or influence diagram), performing sensitivity analysis (discussed in Chapter 6, Section 6.4), and possibly reiterating these steps is called **decision analysis**.

Another example is next.

Example 5.2 *Suppose you are in the same situation as in Example 5.1, except, instead of considering leaving your money in the bank, your other choice is to buy an option on NASDIP. The option costs $1000, and it allows you to buy 500 shares of NASDIP at $11 per share in one month. So if NASDIP is at $5 or $0 per share in one month, you would not exercise your option, and you would*

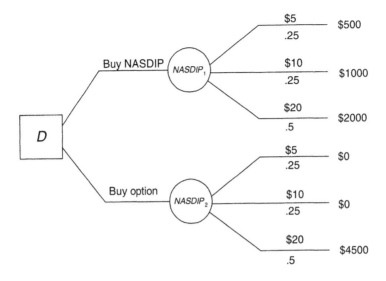

Figure 5.3: The decision tree modeling the investment decision concerning NAS-DIP when the other choice is to buy an option on NASDIP.

lose $1000. However, if NASDIP is at $20 per share in one month, you would exercise your option, and your $1000 investment would be worth

$$500(\$20 - \$11) = \$4500.$$

Figure 5.3 shows a decision tree representing this problem instance. From that tree, we have

$$EU(Buy\ option) = EU(NASDIP_2) = .25(\$0) + .25(\$0) + .5(\$4500)$$
$$= \$2250.$$

Recall that $EU(Buy\ NASDIP)$ is only $1375. So our decision would be to buy the option. It is left as an exercise to show the solved decision tree.

Notice that the decision tree in Figure 5.3 is symmetrical, whereas the one in Figure 5.2 is not. The reason is that we encounter the same uncertain event regardless of which decision is made. Only the utilities of the outcomes are different.

Before proceeding, we address a concern you may have. That is, you may be wondering how an individual could arrive at the probabilities of .25, .5, and .25 in Example 5.1. These probabilities are often not relative frequencies; rather, they are subjective probabilities that represent an individual's reasonable numeric beliefs. The individual arrives at them by a careful analysis of the situation. Methods for assessing subjective beliefs were discussed briefly in Chapter 2, Section 2.3.2 and are discussed in more detail in [Neapolitan, 1996]. Even so, you may argue that the individual surely must believe there are many more possible future values for a share of NASDIP than three. How can the individual claim the only possible values are $5, $10, and $20? You are correct. However, if further refinement of one's beliefs would not affect the decision, then it is not necessary to do so. So if the decision maker feels that a model containing more values would not result in a different decision, it is sufficient to use the model containing only the values $5, $10, and $20.

5.1.2 Solving More Complex Decision Trees

The general algorithm for solving decision trees is quite simple. There is a time ordering from left to right in a decision tree. That is, any node to the right of another node occurs after that node in time. The tree is solved as follows:

> Starting at the right,
>> proceed to the left
>>> passing expected utilities to chance nodes;
>>> passing maximums to decision nodes;
>> until the root is reached.

We now present more complex examples of modeling with decision trees.

Example 5.3 *Suppose Nancy is a high roller, and she is considering buying 10,000 shares of ICK for $10 a share. This number of shares is so high that if she purchases them, it could affect market activity and bring up the price of ICK. She also believes the overall value of the Dow Jones Industrial Average will affect the price of ICK. She feels that in one month the Dow will be at either 10,000 or 11,000, and ICK will be at either $5 or $20 per share. Her other choice is to buy an option on ICK for $100,000. The option will allow her to buy 50,000 shares of ICK for $15 a share in one month. To analyze this problem instance, she constructs the following probabilities:*

$$P(ICK = \$5 | Decision = Buy\ ICK, Dow = 11,000) = .2$$

$$P(ICK = \$5 | Decision = Buy\ ICK, Dow = 10,000) = .5$$

$$P(ICK = \$5 | Decision = Buy\ option, Dow = 11,000) = .3$$

$$P(ICK = \$5 | Decision = Buy\ option, Dow = 10,000) = .6.$$

Furthermore, she assigns

$$P(Dow = 11,000) = .6.$$

This problem instance is represented by the decision tree in Figure 5.4. Next we solve the tree:

$$EU(ICK_1) = (.2)(\$50,000) + (.8)(\$200,000) = \$170,000$$

$$EU(ICK_2) = (.5)(\$50,000) + (.5)(\$200,000) = \$125,000$$

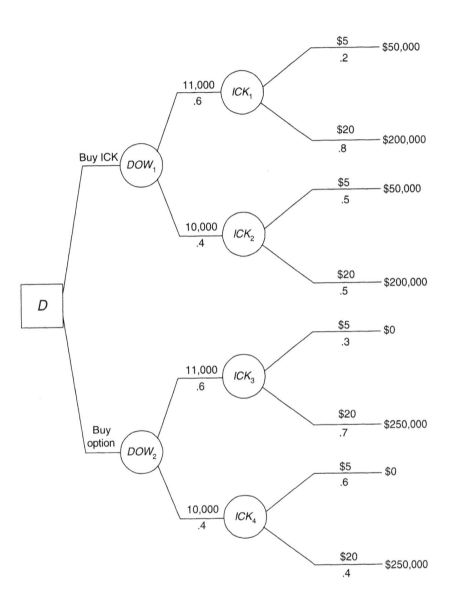

Figure 5.4: A decision tree representing Nancy's decision whether to buy ICK or an option on ICK.

$$EU(\text{Buy ICK}) = EU(DOW_1) = (.6)(\$170{,}000) + (.4)(\$125{,}000) = \$152{,}000$$

$$EU(ICK_3) = (.3)(\$0) + (.7)(\$250{,}000) = \$175{,}000$$

$$EU(ICK_4) = (.6)(\$0) + (.4)(\$250{,}000) = \$100{,}000$$

$$EU(\text{Buy option}) = EU(DOW_2) = (.6)(\$175{,}000) + (.4)(\$100{,}000) = \$145{,}000$$

$$EU(D) = \max(\$152{,}000, \ \$145{,}000) = \$152{,}000.$$

The solved decision tree is shown in Figure 5.5. The decision is to buy ICK.

The previous example illustrates a problem with decision trees. That is, the representation of a problem instance by a decision tree grows exponentially with the size of the instance. Notice that the instance in Example 5.3 only has one more element in it than the instance in Example 5.2; that is, it includes that uncertainty about the Dow. Yet its representation is twice as large. So it is quite difficult to represent a large instance with a decision tree. We will see in the next section that influence diagrams do not have this problem. Before that, we show more examples.

Example 5.4 *Sam has the opportunity to buy a 1996 Spiffycar automobile for $10,000, and he has a prospect who would be willing to pay $11,000 for the auto if it is in excellent mechanical shape. Sam determines that all mechanical parts except for the transmission are in excellent shape. If the transmission is bad, it will cost Sam $3000 to repair it, and he would have to repair it before the prospect would buy it. So he would only end up with $8000 if he bought the vehicle and its transmission was bad. He cannot determine the state of the transmission himself. However, he has a friend who can run a test on the transmission. The test is not absolutely accurate. Rather 30% of the time it judges a good transmission to be bad and 10% of the time it judges a bad transmission to be good. To represent this relationship between the transmission and the test, we define the following random variables:*

Variable	Value	When the Variable Takes This Value
Test	*positive*	Test judges the transmission to be bad.
	negative	Test judges the transmission to be good.
Tran	*good*	Transmission is good.
	bad	Transmission is bad.

The previous discussion implies that we have these conditional probabilities:

$$P(Test = positive | Tran = good) = .3$$

$$P(Test = positive | Tran = bad) = .9.$$

Furthermore, Sam knows that 20% of the 1996 Spiffycars have bad transmissions. That is,

$$P(Tran = good) = .8.$$

Sam is going to have his friend run the test for free, and then he will decide whether to buy the car.

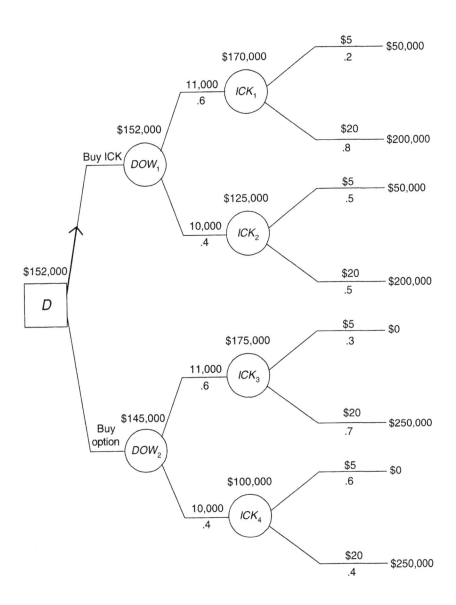

Figure 5.5: The solved decision tree given the decision tree in Figure 5.4.

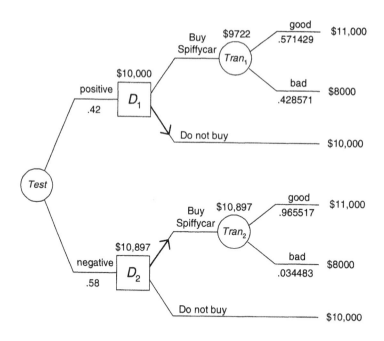

Figure 5.6: The decision tree representing the problem instance in Example 5.4.

This problem instance is represented in the decision tree in Figure 5.6. Notice first that, if he does not buy the vehicle, the outcome is simply $10,000. This is because the point in the future is so near that we can neglect interest as negligible. Note further that the probabilities in that tree are not the ones stated in the example. They must be computed from the stated probabilities. We do that next. The probability on the upper edge emanating from the Test node is the prior probability the test is positive. It is computed as follows (note that we use our abbreviated notation):

$$P(positive) = P(positive|good)P(good) + P(positive|bad)P(bad)$$
$$= (.3)(.8) + (.9)(.2) = .42.$$

The probability on the upper edge emanating from the $Tran_1$ node is the probability the transmission is good given the test is positive. We compute it using Bayes' Theorem as follows:

$$P(good|positive) = \frac{P(positive|good)P(good)}{P(positive)}$$
$$= \frac{(.3)(.8)}{.42} = .571429.$$

It is left as an exercise to determine the remaining probabilities in the tree. Next, we solve the tree:

$$EU(Tran_1) = (.571429)(\$11,000) + (.428571)(\$8000) = \$9714$$

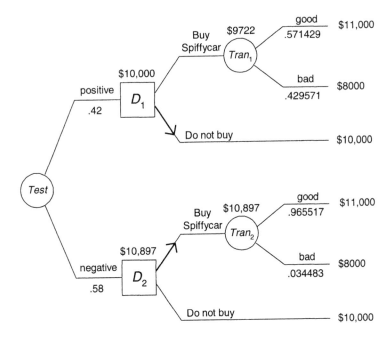

Figure 5.7: The solved decision tree given the decision tree in Figure 5.6.

$$EU(D_1) = \max(\$9714, \$10,000) = \$10,000$$

$$EU(Tran_2) = (.965517)(\$11,000) + (.034483)(\$8000) = \$10,897$$

$$EU(D_2) = \max(\$10,897,\ \$10,000) = \$10,897.$$

We need not compute the expected value of the Test node because there are no decisions to the left of it. The solved decision tree is shown in Figure 5.7. The decision is to not buy the vehicle if the test is positive and to buy it if the test is negative.

The previous example illustrates another problem with decision trees. That is, the probabilities needed in a decision tree are not always the ones that are readily available to us. So we must compute them using the law of total probability and Bayes' Theorem. We will see that influence diagrams do not have this problem either. More examples follow.

Example 5.5 Suppose Sam is in the same situation as in Example 5.4, except that the test is not free. Rather it costs $200. So Sam must decide whether to run the test, buy the car without running the test, or keep his $10,000. The decision tree representing this problem instance is shown in Figure 5.8. Notice that the outcomes when the test is run are all $200 less than their respective outcomes in Example 5.4. This is because it costs $200 to run the test. Note further that, if the vehicle is purchased without running the test, the probability of the transmission being good is simply its prior probability of .8. This is

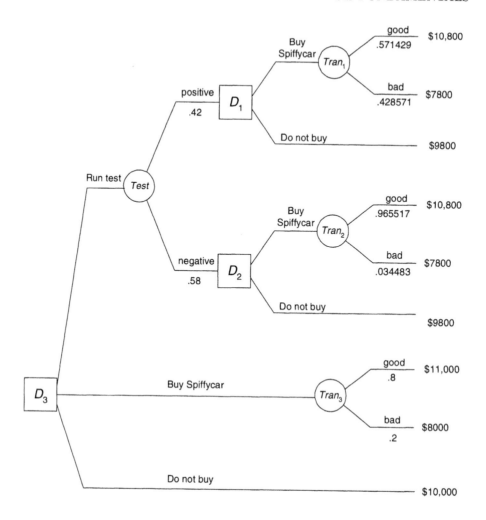

Figure 5.8: The decision tree representing the problem instance in Example 5.5.

because no test was run. So our only information about the transmission is our prior information. Next, we solve the decision tree. It is left as an exercise to show

$$EU(D_1) = \$9800$$

$$EU(D_2) = \$10{,}697.$$

Therefore,

$$EU(Test) = (.42)(\$9800) + (.58)(\$10{,}697) = \$10{,}320.$$

Furthermore,

$$EU(Tran_3) = (.8)(\$11{,}000) + (.2)(\$8000) = \$10{,}400.$$

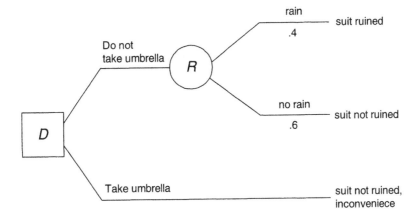

Figure 5.9: The decision tree representing the problem instance in Example 5.6.

Finally,

$$EU(D_3) = \max(\$10{,}320, \ \$10{,}400, \ \$10{,}000) = \$10{,}400.$$

So Sam's decision is to buy the vehicle without running the test. It is left as an exercise to show the solved decision tree.

The next two examples illustrate cases in which the outcomes are not numeric.

Example 5.6 *Suppose Leonardo has just bought a new suit, he is about to leave for work, and it looks like it might rain. Leonardo has a long walk from the train to his office. So he knows if it rains and he does not have his umbrella, his suit will be ruined. His umbrella will definitely protect his suit from the rain. However, he hates the inconvenience of lugging the umbrella around all day. Given that he feels there is a .4 probability it will rain, should he bring his umbrella? A decision tree representing this problem instance is shown in Figure 5.9. We cannot solve this tree yet because its outcomes are not numeric. We can give them numeric utilities as shown next. Clearly, the ordering of the outcomes from worst to best is as follows:*

1. *suit ruined*
2. *suit not ruined, inconvenience*
3. *suit not ruined.*

We assign a utility of 0 to the worst outcome and a utility of 1 to the best outcome. So

$$U(\text{suit ruined}) = 0$$

$$U(\text{suit not ruined}) = 1.$$

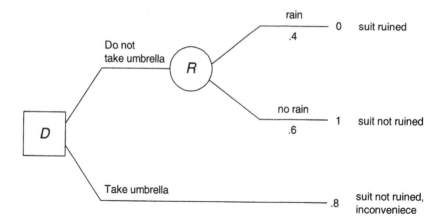

Figure 5.10: The decision tree with numeric values representing the problem instance in Example 5.6.

Then we consider lotteries (chance nodes) L_p in which Leonardo gets the outcome "suit not ruined" with probability p and outcome "suit ruined" with probability $1 - p$. The utility of "suit not ruined, inconvenience" is defined to be the expected utility of the lottery L_p for which Leonardo would be indifferent between lottery L_p and being assured of "suit not ruined, inconvenience." We then have

U(suit not ruined, inconvenience)

$$
\begin{aligned}
&\equiv \quad EU(L_p) \\
&= \quad pU(\text{suit not ruined}) + (1 - p)U(\text{suit ruined}) \\
&= \quad p(1) + (1 - p)0 = p.
\end{aligned}
$$

Let's say Leonardo decides $p = .8$. Then

$$U(\text{suit not ruined, inconvenience}) = .8.$$

The decision tree with these numeric values is shown in Figure 5.10. We solve that decision tree next.

$$EU(R) = (.4)(0) + (.6)(1) = .6$$

$$EU(D) = \max(.6, .8) = .8.$$

So the decision is to take the umbrella.

The method used to obtain numeric values in the previous example easily extends to the case where there are more than three outcomes. For example, suppose there was a fourth outcome "suit goes to cleaners" in between "suit not ruined, inconvenience" and "suit not ruined" in the preference ordering. We consider lotteries L_q in which Leonardo gets outcome "suit not ruined" with probability q and outcome "suit not ruined, inconvenience" with probability $1 - q$. The utility of "suit goes to cleaners" is defined to be the expected utility of the lottery L_q for which Leonardo would be indifferent between lottery L_q and being assured of "suit goes to cleaners." We then have

U(suit goes to cleaners)

$$
\begin{aligned}
&\equiv\ EU(L_q) \\
&=\ qU(\text{suit not ruined}) + (1 - q)U(\text{suit not ruined, inconvenience}) \\
&=\ q(1) + (1 - q)(.8) = .8 + .2q.
\end{aligned}
$$

Let's say Leonardo decides $q = .6$. Then

$$U(\text{suit goes to cleaners}) = .8 + (.2)(.6) = .92.$$

Next, we give an example from the medical domain.[1]

Example 5.7 *Amit, a 15-year-old high school student, has been definitively diagnosed with streptococcal infection, and he is considering having a treatment which is known to reduce the number of days with a sore throat from 4 to 3. He learns, however, that the treatment has a .000003 probability of causing death due to anaphylaxis. Should he have the treatment?*

*You may argue that, if he may die from the treatment, he certainly should not have it. However, the probability of dying is extremely small, and we daily accept small risks of dying in order to obtain something of value to us. For example, many people take a small risk of dying in a car accident in order to arrive at work. We see then that we cannot discount the treatment based solely on that risk. So what should Amit do? Next, we apply decision analysis to recommend a decision to him. Figure 5.11 shows a decision tree representing Amit's decision. To solve this problem instance we need to quantify the outcomes in that tree. We can do this using **quality adjusted life expectancies** (QALE). We ask Amit to determine what one year of life with a sore throat is worth relative to one year of life without one. We will call such years "well years." Let's say he says it is worth .9 well years. That is, for Amit*

> 1 year with sore throat *is equivalent to* .9 well years.

*We then assume a **constant proportional trade-off**. That is, we assume the time trade-off associated with having a sore throat is independent of the time spent with one. The validity of this assumption and alternative models are discussed in [Nease and Owens, 1997]. Given this assumption, for Amit*

> t years with sore throat *is equivalent to* .9t well years.

*The value .9 is called the **time-trade-off quality adjustment** for a sore throat. Another way to look at it is that Amit would give up .1 years of life to*

---

[1]This example is based on an example in [Nease and Owens, 1997]. Although the information is not fictitious, some of it is controversial.

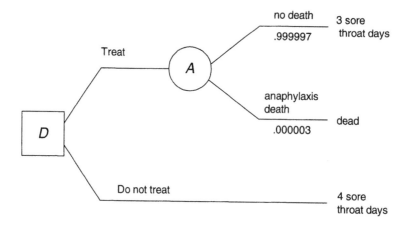

Figure 5.11: A decision tree modeling Amit's decision concerning being treated for streptococcal infection.

*avoid having a sore throat for .9 years of life. Now, if we let t be the amount of time Amit will have a sore throat due to this infection, and l be Amit's remaining life expectancy, we define his quality **QALE** as follows:*

$$QALE(l, t) = (l - t) + .9t.$$

From life expectancy charts, we determine Amit's remaining life expectancy is 60 years. Converting days to years, we have the following:

$$3 \ days = .008219 \ years$$

$$4 \ days = .010959 \ years.$$

Therefore, Amit's QALEs are as follows:

$$
\begin{aligned}
QALE(60 \ years, \ 3 \ sore \ throat \ days) \ &= \ 60 - .008219 + .9(.008219) \\
&= \ 59.999178
\end{aligned}
$$

$$
\begin{aligned}
QALE(60 \ years, \ 4 \ sore \ throat \ days) \ &= \ 60 - .010959 + .9(.010959) \\
&= \ 59.998904.
\end{aligned}
$$

Figure 5.12 shows the decision tree in Figure 5.11 with the actual outcomes augmented with QALEs. Next, we solve that tree.

$$
\begin{aligned}
EU(Treat) = EU(A) \ &= \ (.999993)(59.999178) + (.000003)(0) \\
&= \ 59.998758
\end{aligned}
$$

$$EU(Do \ not \ treat) = 59.998904$$

$$EU(D) = \max(59.998758, \ 59.998904) = 59.998904.$$

So the decision is to not treat, but just barely.

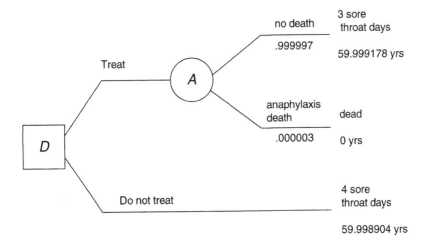

Figure 5.12: The decision tree in Figure 5.11 with the actual outcomes augmented by QALEs.

Example 5.8 *This example is an elaboration of the previous one. Actually, Streptococcus infection can lead to rheumatic heart disease (RHD), which is less probable if the patient is treated. Specifically, if we treat a patient with Streptococcus infection, the probability of rheumatic heart disease is .000013, while if we do not treat the patient, the probability is .000063. Rheumatic heart disease would be for life. So Amit needs to take all this into account. First, he must determine time trade-off quality adjustments both for having rheumatic heart disease alone and for having it along with a sore throat. Suppose he determines the following:*

$$1 \text{ year with RHD} \qquad \text{is equivalent to} \qquad .15 \text{ well years.}$$

$$1 \text{ year with sore throat and RHD} \qquad \text{is equivalent to} \qquad .1 \text{ well years.}$$

We then have

$$QALE(60 \text{ years, RHD, 3 sore throat days}) \;=\; .15\left(60 - \frac{3}{365}\right) + .1\left(\frac{3}{365}\right)$$
$$=\; 8.999589$$

$$QALE(60 \text{ years, RHD, 4 sore throat days}) \;=\; .15\left(60 - \frac{4}{365}\right) + .1\left(\frac{4}{365}\right)$$
$$=\; 8.999452.$$

We have already computed QALEs for 3 or 4 days with only a sore throat in the previous example. Figure 5.13 shows the resultant decision tree. We solve that decision tree next.

$$EU(RHD_1) \;=\; (.000013)(8.999569) + (.999987)(59.999178)$$
$$=\; 59.998515$$

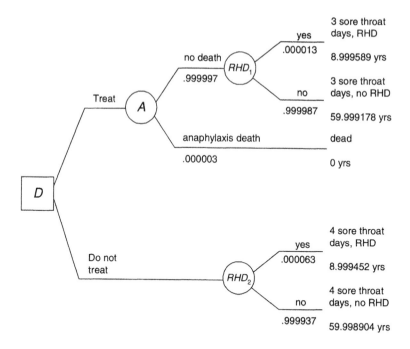

Figure 5.13: A decision tree modeling Amit's decision concerning being treated for streptococcal infection when rheumatic heart disease is considered.

$$EU(\textit{Treat}) = EU(A) \quad = \quad (.999997)(59.998515) + (.000003)(0)$$
$$= \quad 59.998335$$

$$EU(\textit{Do not treat}) \quad = \quad EU(RHD_2)$$
$$= \quad (.000063)(8.999452) + (.999937)(59.998904)$$
$$= \quad 59.995691$$

$$EU(D) = \max(59.998335, \ 59.995691) = 59.998335.$$

So now the decision is to treat, but again just barely.

You may argue that, in the previous two examples, the difference in the expected utilities is negligible because the number of significant digits needed to express it is far more than the number of significant digits in Amit's assessments. This argument is reasonable. However, the utilities of the decisions are so close because the probabilities of both anaphylaxis death and rheumatic heart disease are so small. In general, this situation is not always the case. It is left as an exercise to rework the previous example with the probability of rheumatic heart disease being .13 instead of .000063.

Another consideration in medical decision making is the financial cost of the treatments. In this case, the value of an outcome is a function of both the QALE and the financial cost associated with the outcome.

5.2 Influence Diagrams

In Section 5.1, we noted two difficulties with decision trees. First, the representation of a problem instance by a decision tree grows exponentially with the size of the instance. Second, the probabilities needed in a decision tree are not always the ones that are readily available to us. Next, we present an alternative representation of decision problem instances, namely influence diagrams, which do not have either of these difficulties. First, we only discuss representing problem instances with influence diagrams. Then in Section 5.2.2 we discuss solving influence diagrams.

5.2.1 Representing with Influence Diagrams

An **influence diagram** contains three kinds of nodes: **chance (or uncertainty) nodes** representing random variables; **decision nodes** representing decisions to be made; and one **utility node**, which is a random variable whose possible values are the utilities of the outcomes. We depict these nodes as follows:

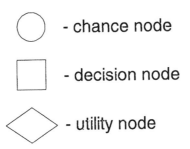

The edges in an influence diagram have the following meaning:

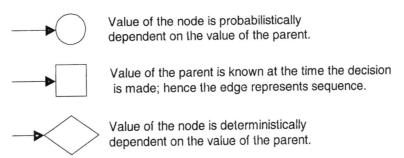

The chance nodes in an influence diagram satisfy the Markov condition with the probability distribution. That is, each chance node X is conditionally independent of the set of all its nondescendents given the set of all its parents. So an influence diagram is actually a Bayesian network augmented with decision nodes and a utility node. There must be an ordering of the decision nodes in an influence diagram based on the order in which the decisions are made. The

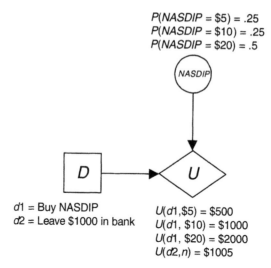

$P(NASDIP = \$5) = .25$
$P(NASDIP = \$10) = .25$
$P(NASDIP = \$20) = .5$

d1 = Buy NASDIP
d2 = Leave $1000 in bank

$U(d1, \$5) = \$500$
$U(d1, \$10) = \$1000$
$U(d1, \$20) = \$2000$
$U(d2, n) = \$1005$

Figure 5.14: An influence diagram modeling your decision whether to buy NAS-DIP.

order is specified using the edges between the decision nodes. For example, if we have the order

$$D_1, D_2, D_3,$$

then there are edges from D_1 to D_2 and D_3 and an edge from D_2 to D_3.

To illustrate influence diagrams, we next represent the problem instances, in the examples in the section on decision trees by influence diagrams.

Example 5.9 *Recall Example 5.1 in which you felt there was a .25 probability NASDIP will be at $5 at month's end, a .5 probability it will be at $20, and a .25 probability it will be at $10. Your decision is whether to buy 100 shares of NASDIP for $1000 or to leave the $1000 in the bank where it will earn .005 interest. Figure 5.14 shows an influence diagram representing this problem instance. Notice a few things about that diagram. There is no edge from D to NASDIP because your decision as to whether to buy NASDIP has no affect on its performance. (We assume your 100 shares are not enough to affect market activity.) There is no edge from NASDIP to D because at the time you make your decision you do not know NASDIP's value in one month. There are edges from both NASDIP and D to U because your utility depends both on whether NASDIP goes up and whether you buy it. Notice that if you do not buy it, the utility is the same regardless of what happens to NASDIP. This is why we write $U(d2, n) = \$1005$. The variable n represents any possible value of NASDIP.*

Example 5.10 *Recall Example 5.2 which concerned the same situation as in Example 5.1, except that your choices were either to buy NASDIP or to buy an option on NASDIP. Recall further that if NASDIP was at $5 or $0 per share*

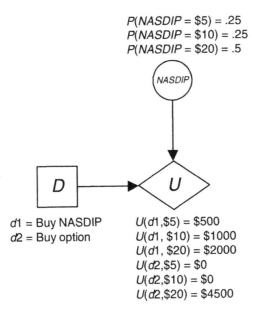

$P(NASDIP = \$5) = .25$
$P(NASDIP = \$10) = .25$
$P(NASDIP = \$20) = .5$

NASDIP

D   U

$d1$ = Buy NASDIP
$d2$ = Buy option

$U(d1,\$5) = \$500$
$U(d1, \$10) = \$1000$
$U(d1, \$20) = \$2000$
$U(d2,\$5) = \$0$
$U(d2,\$10) = \$0$
$U(d2,\$20) = \$4500$

Figure 5.15: An influence diagram modeling your decision whether to buy NAS-DIP when the other choice is to buy an option.

in one month, you would not exercise your option and you would lose your $1000; and if NASDIP was at $20 per share in one month, you would exercise your option and your $1000 investment would be worth $4500. Figure 5.15 shows an influence diagram representing this problem instance. Recall that when we represented this instance with a decision tree (Figure 5.3) that tree was symmetrical because we encountered the same uncertain event regardless of which decision was made. This symmetry manifests itself in the influence diagram in that the value of the utility node U depends on the value of the chance node *NASDIP* regardless of the value of the decision node D.

Example 5.11 *Recall Example 5.3 in which Nancy was considering either buying 10,000 shares of ICK for $10 a share or an option on ICK for $100,000 which would allow her to buy 50,000 shares of ICK for $15 a share in one month. Recall further that she believed that in one month the Dow would be either at 10,000 or at 11,000, and ICK would be either at $5 or at $20 per share. Finally, recall that she assigned the following probabilities:*

$$P(ICK = \$5 | Dow = 11,000, Decision = Buy\ ICK) = .2$$

$$P(ICK = \$5 | Dow = 11,000, Decision = Buy\ option) = .3$$

$$P(ICK = \$5 | Dow = 10,000, Decision = buy\ ICK) = .5$$

$$P(ICK = \$5 | Dow = 10,000, Decision = Buy\ option) = .6$$

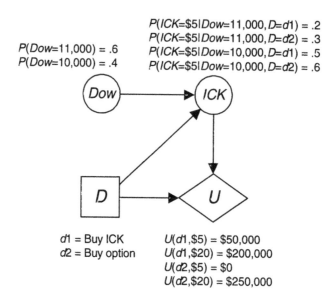

Figure 5.16: An influence diagram modeling Nancy's decision concerning buying ICK or an option on ICK.

$$P(Dow = \$11{,}000) = .6.$$

Figure 5.16 shows an influence diagram representing this problem instance. Notice that the value of ICK depends not only on the value of the Dow, but also on the decision D. This is because Nancy's purchase can affect market activity. Note further that this instance has one more component than the instance in Example 5.10, and we needed to add only one more node to represent it with an influence diagram. So the representation grew linearly with the size of the instance. By contrast, recall that when we represented the instances with decision trees, the representation grew exponentially.

Example 5.12 *Recall Example 5.4 in which Sam had the opportunity to buy a 1996 Spiffycar automobile for $10,000, and he had a prospect who would be willing to pay $11,000 for the auto if it were in excellent mechanical shape. Recall further that if the transmission were bad, Sam would have to spend $3000 to repair it before he could sell the vehicle. So he would only end up with $8000 if he bought the vehicle and its transmission was bad. Finally, recall he had a friend who could run a test on the transmission, and we had the following:*

$$P(Test = positive|Tran = good) = .3$$

$$P(Test = positive|Tran = bad) = .9$$

$$P(Tran = good) = .8.$$

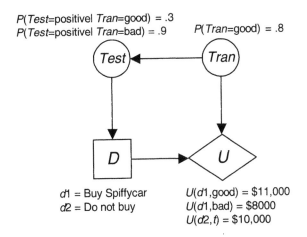

P(Test=positivel Tran=good) = .3
P(Test=positivel Tran=bad) = .9   P(Tran=good) = .8

d1 = Buy Spiffycar
d2 = Do not buy

U(d1,good) = \$11,000
U(d1,bad) = \$8000
U(d2,t) = \$10,000

Figure 5.17: An influence diagram modeling Sam's decision concerning buying the Spiffycar.

Figure 5.17 shows an influence diagram representing this problem instance. Notice that there is an arrow from $Tran$ to $Test$ because the value of the test is probabilistically dependent on the state of the transmission, and there is an arrow from $Test$ to D because the outcome of the test will be known at the time the decision is made. That is, D follows $Test$ in sequence. Note further that the probabilities in the influence diagram are the ones we know. We did not need to use the law of total probability and Bayes' Theorem to compute them, as we did when we represented the instance with a decision tree.

Example 5.13 *Recall Example 5.5 in which Sam was in the same situation as in Example 5.4 except that the test was not free. Rather it costs \$200. So Sam had to decide whether to run the test, buy the car without running the test, or keep his \$10,000. Figure 5.18 shows an influence diagram representing this problem instance. Notice that there is an edge from R to D because decision R is made before decision D. Note further that there is an edge from D to T because the test T is run only if we make decision $r1$.*

Next, we show a more complex instance, which we did not represent with a decision tree.

Example 5.14 *Suppose Sam is in the same situation as in Example 5.13, but with the following modifications. First, Sam knows that 20% of the Spiffycars were manufactured in a plant that produced lemons and 80% of them were manufactured in a plant that produced peaches. Furthermore, he knows 40% of the lemons have good transmissions and 90% of the peaches have good transmissions. Also, 5% of the lemons have fine alternators, and 80% of the peaches have fine alternators. If the alternator is faulty (not fine), it will cost Sam \$600*

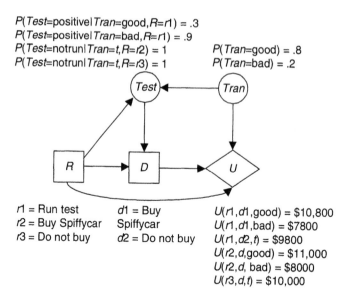

P(Test=positive|Tran=good,R=r1) = .3
P(Test=positive|Tran=bad,R=r1) = .9
P(Test=notrun|Tran=t,R=r2) = 1
P(Test=notrun|Tran=t,R=r3) = 1   P(Tran=good) = .8
                                       P(Tran=bad) = .2

r1 = Run test   d1 = Buy   U(r1,d1,good) = $10,800
r2 = Buy Spiffycar   Spiffycar   U(r1,d1,bad) = $7800
r3 = Do not buy   d2 = Do not buy   U(r1,d2,t) = $9800
                                       U(r2,d,good) = $11,000
                                       U(r2,d, bad) = $8000
                                       U(r3,d,t) = $10,000

Figure 5.18: An influence diagram modeling Sam's decision concerning buying the Spiffycar when he must pay for the test.

to repair it before he can sell the vehicle. Figure 5.19 shows an influence diagram representing this problem instance. Notice that the set of chance nodes in the influence diagram constitutes a Bayesian network. For example, Tran and Alt are not independent, but they are conditionally independent given Car.

We close with a large problem instance in the medical domain.

Example 5.15 This example is taken from [Nease and Owens, 1997]. Suppose a patient has a non-small-cell carcinoma of the lung. The primary tumor is 1 cm in diameter, a chest X-ray indicates the tumor does not abut the chest wall or mediastinum, and additional workup shows no evidence of distant metastases. The preferred treatment in this situation is a thoracotomy. The alternative treatment is radiation. Of fundamental importance in the decision to perform a thoracotomy is the likelihood of mediastinal metastases. If mediastinal metastases are present, a thoracotomy would be contraindicated because it subjects the patient to a risk of death with no health benefit. If mediastinal metastases are absent, a thoracotomy offers a substantial survival advantage as long as the primary tumor has not metastasized to distant organs.

We have two tests available for assessing the involvement of the mediastinum. They are computed tomography (CT scan) and mediastinoscopy. This problem instance involves three decisions. First, should the patient undergo a CT scan? Second, given this decision and any CT results, should the patient undergo mediastinoscopy? Third, given these decisions and any test results, should the patient undergo a thoracotomy?

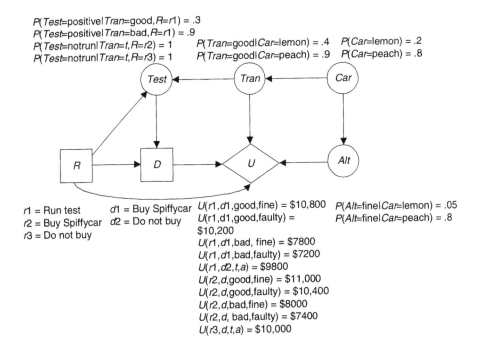

$P(Test=positive|Tran=good,R=r1) = .3$
$P(Test=positive|Tran=bad,R=r1) = .9$
$P(Test=notrun|Tran=t,R=r2) = 1$ $P(Tran=good|Car=lemon) = .4$ $P(Car=lemon) = .2$
$P(Test=notrun|Tran=t,R=r3) = 1$ $P(Tran=good|Car=peach) = .9$ $P(Car=peach) = .8$

r1 = Run test   d1 = Buy Spiffycar $U(r1,d1,good,fine) = \$10,800$ $P(Alt=fine|Car=lemon) = .05$
r2 = Buy Spiffycar   d2 = Do not buy $U(r1,d1,good,faulty) =$ $P(Alt=fine|Car=peach) = .8$
r3 = Do not buy $\$10,200$
                                           $U(r1,d1,bad, fine) = \$7800$
                                           $U(r1,d1,bad,faulty) = \$7200$
                                           $U(r1,d2,t,a) = \$9800$
                                           $U(r2,d,good,fine) = \$11,000$
                                           $U(r2,d,good,faulty) = \$10,400$
                                           $U(r2,d,bad,fine) = \$8000$
                                           $U(r2,d, bad,faulty) = \$7400$
                                           $U(r3,d,t,a) = \$10,000$

Figure 5.19: An influence diagram modeling Sam's decision concerning buying the Spiffycar when the alternator may be faulty.

The CT scan can detect mediastinal metastases. The test is not absolutely accurate. Rather, if we let $MedMet$ be a variable whose values are *present* and *absent* depending on whether or not mediastinal metastases are present and $CTest$ be a variable whose values are *cpos* and *cneg* depending on whether or not the CT scan is positive, we have

$$P(CTest = cpos|MedMet = present) = .82$$

$$P(CTest = cpos|MedMet = absent) = .19.$$

The mediastinoscopy is an invasive test of mediastinal lymph nodes for determining whether the tumor has spread to those nodes. If we let $Mtest$ be a variable whose values are *mpos* and *mneg* depending on whether or not the mediastinoscopy is positive, we have

$$P(MTest = mpos|MedMet = present) = .82$$

$$P(MTest = mpos|MedMet = absent) = .005.$$

The mediastinoscopy can cause death. If we let E be the decision concerning whether to have the mediastinoscopy, $e1$ be the choice to have it, $e2$ be the choice not to have it, and $MedDeath$ be a variable whose values are *mdie* and *mlive* depending on whether the patient dies from the mediastinoscopy, we have

$$P(MedDeath = mdie|E = e1) = .005$$

$$P(MedDeath = mdie|E = e2) = 0.$$

The thoracotomy has a greater chance of causing death than the alternative treatment of radiation. If we let T be the decision concerning which treatment to have, $t1$ be the choice to undergo a thoracotomy, $t2$ be the choice to undergo radiation, and $Thordeath$ be a variable whose values are $tdie$ and $tlive$ depending on whether the patient dies from the treatment, we have

$$P(ThorDeath = tdie|T = t1) = .037$$

$$P(ThorDeath = tdie|T = t2) = .002.$$

Finally, we need the prior probability that mediastinal metastases are present. We have

$$P(MedMet = present) = .46.$$

Figure 5.20 shows an influence diagram representing this problem instance. Note that we considered quality adjustments to life expectancy (QALE) and financial costs to be insignificant in this example. The value node is only in terms of life expectancy.

5.2.2   Solving Influence Diagrams

We first illustrate how influence diagrams can be solved by presenting some examples. Then we show solving influence diagrams using the package Netica.

5.2.3   Techniques for Solving Influence Diagrams⋆

Next, we show how to solve influence diagrams. If you are only interested in seeing them solved using packages, you can skip to the next subsection.

Example 5.16 *Consider the influence diagram in Figure 5.14, which was developed in Example 5.9. To solve the influence diagram, we need to determine which decision choice has the largest expected utility. The expected utility of a decision choice is the expected value E of U given the choice is made. We have*

$$
\begin{aligned}
EU(d1) &= E(U|d1) \\
&= P(\$5|d1)U(d1,\$5) + P(\$10|d1)U(d1,\$10) + P(\$20|d1)U(d1,\$20) \\
&= (.25)(\$500) + (.25)(\$1000) + (.5)(\$2000) \\
&= \$1375
\end{aligned}
$$

$$
\begin{aligned}
EU(d2) &= E(U|d2) \\
&= P(\$5|d2)U(d2,\$5) + P(\$10|d2)U(d2,\$10) + P(\$20|d2)U(d2,\$20) \\
&= (.25)(\$1005) + (.25)(\$1005) + (.5)(\$1005) \\
&= \$1005.
\end{aligned}
$$

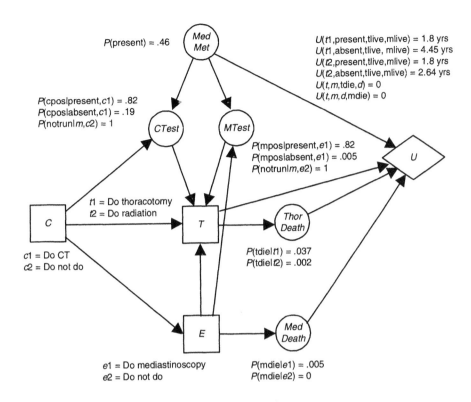

P(present) = .46

U(t1,present,tlive,mlive) = 1.8 yrs
U(t1,absent,tlive, mlive) = 4.45 yrs
U(t2,present,tlive,mlive) = 1.8 yrs
U(t2,absent,tlive,mlive) = 2.64 yrs
U(t,m,tdie,d) = 0
U(t,m,d,mdie) = 0

P(cpos|present,c1) = .82
P(cpos|absent,c1) = .19
P(notrun|m,c2) = 1

P(mpos|present,e1) = .82
P(mpos|absent,e1) = .005
P(notrun|m,e2) = 1

t1 = Do thoracotomy
t2 = Do radiation

c1 = Do CT
c2 = Do not do

P(tdie|t1) = .037
P(tdie|t2) = .002

e1 = Do mediastinoscopy
e2 = Do not do

P(mdie|e1) = .005
P(mdie|e2) = 0

Figure 5.20: An influence diagram modeling the decision of whether to be treated with a thoracotomy.

The utility of our decision is therefore

$$EU(D) = \max(EU(d1), EU(d2))$$
$$= \max(\$1375, \ \$1005) = \$1375,$$

and our decision choice is d1.

Notice in the previous example that the probabilities do not depend on the decision choice. This is because there is no edge from D to $NASDIP$. In general, this is not always the case, as the next example illustrates.

Example 5.17 *Consider the influence diagram in Figure 5.16, which was developed in Example 5.11. We have*

$$EU(d1) = E(U|d1)$$
$$= P(\$5|d1)U(d1,\$5) + P(\$20|d1)U(d1,\$20)$$
$$= (.32)(\$50,000) + (.68)(\$200,000)$$
$$= \$152,000$$

$$
\begin{aligned}
EU(d2) &= E(U|d2) \\
&= P(\$5|d2)U(d2, \$5) + P(\$20|d2)U(d2, \$20) \\
&= (.42)(\$0) + (.58)(\$250,000) \\
&= \$145,000
\end{aligned}
$$

$$
\begin{aligned}
EU(D) &= \max(EU(d1), EU(d2)) \\
&= \max(\$152,000,\ \$145,000) = \$152,000,
\end{aligned}
$$

and our decision choice is $d1$. You may wonder where we obtained the values of $P(\$5|d1)$ and $P(\$5|d2)$. Once we instantiate the decision node, the chance nodes comprise a Bayesian network. We then call a Bayesian network inference algorithm to compute the needed conditional probabilities. For example, that algorithm would do the following computation:

$$
\begin{aligned}
P(\$5|d1) &= P(\$5|11,000, d1)P(11,000) + P(\$5|10,000, d1)P(10,000) \\
&= (.2)(.6) + (.5)(.4) = .32.
\end{aligned}
$$

Henceforth, we will not usually show the computations done by the Bayesian network inference algorithm. We will only show the results.

Example 5.18 Consider the influence diagram in Figure 5.17, which was developed in Example 5.12. Since there is an arrow from $Test$ to D, the value of $Test$ will be known when the decision is made. So we need to determine the expected value of U given each value of $Test$. We have

$$
\begin{aligned}
EU(d1|positive) &= E(U|d1, positive) \\
&= P(good|d1, positive)U(d1, good) + \\
&\quad P(bad|d1, positive)U(d1, bad) \\
&= (.571429)(\$11,000) + (.428571)(\$8000) \\
&= \$9714
\end{aligned}
$$

$$
\begin{aligned}
EU(d2|positive) &= E(U|d2, positive) \\
&= P(good|d2, positive)U(d2, good) + \\
&\quad P(bad|d2, positive)U(d2, bad) \\
&= (.571429)(\$10,000) + (.428571)(\$10,000) \\
&= \$10,000
\end{aligned}
$$

$$
\begin{aligned}
EU(D|positive) &= \max(EU(d1|positive), EU(d2|positive)) \\
&= \max(\$9714,\ \$10,000) = \$10,000,
\end{aligned}
$$

and our decision choice is $d2$. As in the previous example, the needed conditional probabilities are obtained from a Bayesian network inference algorithm.

It is left as an exercise to compute $EU(D|negative)$.

Example 5.19 *Consider the influence diagram in Figure 5.18, which was developed in Example 5.13. Now we have two decisions: R and D. Since there is an edge from R to D, decision R is made first, and the EU of this decision is the one we need to compute. We have*

$$
\begin{aligned}
EU(r1) &= E(U|r1) \\
&= P(d1, good|r1)U(r1, d1, good) + P(d1, bad|r1)U(r1, d1, bad) + \\
&\quad P(d2, good|r1)U(r1, d2, good) + P(d2, bad|r1)U(r1, d2, bad).
\end{aligned}
$$

We need to compute the conditional probabilities in this expression. Since D and $Tran$ are not dependent on R (decision R only determines the value of decision D in the sense that decision D does not take place for some values of R), we no longer show $r1$ to the right of the conditioning bar. We have

$$
\begin{aligned}
P(d1, good) &= P(d1|good)P(good) \\
&= [P(d1|positive)P(positive|good) + \\
&\quad P(d1|negative)P(negative|good)]P(good) \\
&= [(0)P(positive|good) + (1)P(negative|good)]P(good) \\
&= P(negative|good)P(good) \\
&= (.7)(.8) = .56.
\end{aligned}
$$

The second equality above is obtained because D and $Tran$ are independent conditional on $Test$. The values of $P(d1|positive)$ and $P(d1|negative)$ were obtained by first computing expected utilities as in Example 5.18 and then by setting the conditional probability to 1 if the decision choice is the one that maximizes expected utility and to 0 otherwise. It is left as an exercise to show that the other three probabilities are .02, .24, and .18, respectively. We therefore have

$$
\begin{aligned}
EU(r1) &= E(U|r1) \\
&= P(d1, good)U(r1, d1, good) + P(d1, bad)U(r1, d1, bad) + \\
&\quad P(d2, good)U(r1, d2, good) + P(d2, bad)U(r1, d2, bad) \\
&= (.56)(\$10{,}800) + (.02)(\$7800) + (.24)(\$9800) + (.18)(\$9800) \\
&= \$10{,}320.
\end{aligned}
$$

It is left as an exercise to show

$$
EU(r2) = \$10{,}400
$$

$$
EU(r3) = \$10{,}000.
$$

So

$$
\begin{aligned}
EU(R) &= \max(EU(r1), EU(r2), EU(r2)) \\
&= \max(\$10{,}320,\ \$10{,}400,\ \$10{,}000) = \$10{,}500,
\end{aligned}
$$

and our decision choice is $r2$.

Example 5.20 Next, we show another method for solving the influence diagram in Figure 5.18 which, although it may be less elegant than the previous method, corresponds more to the way decision trees are solved. In this method, with decision R fixed at each of its choices, we solve the resultant influence diagram for decision D, and then we use these results to solve R.

First, fixing R at $r1$, we solve the influence diagram for D. The steps are the same as those in Example 5.18. That is, since there is an arrow from $Test$ to D, the value of $Test$ will be known when the decision is made. So we need to determine the expected value of U given each value of $Test$. We have

$$
\begin{aligned}
EU(d1|r1, positive) &= E(U|r1, d1, positive) \\
&= P(good|positive)U(r1, d1, good) + \\
&\quad P(bad|positive)U(r1, d1, bad) \\
&= (.571429)(\$11{,}000) + (.429571)(\$8000) \\
&= \$9522
\end{aligned}
$$

$$
\begin{aligned}
EU(d2|r1, positive) &= E(U|r1, d2, positive) \\
&= P(good|positive)U(r1, d2, good) + \\
&\quad P(bad|positive)U(r1, d2, bad) \\
&= (.571429)(\$9800) + (.429571)(\$9800) \\
&= \$9800
\end{aligned}
$$

$$
\begin{aligned}
EU(D|r1, positive) &= \max(EU(d1|r1, positive), EU(d2|r1, positive)) \\
&= \max(\$9522, \ \$9800) = \$9800
\end{aligned}
$$

$$
\begin{aligned}
EU(d1|r1, negative) &= E(U|r1, d1, negative) \\
&= P(good|negative)U(r1, d1, good) + \\
&\quad P(bad|negative)U(r1, d1, bad) \\
&= (.965517)(\$10{,}800) + (.034483)(\$7800) \\
&= \$10{,}697
\end{aligned}
$$

$$
\begin{aligned}
EU(d2|r1, negative) &= E(U|r1, d2, negative) \\
&= P(good|negative)U(r1, d2, good) + \\
&\quad P(bad|negative)U(r1, d2, bad) \\
&= (.965517)(\$9800) + (.034483)(\$9800) \\
&= \$9800
\end{aligned}
$$

$$
\begin{aligned}
EU(D|r1, negative) &= \max(EU(d1|r1, negative), EU(d2|r1, negative)) \\
&= \max(\$10{,}697, \ \$9800) = \$10{,}697.
\end{aligned}
$$

As before, the conditional probabilities are obtained from a Bayesian network inference algorithm. Once we have the expected utilities of D, we can compute the expected utility of R as follows:

$$
\begin{aligned}
EU(r1) &= EU(D|r1, positive)P(positive) + EU(D|r1, negative)P(negative) \\
&= \$9800(.42) + \$10{,}697(.58) \\
&= \$10{,}320.
\end{aligned}
$$

Note that this is the same value we obtained using the other method. We next proceed to compute $EU(r2)$ and $EU(r3)$ in the same way. It is left as an exercise to do so.

The second method illustrated in the previous example extends readily to an algorithm for solving influence diagrams. The algorithm solves the influence diagram by converting it to the decision tree corresponding to the influence diagram. For example, if we had three decision nodes D, E, and F in that order, we would first instantiate D to its first decision choice $d1$. This amounts to focusing on the subtree (of the corresponding decision tree) emanating from decision $d1$. Then, we'd instantiate E to its first decision choice $e1$. This amounts to focusing on the subtree emanating from decision $e1$. Since F is our last decision, we'd then solve the influence diagram for decision F. Next, we'd compute the expected utility of E's first decision choice $e1$. After doing this for all of E's decision choices, we'd solve the influence diagram for decision E. We'd then compute the expected utility of D's first decision choice. This process would be repeated for each of D's decision choices. It is left as an exercise to write an algorithm that implements this method.

Olmsted [1983] developed a way to evaluate an influence diagram without transforming it to a decision tree. The method operates directly on the influence diagram by performing arc reversal/node reduction operations. These operations successively transform the diagram, ending with a diagram with only one utility node that holds the utility of the optimal decision. The method appears in [Shachter, 1986]. Tatman and Schachter [1990] use supervalue nodes, which simplify the construction of influence diagrams and subsequent sensitivity analysis. Another method for evaluating influence diagrams is to use variable elimination, which is described in Jensen [2001].

5.2.4 Solving Influence Diagrams Using Netica

In Chapter 3, Section 3.4.3, we showed how to do inference in a Bayesian network using the software package Netica. Next we show how to solve an influence diagram using it.

Example 5.21 *Recall that Figure 5.16 showed an influence diagram representing the problem instance in Example 5.11. Figure 5.21 shows that influence diagram developed using Netica. Recall from Chapter 3, Section 3.4.3, that Netica computes and shows the prior probabilities of the variables rather than showing the conditional probability distributions, and probabilities are shown*

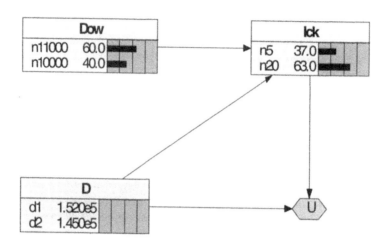

Figure 5.21: The influence diagram in Figure 5.16 developed using Netica.

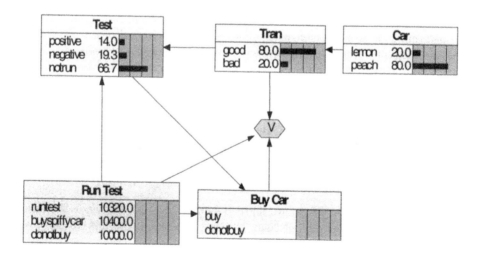

Figure 5.22: The influence diagram in Figure 5.18 developed using Netica.

as percentages. A peculiarity of Netica is that node values must start with a letter. So we placed an "n" before numeric values. Another unfortunate feature is that both chance and decision nodes are depicted as rectangles.

The values shown at the decision node D are the expected values of the decision alternatives. We see that

$$E(d1) = 1.520 \times 10^5 = 152,000$$

$$E(d2) = 1.450 \times 10^5 = 145,000.$$

So the decision alternative that maximizes expected value is $d1$.

Example 5.22 *Recall that Figure 5.18 showed an influence diagram representing the problem instance in Example 5.13. Figure 5.22 shows that influence diagram developed using Netica. We see that the decision alternative of "Run Test" that maximizes expected utility is to buy the Spiffycar without running the test.*

Example 5.23 *Recall that Figure 5.19 showed an influence diagram representing the problem instance in Example 5.14. Figure 5.23 (a) shows that influence diagram developed using Netica. We see that the decision alternative of "Run Test" that maximizes expected utility is to run the test. After running the test, the test will come back either positive or negative, and we must then decide whether or not to buy the car. The influence diagram updated to running the test and the test coming back positive appears in Figure 5.23 (b). We see that in this case the decision alternative that maximizes expected utility is to not buy the car. The influence diagram updated to running the test and the test coming back negative appears in Figure 5.23 (c). We see that in this case the decision alternative that maximizes expected utility is to buy the car.*

Example 5.24 *Recall that Figure 5.20 showed an influence diagram representing the problem instance in Example 5.15. Figure 5.24 (a) shows that influence diagram developed using Netica. We see that the decision alternative of "CT Scan" that maximizes expected utility is $c1$, which is to do the scan. After doing the scan, the scan will come back either positive or negative, and we must then decide whether or not to do the mediastinoscopy. The influence diagram updated to doing the CT scan and the scan coming back positive appears in Figure 5.24 (b). We see that in this case the decision alternative that maximizes expected utility is to do the mediastinoscopy. The influence diagram updated to then doing the mediastinoscopy and the test coming back negative appears in Figure 5.24 (c). We see that in this case the decision alternative that maximizes expected utility is to do the thoracotomy.*

In the previous example, it is not surprising that the decision alternative that maximizes expected utility is to do the CT scan since that scan has no cost. Suppose instead that there is a financial cost of $1000 involved in doing the scan. Since the utility function is in terms of years of life, to perform a decision analysis we must convert the $1000 to units of years of life (or vice

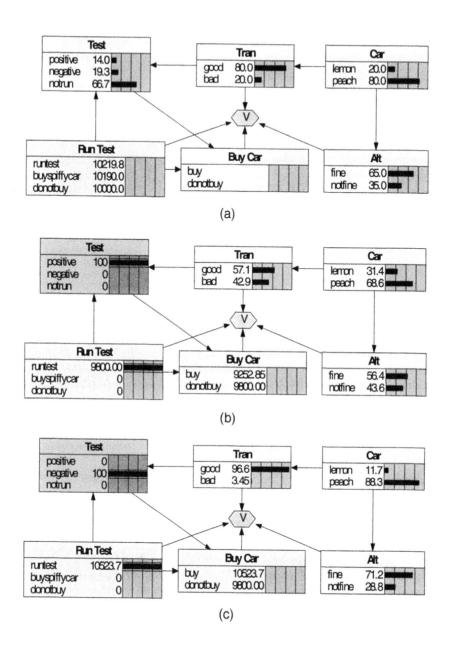

Figure 5.23: The influence diagram in Figure 5.19 developed using Netica appears in (a). The influence diagram updated to running the test and the test coming back positive appears in (b). The influence diagram updated to running the test and the test coming back negative appears in (c).

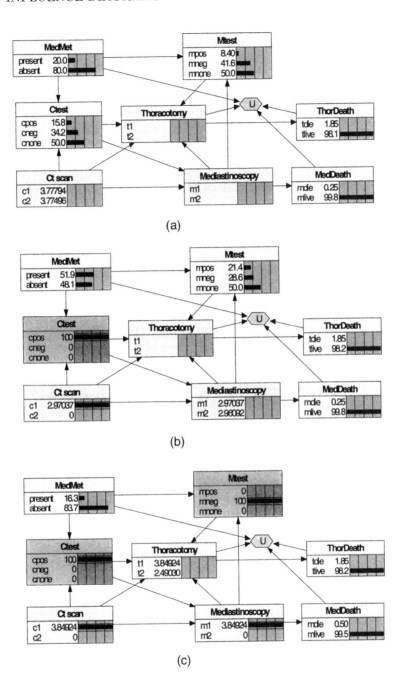

Figure 5.24: The influence diagram in Figure 5.20 developed using Netica appears in (a). The influence diagram updated to doing the CT scan and the scan coming back positive appears in (b). The influence diagram updated to then doing the mediastinoscopy and the test coming back negative appears in (c).

versa). Let's say the decision maker decides $1000 is equivalent to .01 years of life. It is left as an exercise to determine whether in this case the decision alternative that maximizes expected utility is still to do the CT scan.

5.3   Dynamic Networks*

Dynamic Bayesian networks model relationships among random variables that change over time. After introducing dynamic Bayesian networks, we discuss dynamic influence diagrams. This section is somewhat technical and can be omitted without affecting your understanding of the rest of the book.

5.3.1   Dynamic Bayesian Networks

After developing the theory, we present an example.

Formulation of the Theory

Bayesian networks do not model temporal relationships among variables. That is, a Bayesian network only represents the probabilistic relationships among a set of variables at some point in time. It does not represent how the value of some variable may be related to its value and the values of other variables at previous points in time. In many problems, however, the ability to model temporal relationships is very important. For example, in medicine it is often important to represent and reason about time in tasks such as diagnosis, prognosis, and treatment options. Capturing the dynamic (temporal) aspects of the problem is also important in artificial intelligence, economics, and biology. Next, we discuss dynamic Bayesian networks, which do model the temporal aspects of a problem.

We start by defining a random vector. Given random variables $X_1, \ldots, X_n$, the column vector

$$\mathbf{X} = \begin{pmatrix} X_1 \\ \vdots \\ X_n \end{pmatrix}$$

is called a **random vector**. A **random matrix** is defined in the same manner. We use $\mathbf{X}$ to denote both a random vector and the set of random variables which comprises $\mathbf{X}$. Similarly, we use $\mathbf{x}$ to denote both a vector value of $\mathbf{X}$ and the set of values which comprises $\mathbf{x}$. The meaning is clear from the context. Given this convention and a random vector $\mathbf{X}$ with dimension n, $P(\mathbf{x})$ denotes the joint probability distribution $P(x_1, \ldots, x_n)$. Random vectors are called **independent** if the sets of variables that comprise them are independent. A similar definition holds for conditional independence.

Now we can define a dynamic Bayesian network, which extends the Bayesian network to model temporal processes. We assume that changes occur between discrete time points, which are indexed by the non-negative integers, and we have some finite number T of points in time. Let $\{X_1, \ldots, X_n\}$ be the set of features whose values change over time, let $X_i[t]$ be a random variable representing the value of X_i at time t for $0 \leq t \leq T$, and let

$$\mathbf{X}[t] = \begin{pmatrix} X_1[t] \\ \vdots \\ X_n[t] \end{pmatrix}.$$

For all t, each $X_i[t]$ has the same space which depends on i, and we call it the space of X_i. A **dynamic Bayesian network** is a Bayesian network containing the variables that comprise the T random vectors $\mathbf{X}[t]$ and is determined by the following specifications:

1. An initial Bayesian network consisting of (a) an initial DAG $\mathbb{G}_0$ containing the variables in $\mathbf{X}[0]$ and (b) an initial probability distribution P_0 of these variables.

2. A transition Bayesian network which is a template consisting of (a) a transition DAG $\mathbb{G}_\rightarrow$ containing the variables in $\mathbf{X}[t] \cup \mathbf{X}[t+1]$ and (b) a transition probability distribution $P_\rightarrow$ which assigns a conditional probability to every value of $\mathbf{X}[t+1]$ given every value $\mathbf{X}[t]$. That is, for every value $\mathbf{x}[t+1]$ of $\mathbf{X}[t+1]$ and value $\mathbf{x}[t]$ of $\mathbf{X}[t]$, we specify

$$P_\rightarrow(\mathbf{X}[t+1] = \mathbf{x}[t+1]|\mathbf{X}[t] = \mathbf{x}[t]).$$

Since for all t each X_i has the same space, the vectors $\mathbf{x}[t+1]$ and $\mathbf{x}[t]$ each represent values from the same set of spaces. The index in each indicates the random variable which has the value. We showed the random variables above; henceforth we do not show them.

3. The dynamic Bayesian network consisting of (a) the DAG composed of the DAG $\mathbb{G}_0$ and for $0 \leq t \leq T-1$ the DAG $\mathbb{G}_\rightarrow$ evaluated at t and (b) the following joint probability distribution:

$$P(\mathbf{x}[0], \dots \mathbf{x}[T]) = P_0(\mathbf{x}[0]) \prod_{t=0}^{T-1} P_\rightarrow(\mathbf{x}[t+1]|\mathbf{x}[t]). \qquad (5.1)$$

Figure 5.25 shows an example. The transition probability distribution entailed by the network in Figure 5.25 is

$$P_\rightarrow(\mathbf{x}[t+1]|\mathbf{x}[t]) = \prod_{i=0}^{n} P_\rightarrow(x_i[t+1]|\mathsf{pa}_i[t+1]),$$

where $\mathsf{pa}_i[t+1]$ denotes the values of the parents of $X_i[t+1]$. Note that there are parents in both $\mathbf{X}[t]$ and $\mathbf{X}[t+1]$.

Owing to Equality 5.1, for all t and for all $\mathbf{x}$,

$$P(\mathbf{x}[t+1]|\mathbf{x}[0], \dots \mathbf{x}[t]) = P(\mathbf{x}[t+1]|\mathbf{x}[t]).$$

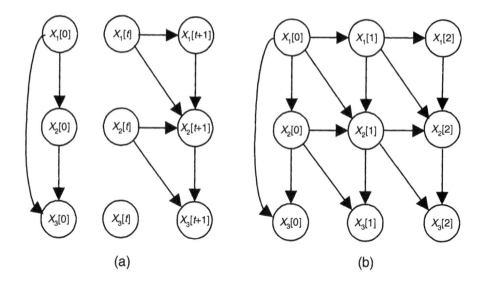

Figure 5.25: Prior and transition Bayesian networks are in (a). The resultant dynamic Bayesian network for $T = 2$ is in (b). Note that the probability distributions are not shown.

That is, all the information needed to predict a world state at time t is contained in the description of the world at time $t-1$. No information about earlier times is needed. Owing to this feature, we say the process has the **Markov** property. Furthermore, the process is **stationary**. That is, $P(\mathbf{x}[t + 1]|\mathbf{x}[t])$ is the same for all t. In general, it is not necessary for a dynamic Bayesian network to have either of these properties. However, they reduce the complexity of representing and evaluating the networks, and they are reasonable assumptions in many applications. The process need not stop at a particular time T. However, in practice, we reason only about some finite amount of time. Furthermore, we need a terminal time value to properly specify a Bayesian network.

Probabilistic inference in a dynamic Bayesian network can be done using the standard algorithms discussed in Chapter 3. However, since the size of a dynamic Bayesian network can become enormous when the process continues for a long time, the algorithms can be quite inefficient. There is a special subclass of dynamic Bayesian networks in which this computation can be done more efficiently. This subclass includes Bayesian networks in which the networks in different time steps are connected only through non-evidence variables. An example of such a network is shown in Figure 5.26. The variables labeled with an E are the evidence variables and are instantiated in each time step. We lightly shade nodes representing them. An application which uses such a dynamic Bayesian network is shown in the next subsection. Presently, we illustrate how updating can be done effectively in such networks.

Let $\mathbf{e}[t]$ be the set of values of the evidence variables at time step t and let

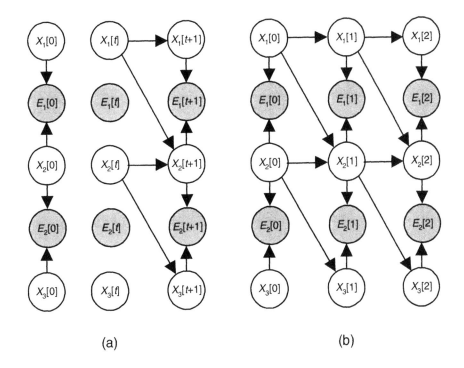

Figure 5.26: Prior and transition Bayesian networks in the case where the networks in different time slots are connected only through non-evidence variables, are in (a). The resultant dynamic Bayesian network for $T = 2$ is in (b).

$\mathsf{f}[t]$ be the set of values of the evidence variables up to and including time step t. Suppose for each value $\mathbf{x}[t]$ of $\mathbf{X}[t]$ we know

$$P(\mathbf{x}[t]|\mathsf{f}[t]).$$

We want to now compute $P(\mathbf{x}[t+1]|\mathsf{f}[t+1])$. First, we have

$$
\begin{aligned}
P(\mathbf{x}[t+1]|\mathsf{f}[t]) &= \sum_{\mathbf{x}[t]} P(\mathbf{x}[t+1]|\mathbf{x}[t],\mathsf{f}[t])P(\mathbf{x}[t]|\mathsf{f}[t]). \\
&= \sum_{\mathbf{x}[t]} P(\mathbf{x}[t+1]|\mathbf{x}[t])P(\mathbf{x}[t]|\mathsf{f}[t]). \qquad (5.2)
\end{aligned}
$$

Using Bayes' Theorem, we then have

$$
\begin{aligned}
P(\mathbf{x}[t+1]|\mathsf{f}[t+1]) &= P(\mathbf{x}[t+1]|\mathsf{f}[t],\mathsf{e}[t+1]) \\
&= \alpha P(\mathsf{e}[t+1]|\mathbf{x}[t+1],\mathsf{f}[t])P(\mathbf{x}[t+1]|\mathsf{f}[t]) \\
&= \alpha P(\mathsf{e}[t+1]|\mathbf{x}[t+1])P(\mathbf{x}[t+1]|\mathsf{f}[t]), \qquad (5.3)
\end{aligned}
$$

where α is a normalizing constant. The value of $P(\mathsf{e}[t+1]|\mathbf{x}[t+1])$ can be computed using an inference algorithm for Bayesian networks. We start the

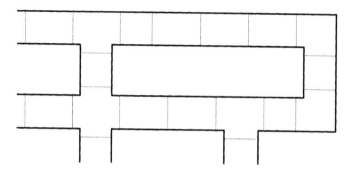

Figure 5.27: Tessellation of corridor layout.

process by computing $P(\mathbf{x}[0]|\mathbf{f}[0]) = P(\mathbf{x}[0]|\mathbf{e}[0])$. Then at each time step $t+1$ we compute $P(\mathbf{x}[t+1]|\mathbf{f}[t+1])$ using Equalities 5.2 and 5.3 in sequence. Note that to update the probability for the current time step we only need values computed at the previous time step and the evidence at the current time step. We can throw out all previous time steps, which means we need only keep enough network structure to represent two time steps.

A simple way to view the process is shown next. We define

$$P'(\mathbf{x}[t+1]) \equiv P(\mathbf{x}[t+1]|\mathbf{f}[t]),$$

which is the probability distribution of $\mathbf{X}[t+1]$ given the evidence in the first t time steps. We determine this distribution at the beginning of time step $t+1$ using Equality 5.2, and then we discard all previous information. Next we obtain the evidence in the time step $t+1$ and update P' using Equality 5.3.

An Example: Mobile Target Localization

We show an application of dynamic Bayesian networks to mobile target localization, which was developed by Basye et al. [1993]. The **mobile target localization problem** concerns tracking a target while maintaining knowledge of one's own location. Basye et al. [1993] developed a world in which a target and a robot reside. The robot is supplied with a map of the world, which is divided into corridors and junctions. Figure 5.27 shows a portion of one such world tessellated according to this scheme. Each rectangle in that figure is a different region. The state space for the location of the target is the set of all the regions shown in the figure, and the state space for the location of the robot is the set of all these regions augmented with four quadrants to represent the directions the robot can face. Let L_R and L_A be random variables whose values are the locations of the robot and the target, respectively.

Both the target and the robot are mobile, and the robot has sensors it uses to maintain knowledge of its own location and to track the target's location. Specifically, the robot has a sonar ring consisting of eight sonar transducers, configured in pairs pointing forward, backward, and to each side of the robot.

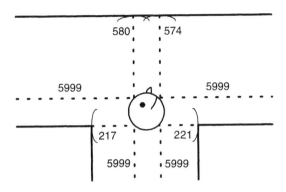

Figure 5.28: Sonar readings upon entering a T-junction.

Each sonar gives a reading between 30 and 6000 mm, where 6000 means 6000 or more. Figure 5.28 shows one set of readings obtained from the sonars upon entering a T-junction. We want the sensors to tell us what kind of region we are in. So we need a mapping from the raw sensor data to an abstract sensor space consisting of the following: corridor, T-junction, L-junction, dead-end, open space, and crossing. This mapping could be deterministic or probabilistic. Basye et al. [1993] discuss methods for developing it. Sonar data are notoriously noisy and difficult to disambiguate. A sonar which happens to be pointed at an angle of greater than 70 degrees to a wall will likely not see the wall. So we will assume the relationship is probabilistic. The robot also has a forward-pointing camera to identify the presence of its target. The camera can detect the presence of a blob identified to be the target. If it does not detect a suitable blob, this evidence is reported. If it does find a suitable blob, the size of the blob is used to estimate its distance from the robot, which is reported in rather gross units, i.e., within 1 meter, between 2 and 3 meters, etc. The detection of a blob at a given distance is only probabilistically dependent on the actual presence of the target at that distance. Let E_R be a random variable whose value is the sonar reading, which tells the robot something about its own location, and E_A be a random variable whose value is the camera reading, which tells the robot something about the target's location relative to the robot. It follows from the previous discussion that E_R is probabilistically dependent on L_R and E_A is probabilistically dependent on both L_R and L_A. At each time step, the robot obtains readings from its sonar ring and camera. For example, it may obtain the sonar readings in Figure 5.28, and its camera may inform it that the target is visible at a certain distance.

The actions available to the robot and the target are as follows: travel down the corridor the length of one region, turn left around the corner, turn around, etc. In the dynamic Bayesian network model, these actions are simply performed in some pre-programmed probabilistic way, which is not related to the sensor data. So the location of the robot at time $t + 1$ is a probabilistic function of its location at time t. When we model the problem with a dynamic

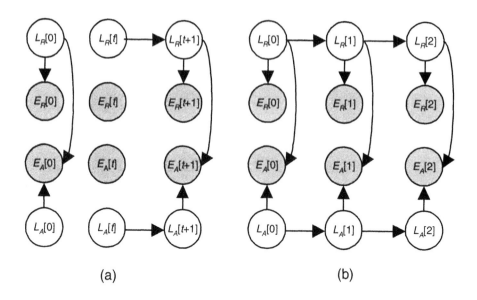

(a)   (b)

Figure 5.29: The prior and transition Bayesian networks for the mobile target mobilization problem are in (a). The resultant dynamic Bayesian network for $T = 2$ is in (b).

influence diagram in Section 5.3.2, the robot will decide on its action based on the sensor data. The target's movement could be determined by a person or be pre-programmed probabilistically.

In summary, the random variables in the problem are as follows:

Variable	What the Variable Represents
L_R	Location of the robot
L_A	Location of the target
E_R	Sensor reading regarding location of robot
E_A	Camera reading regarding location of target relative to robot

Figure 5.29 shows a dynamic Bayesian network which models this problem (without showing any actual probability distributions). The prior probabilities in the prior network represent information initially known about the location of the robot and the target. The conditional probabilities in the transition Bayesian network can be obtained from data. For example, $P(e_A|l_R, l_A)$ can be obtained by repeatedly putting the robot and the target in positions l_R and l_A, respectively, and seeing how often reading e_A is obtained.

Note that although the robot can sometimes view the target, the robot makes no effort to track the target. That is, the robot moves probabilistically according to some scheme. Our goal is for the robot to track the target. However, to do this it must decide on where to move next based on the sensor data and camera reading. As mentioned above, we need dynamic influence diagrams to produce such a robot. They are discussed next.

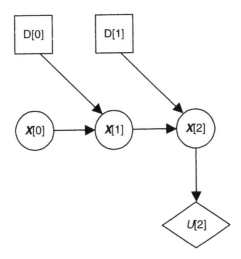

Figure 5.30: The high-level structure of a dynamic influence diagram.

5.3.2 Dynamic Influence Diagrams

Again, we first develop the theory, and then we give an example.

Formulation of the Theory

To create a **dynamic influence diagram** from a dynamic Bayesian network, we need only add decision nodes and a value node. Figure 5.30 shows the high-level structure of such a network for $T = 2$. The chance node at each time step in that figure represents the entire DAG at that time step, and so the edges represent sets of edges. There is an edge from the decision node at time t to the chance nodes at time $t + 1$ because the decision made at time t can affect the state of the system at time $t + 1$. The problem is to determine the decision at each time step which maximizes expected utility at some point in the future. Figure 5.30 represents the situation where we are determining the decision at time 0 which maximizes expected utility at time 2. The final utility could, in general, be based on the earlier chance nodes and even the decision nodes. However, we do not show such edges to simplify the diagram. Furthermore, the final expected utility is often a weighted sum of expected utilities independently computed for each time step up to the point in the future we are considering. Such a utility function is called **time-separable**.

In general, dynamic influence diagrams can be solved using the algorithm presented in Section 5.2.2. The next section contains an example.

An Example: Mobile Target Localization Revisited

After we present the model, we show some results concerning a robot constructed according to the model.

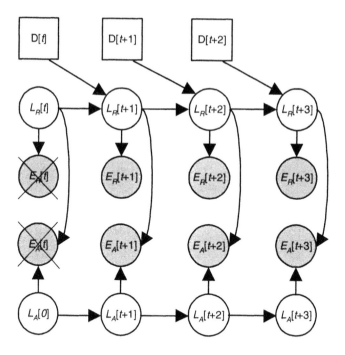

Figure 5.31: The dynamic influence diagram modeling the robot's decision of which action to take at time t.

The Model Recall the robot discussed in Section 5.3.1. Our goal is for the robot to track the target by deciding on its move at time t based on its evidence at time t. So now we allow the robot to make a decision $D[t]$ at time t of which action it will take, where the value of $D[t]$ is a result of maximizing some expected utility function based on the evidence in time step t. We assume there is error in the robot's movement. So the location of the robot at time $t + 1$ is a probabilistic function of its location at the previous time step and the action taken. The conditional probability distribution of L_R is obtained from data, as discussed at the end of Section 5.3.1. That is, we repeatedly place the robot in a location, perform an action, and then observe its new location.

The dynamic influence diagram, which represents the decision at time t and in which the robot is looking three time steps into the future, is shown in Figure 5.31. Note that there are crosses through the evidence variable at time t to indicate their values are already known. We need to maximize expected utility using the probability distribution conditional on these values and the values of all previous evidence variables. Recall that at the end of Section 5.3.1 we called this probability distribution P', and we discussed how it can be obtained. First, we need to define a utility function. Suppose we decide to determine the decision at time t by looking M time steps into the future. Let

$$\mathsf{d}_M = \{d[t], d[t+1], \ldots d[t+M-1]\}$$

be a set of values of the next M decisions including the current one, and let

$$\mathsf{f}_M = \{e_R[t+1], e_A[t+1], e_R[t+2], e_A[t+2], \dots e_R[t+M], e_A[t+M]\}$$

be a set of values of the evidence variables observed after the decisions are made. For $1 \le k \le M$ let d_k and f_k, respectively, be the first k decisions and evidence pairs in each of these sets. Define

$$U_k(\mathsf{f}_k, \mathsf{d}_k) = -\min_u \sum_v dist(u,v) P'(L_A[t+k] = v)|\mathsf{f}_k, \mathsf{d}_k), \qquad (5.4)$$

where $dist$ is the Euclidean distance, the sum is over all values v in the space of L_A, and the minimum is over all values u in the space of L_A. Recall from the beginning of Section 5.3.1 that the robot is supplied with a map of the world. It uses this map to find every element in the space of L_A. The idea is that if we make these decisions and obtain these observations at time $t+k$, the sum in Equality 5.4 is the expected value of the distance between the target and a given location u. The smaller this expected value is, the more likely it is that the target is close to u. The location $\breve{u}$ which has the minimum expected value is then our best guess at where the target is if we make these decisions and obtain these observations. So the utility of the decisions and the observations is the expected value for $\breve{u}$. The minus sign occurs because we maximize expected utility.

We then have

$$EU_k(\mathsf{d}_k) = \sum_{\mathsf{f}_k} U_k(\mathsf{f}_k, \mathsf{d}_k) P'(\mathsf{f}_k|\mathsf{d}_k). \qquad (5.5)$$

This expected utility only concerns the situation k time steps into the future. To take into account all time steps up to and including time $t+M$, we use a utility function which is a weighted sum of the utilities at each time step. We then have

$$EU(\mathsf{d}_M) = \sum_{k=1}^{M} \gamma_k EU_k(\mathsf{d}_k), \qquad (5.6)$$

where γ_k decreases with k to discount the impact of future consequences. Note that implicitly $\gamma_k = 0$ for $k > M$. Note further that we have a time-separable utility function. We choose the decision sequence which maximizes this expected utility in Equality 5.6, and we then make the first decision in this sequence at time step t.

In summary, the process proceeds as follows: In time step t the robot updates its probability distribution based on the evidence (sensor and camera readings) obtained in that step. Then the expected utility of a sequence of decisions (actions) is evaluated. This is repeated for other decision sequences, and the one that maximizes expected utility is chosen. The first decision (action) in that sequence is executed, the sensor and camera readings in time step $t+1$ are obtained, and the process repeats.

The computation of $P'(\mathsf{f}_k|\mathsf{d}_k)$ in Equality 5.5 for all values of f can be quite expensive. Dean and Wellman [1991] discuss ways to reduce the complexity of the decision evaluation.

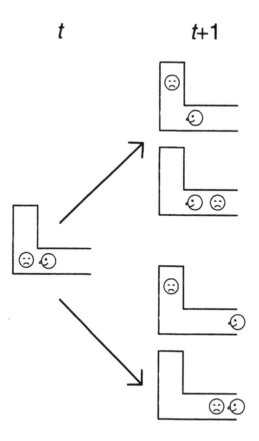

Figure 5.32: Staying close to the target may not be optimal.

Result: Emergent Behavior Basye et al. [1993] developed a robot using the model just described, and they observed some interesting, unanticipated emergent behavior. By **emergent behavior** we mean behavior that is not purposefully programmed into the robot, but emerges as a consequence of the model. For example, when the target moves towards a fork, the robot stays close behind it, since this will enable it to determine which branch the target takes. However, when the target moves toward a cul-de-sac, the robot keeps fairly far away. Basye et al. [1993] expected it to remain close behind. By analyzing the probability distributions and results of the value function, they discovered that the model allows for the possibility that the target might slip behind the robot, leaving the robot unable to determine the location of the target without additional actions. If the robot stays some distance away, regardless of what action the target takes, the observations made by the robot are sufficient to determine the target's location. Figure 5.32 illustrates the situation. In time step t the robot is close to the target as the target is about to enter the cul-de-sac. If the robot stays close, as illustrated by the top path, in time step $t + 1$ it is just as likely that the target will slip behind the robot as it is that the target

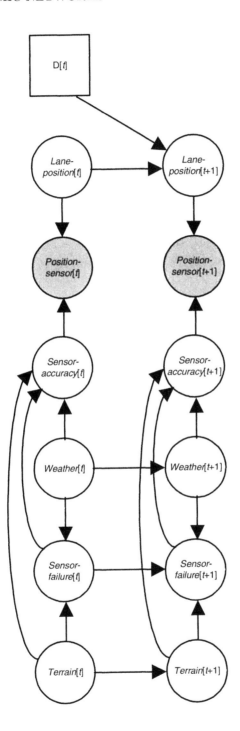

Figure 5.33: Two time steps in a dynamic influence diagram, which models the decision faced by an autonomous vehicle.

will move up the cul-de-sac. If the target does slip behind the robot, it will no longer be visible. However, if the robot backs off, as illustrated by the bottom path, the robot will be able to determine the location of the target regardless of what the target does. When considering its possible observations in time step $t+1$, the observation "target not visible" would not give the robot a good idea as to the target's location. So the move to stay put is less valued than the move to back off.

Large-Scale Systems   The method used to control our robot could be used in a more complex system. For example, an autonomous vehicle could use a vision-based, lane-position sensor to keep it in the center of its lane. The position sensor's accuracy would be affected by rain and an uneven road surface. Also, both rain and a bumpy road could cause the position sensor to fail. Clearly, sensor failure would affect the sensor's accuracy. A dynamic influence diagram that models this situation appears in Figure 5.33. This example was taken from [Russell and Norvig, 1995].

Other Applications

Applications of dynamic Bayesian networks and influence diagrams include planning under uncertainty (e.g., our robot) [Dean and Wellman, 1991], analysis of freeway traffic using computer vision [Huang et al., 1994], modeling the stepping patterns of the elderly to diagnose falls [Nicholson, 1996], and audio-visual speech recognition [Nefian et al., 2002].

EXERCISES

Section 5.1

Exercise 5.1 Solve the decision tree in Figure 5.34.

Exercise 5.2 Solve the decision tree in Figure 5.35.

Exercise 5.3 Show the solved decision tree given the decision tree in Figure 5.3.

Exercise 5.4 Compute the conditional probabilities in the decision tree in Figure 5.6 from the conditional probabilities given in Example 5.4.

Exercise 5.5 Show $EU(D_1) = \$9800$ and $EU(D_2) = \$10{,}697$ for the decision tree in Figure 5.8.

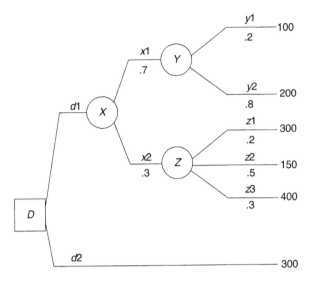

Figure 5.34: A decision tree.

Exercise 5.6 Consider Example 5.6. Suppose Leonardo has the opportunity to consult the weather forecast before deciding on whether to take his umbrella. Suppose further that the weather forecast says it will rain on 90% of the days it actually does rain and on 20% of the days it does not rain. That is,

$$P(Forecast = \text{rain}|R = \text{rain}) = .9$$

$$P(Forecast = \text{rain}|R = \text{no rain}) = .2.$$

As before, suppose Leonardo judges that

$$P(R = \text{rain}) = .4.$$

Show the decision tree representing this problem instance assuming the utilities in Example 5.6. Solve that decision tree.

Exercise 5.7 Consider again Example 5.6. Assume that if it rains there is a .7 probability the suit will only need to go to the cleaners, and a .3 probability it will rain. Assume again that

$$P(R = \text{rain}) = .4.$$

Assess your own utilities for this situation, show the resultant decision tree, and solve that decision tree.

Exercise 5.8 Consider Example 5.8. Assume that your life expectancy from birth is 75 years. Assess your own QALEs for the situation described in that example, show the resultant decision tree, and solve that decision tree.

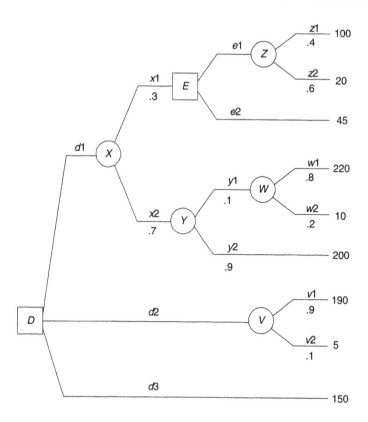

Figure 5.35: A decision tree with two decisions.

Exercise 5.9 Suppose Jennifer is a young, potential capitalist with $1000 to invest. She has heard glorious tales of many who have made fortunes in the stock market. So she decides to do one of three things with her $1000. (1) She could buy an option on Techjunk which would allow her to buy 1000 shares of Techjunk for $22 a share in one month. (2) She could use the $1000 to buy shares of Techjunk. (3) She could leave the $1000 in the bank earning .07 annually. Currently, Techjunk is selling for $20 a share. Suppose further she feels there is a .5 chance the NASDAQ will be at 2000 in two months and a .5 chance it will be at 2500. If it is at 2000, she feels there is a .3 chance Techjunk will be at $23 a share and a .7 chance it will be at $15 a share. If the NASDAQ is at 2500, she feels there is a .7 chance Techjunk will be at $26 a share and a .3 chance it will be $20 a share. Show a decision tree that represents this decision, and solve that decision tree.

Let $P(NASDAQ = 2000) = p$ and $P(NASDAQ = 2500) = 1 - p$. Determine the maximal value of p for which the decision would be to buy the option. Is there any value of p for which the decision would be to buy the stock?

Exercise 5.10 This exercise is based on an example in [Clemen, 1996]. In

1984, Penzoil and Getty Oil agreed to a merger. However, before the deal was closed, Texaco offered Getty a better price. So Gordon Getty backed out of the Penzoil deal and sold to Texaco. Penzoil immediately sued, won the case, and was awarded $11.1 billion. A court order reduced the judgment to $2 billion, but interest and penalties drove the total back up to $10.3 billion. James Kinnear, Texaco's chief executive officer, said he would fight the case all the way up to the U.S. Supreme Court because, he argued, Penzoil had not followed Security and Exchange Commission regulations when negotiating with Getty. In 1987, just before Penzoil was to begin filing liens against Texaco, Texaco offered to give Penzoil $2 billion to settle the entire case. Hugh Liedke, chairman of Penzoil, indicated that his advisors told him a settlement between $3 billion and $5 billion would be fair.

What should Liedke do? Two obvious choices are (1) he could accept the $2 billion or (2) he could turn it down. Let's say that he is also considering counteroffering $5 billion. If he does, he judges that Texaco will either accept the counteroffer with probability .17, refuse the counteroffer with probability .5, or counter back in the amount of $3 billion with probability .33. If Texaco does counter back, Liedke will then have the decision of whether to refuse or accept the counteroffer. Liedke assumes that if he simply turns down the $2 billion with no counteroffer, if Texaco refuses his counteroffer, or if he refuses their return counteroffer, the matter will end up in court. If it does go to court, he judges that there is .2 probability Penzoil will be awarded $10.3 billion, a .5 probability they will be awarded $5 billion, and a .3 probability they will get nothing.

Show a decision tree that represents this decision, and solve that decision tree.

What finally happened? Liedke simply refused the $2 billion. Just before Penzoil began to file liens on Texaco's assets, Texaco filed for protection from creditors under Chapter 11 of the federal bankruptcy code. Penzoil then submitted a financial reorganization plan on Texaco's behalf. Under the plan, Penzoil would receive about $4.1 billion. Finally, the two companies agreed on $3 billion as part of Texaco's financial reorganization.

Section 5.2

Exercise 5.11 Represent the problem instance in Exercise 5.6 with an influence diagram. Hand solve the influence diagram. Using Netica or some other software package, construct and solve the influence diagram.

Exercise 5.12 Represent the problem instance in Exercise 5.7 with an influence diagram. Hand solve the influence diagram. Using Netica or some other software package, construct and solve the influence diagram.

Exercise 5.13 Represent the problem instance in Exercise 5.9 with an influence diagram. Hand solve the influence diagram. Using Netica or some other software package, construct and solve the influence diagram.

Exercise 5.14 Represent the problem instance in Exercise 5.10 with an influence diagram. Hand solve the influence diagram. Using Netica or some other software package, construct and solve the influence diagram.

Exercise 5.15 After Example 5.24, we noted that it was not surprising that the decision alternative that maximizes expected utility is to do the CT scan since that scan has no cost. Suppose instead that there is a financial cost of $1000 involved in doing the scan. Let's say the decision maker decides $1000 is equivalent to .01 years of life. Using Netica or some other software package, construct an influence diagram representing this problem instance, and determine whether in this case the decision alternative that maximizes expected utility is still to do the CT scan.

Section 5.3

Exercise 5.16 Assign parameter values to the dynamic Bayesian network in Figure 5.29, and compute the conditional probability of the locations of the robot and the target at time 1 given some evidence at times 0 and 1.

Exercise 5.17 Assign parameter values to the dynamic influence diagram in Figure 5.31, and determine the decision at time 0 based on some evidence at time 0 and by looking 1 time into the future.

Chapter 6

Further Techniques in Decision Analysis

The previous chapter presented the fundamentals of decision analysis. Here, we present further techniques in the use of decision analysis. Most individuals would not make a monetary decision by simply maximizing expected values if the amounts of money involved were large compared to their total wealth. That is, most individuals are risk averse. So, in general, we need to model an individual's attitude toward risk when using decision analysis to recommend a decision. Section 6.1 shows how to do this using a personal utility function. Rather than assess a utility function, a decision maker may prefer to analyze the risk directly. In Section 6.2 we discuss risk profiles, which enable the decision maker to do this. Some decisions do not require the use of utility functions or risk profiles because one decision alternative dominates the other for all decision makers. In Section 6.3 we present examples of such decisions. Both influence diagrams and decision trees require that we assess probabilities and outcomes. Sometimes assessing these values precisely can be a difficult and laborious task.

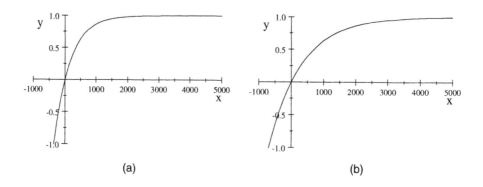

Figure 6.1: The $U_{500}(x) = 1 - e^{-x/500}$ function is in (a), while the $U_{1000}(x) = 1 - e^{-x/1000}$ function is in (b).

For example, it would be difficult and time consuming to determine whether the probability that the S&P 500 will be above 1500 in January 2008 is .3 or .35. Sometimes further refinement of these values would not affect our decision anyway. Section 6.4 shows how to measure the sensitivity of our decisions to the values of outcomes and probabilities. Often, before making a decision we have access to information, but at a cost. For example, before deciding to buy a stock we may be able to purchase the advice of an investment analyst. In Section 6.5 we illustrate how to compute the value of information, which enables us to determine whether the information is worth the cost.

6.1 Modeling Risk Preferences

Recall that in Chapter 5, Example 5.1, we chose the alternative with the largest expected value. Surely, a person who is very risk-averse might prefer the sure $1005 over the possibility of ending up with only $500. However, many people maximize expected value when the amount of money is small relative to their total wealth. The idea is that in the long run they will end up better off by so doing. When an individual maximizes expected value to reach a decision, the individual is called an **expected value maximizer**. On the other hand, given the situation discussed in Chapter 5, Example 5.1, most people would not invest $100,000 in NASDIP because that is too much money relative to their total wealth. In the case of decisions in which an individual would not maximize expected value, we need to model the individual's attitude toward risk in order to use decision analysis to recommend a decision. One way to do this is to use a **utility function**, which is a function that maps dollar amounts to utilities. We discuss such functions next.

6.1.1 The Exponential Utility Function

The **exponential utility function** is given by

$$U_r(x) = 1 - e^{-x/r}.$$

In this function the parameter r, called the **risk tolerance**, determines the degree of risk-aversion modeled by the function. As r becomes smaller, the function models more risk-averse behavior. Figure 6.1 (a) shows $U_{500}(x)$, while Figure 6.1 (b) shows $U_{1000}(x)$. Notice that both functions are concave (opening downward), and the one in Figure 6.1 (b) is closer to being a straight line. The more concave the function is, the more risk-averse is the behavior modeled by the function. To model risk-neutrality (i.e., simply being an expected value maximizer), we would use a straight line instead of the exponential utility function, and to model risk-seeking behavior we would use a convex (opening upward) function. Chapter 5 showed many examples of modeling risk-neutrality. Here, we concentrate on modeling risk-averse behavior.

Example 6.1 *Suppose Sam is making the decision in Chapter 5, Example 5.1, and Sam decides his risk tolerance r is equal to 500. Then for Sam*

EU(Buy NASDIP)

$$
\begin{aligned}
&= \ EU(NASDIP) \\
&= \ .25U_{500}(\$500) + .25U_{500}(\$1000) + .5U_{500}(\$2000) \\
&= \ .25\left(1 - e^{-500/500}\right) + .25\left(1 - e^{-1000/500}\right) + .5\left(1 - e^{-2000/500}\right) \\
&= \ .86504.
\end{aligned}
$$

$$EU(\text{Leave \$1000 in bank}) = U_{500}(\$1005) = 1 - e^{-1005/500} = .86601.$$

So Sam decides to leave the money in the bank.

Example 6.2 *Suppose Sue is less risk averse than Sam, and she decides that her risk tolerance r equals 1000. If Sue is making the decision in Chapter 5, Example 5.1, for Sue*

EU(Buy NASDIP)

$$
\begin{aligned}
&= \ EU(NASDIP) \\
&= \ .25U_{1000}(\$500) + .25U_{1000}(\$1000) + .5U_{1000}(\$2000) \\
&= \ .25\left(1 - e^{-500/1000}\right) + .25\left(1 - e^{-1000/1000}\right) + .5\left(1 - e^{-2000/1000}\right) \\
&= \ .68873.
\end{aligned}
$$

$$EU(\text{Leave \$1000 in bank}) = U_{1000}(\$1005) = 1 - e^{-1005/1000} = .63396.$$

So Sue decides to buy NASDIP.

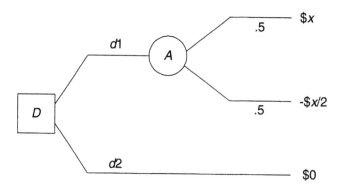

Figure 6.2: You can assess the risk tolerance r by determining the largest value of x for which you would be indifferent between $d1$ and $d2$.

Assessing r

In the previous examples we simply assigned risk tolerances to Sam and Sue. You should be wondering how an individual arrives at her or his personal risk tolerance. Next, we show a method for assessing it.

One way to determine your personal value of r in the exponential utility function is to consider a gamble in which you will win $\$x$ with probability .5 and lose $-\$x/2$ with probability .5. Your value of r is the largest value of x for which you would choose the lottery over obtaining nothing. This is illustrated in Figure 6.2.

Example 6.3 *Suppose we are about to toss a fair coin. I would certainly like the gamble in which I win $\$10$ if a heads occurs and lose $\$5$ if a tails occurs. If we increased the amounts to $\$100$ and $\$50$, or even to $\$1000$ and $\$500$, I would still like the gamble. However, if we increased the amounts to $\$1,000,000$ and $\$500,000$, I would no longer like the gamble because I cannot afford a 50% chance of losing $\$500,000$. By going back and forth like this (similar to a binary cut), I can assess my personal value of r. For me r is about equal to 20,000. (Professors do not make all that much money.)*

You may inquire as to the justification for using this gamble to assess r. Notice that for any r

$$.5\left(1 - e^{-r/r}\right) + .5\left(1 - e^{-(-r/2)/r}\right) = .0083,$$

and

$$1 - e^{-0/r} = 0.$$

We see that for a given value of the risk tolerance r, the gamble in which one wins $\$r$ with probability .5 and loses $-\$r/2$ with probability .5 has about the same utility as receiving $\$0$ for certain. We can use this fact and then work in reverse to assess r. That is, we determine the value of r for which we are indifferent between this gamble and obtaining nothing.

Constant Risk Aversion

Another way to model a decision problem involving money is to consider one's total wealth after the decision is made and the outcomes occur. The next example illustrates this.

Example 6.4 *Suppose Joe has an investment opportunity that entails a .4 probability of gaining $4000 and a .6 probability of losing $2500. If we let d1 be the decision alternative to take the investment opportunity and d2 be the decision alternative to reject it (i.e., he receives $0 for certain), then*

$$E(d1) = .4(\$4000) + .6(-\$2500) = \$100$$

$$E(d2) = \$0.$$

So if Joe were an expected value maximizer, he would choose the investment opportunity.

Suppose next that Joe carefully analyzes his risk tolerance, and he decides that for him $r = \$5000$. Then

$$EU(d1) = .4\left(1 - e^{-4000/5000}\right) + .6\left(1 - e^{-(-2500)/5000}\right) = -.1690$$

$$E(d2) = 1 - e^{-0/5000} = 0.$$

The solved decision tree is shown in Figure 6.3 (a). So given Joe's risk tolerance, he would not choose this risky investment.

Example 6.5 *Suppose Joe's current wealth is $10,000, and he has the same investment opportunity as in the previous example. Joe might reason that what really matters is his current wealth after he makes his decision and the outcome is realized. Therefore, he models the problem instance in terms of his final wealth rather than simply the gain or loss from the investment opportunity. Doing this, we have*

$$EU(d1) = .4\left(1 - e^{-(10,000+4000)/5000}\right) + .6\left(1 - e^{-(10,000-2500)/5000}\right) = .8418$$

$$E(d2) = 1 - e^{-10,000/5000} = .8647.$$

The solved decision tree is shown in Figure 6.3 (b). The fact that his current wealth is $10,000 does not affect his decision. The decision alternative to do nothing still has greater utility than choosing the investment opportunity.

Example 6.6 *Suppose next that Joe rejects the investment opportunity. However, he does well in other investments during the following year, and his total wealth becomes $100,000. Further suppose that he has the exact same investment opportunity he had a year ago. That is, Joe has an investment opportunity that entails a .4 probability of gaining $4000 and a .6 probability of losing $2500. He again models the problem in terms of his final wealth. We then have*

$$EU(d1) = .4\left(1 - e^{-(100,000+4000)/5000}\right) + .6\left(1 - e^{-(100,000-2500)/5000}\right)$$

$$= .9999999976$$

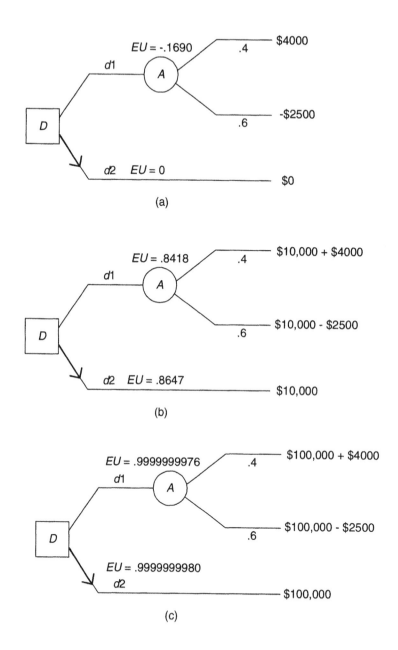

Figure 6.3: The solved decision tree for Example 6.4 is shown in (a). Solved decision trees for that same example when we model in terms of total wealth are shown in (b) and (c). The total wealth in (b) is $10,000, whereas in (c) it is $100,000.

$$E(d2) = 1 - e^{-100,000/5000} = .9999999980.$$

The solved decision tree is shown in Figure 6.3 (c). Although the utility of the investment opportunity is now very close to that of doing nothing, it is still smaller, and he still should choose to do nothing.

It is a property of the exponential utility function that an individual's total wealth cannot affect the decision obtained using the function. A function such as this is called a **constant risk-averse utility function**. If one uses such a function to model one's risk preferences, one must reevaluate the parameters in the function when one's wealth changes significantly. For example, Joe should reevaluate his risk tolerance r when his total wealth changes from \$10,000 to \$100,000.

The reason the exponential utility function displays constant risk aversion is that the term for total wealth cancels out of an inequality comparing two utilities. For example, consider again Joe's investment opportunity that entails a .4 probability of gaining \$4000 and a .6 probability of losing \$2500. Let w be Joe's total wealth. The first inequality in the following sequence of inequalities compares the utility of choosing the investment opportunity to doing nothing when we consider the total wealth w, while the last inequality compares the utility of choosing the investment opportunity to doing nothing when we do not consider total wealth. If you follow the inequalities in sequence, you will see that they are all equivalent to each other. Therefore, consideration of total wealth cannot affect the decision.

$$.4\left(1 - e^{-(w+4000)/5000}\right) + .6\left(1 - e^{-(w-2500)/5000}\right) < \left(1 - e^{-w/5000}\right)$$

$$.4\left(1 - e^{-w/5000}e^{-4000/5000}\right) + .6\left(1 - e^{-w/5000}e^{-(-2500)/5000}\right) < 1 - e^{-w/5000}$$

$$1 - .4\left(e^{-w/5000}e^{-4000/5000}\right) - .6\left(e^{-w/5000}e^{-(-2500)/5000}\right) < 1 - e^{-w/5000}$$

$$-.4\left(e^{-w/5000}e^{-4000/5000}\right) - .6\left(e^{-w/5000}e^{-(-2500)/5000}\right) < -e^{-w/5000}$$

$$-.4\left(e^{-4000/5000}\right) - .6\left(e^{-(-2500)/5000}\right) < -1$$

$$.4\left(1 - e^{-4000/5000}\right) + .6\left(1 - e^{-(-2500)/5000}\right) < 1 - 1$$

$$.4\left(1 - e^{-4000/5000}\right) + .6\left(1 - e^{-(-2500)/5000}\right) < 1 - e^{-0/5000}.$$

6.1.2 A Decreasing Risk-Averse Utility Function

If a change in total wealth can change the decision obtained using a risk-averse utility function, then the function is called a **decreasing risk-averse utility function**. An example of such a function is the logarithm function. We show this by using this function to model Joe's risk preferences.

Example 6.7 *As in Example 6.4, suppose Joe has an investment opportunity that entails a .6 probability of gaining $4000 and a .4 probability of losing $2500. Again let d1 be the decision alternative to take the investment opportunity and d2 be the decision alternative to reject it. Suppose Joe's risk preferences can be modeled using* $\ln(x)$. *First, let's model the problem instance when Joe has a total wealth of $10,000. We then have that*

$$EU(d1) = .4\ln(10{,}000 + 4000) + .6\ln(10{,}000 - 2500) = 9.1723$$

$$EU(d2) = \ln 10{,}000 = 9.2103.$$

So the decision is to reject the investment opportunity and do nothing.

Next let's model the problem instance when Joe has a total wealth of $100,000. We then have that

$$EU(d1) = .4\ln(100{,}000 + 4000) + .6\ln(100{,}000 - 2500) = 11.5134$$

$$EU(d2) = \ln 100{,}000 = 11.5129.$$

So now the decision is to take the investment opportunity.

Modeling risk attitudes is discussed much more in [Clemen, 1996].

6.2   Analyzing Risk Directly

Some decision makers may not be comfortable assessing personal utility functions and making decisions based on such functions. Rather, they may want to directly analyze the risk inherent in a decision alternative. One way to do this is to use the variance as a measure of spread from the expected value. Another way is to develop risk profiles. We discuss each technique in turn.

6.2.1   Using the Variance to Measure Risk

We start with an example.

Example 6.8 *Suppose Patricia is going to make the decision modeled by the decision tree in Figure 6.4. If Patricia simply maximizes expected value, it is left as an exercise to show*

$$
\begin{aligned}
E(d1) &= \$1220 \\
E(d2) &= \$1200.
\end{aligned}
$$

So d1 is the decision alternative that maximizes expected value. However, the expected values by themselves tell us nothing of the risk involved in the alternatives. Let's also compute the variance of each decision alternative. If we choose alternative d1, then

$$
\begin{aligned}
P(2000) &= .8 \times .7 = .56 \\
P(1000) &= .1 \\
P(0) &= .8 \times .3 + .1 = .34.
\end{aligned}
$$

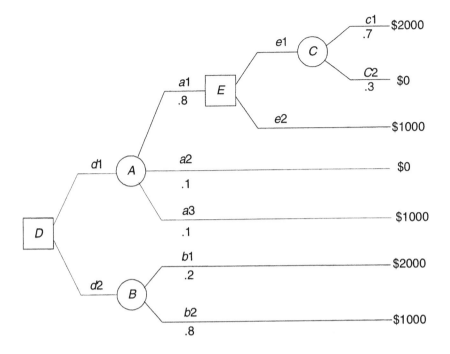

Figure 6.4: The decision tree discussed in Example 6.8.

Notice that there are two ways $0 could be obtained. That is, outcomes $a1$ and $c1$ could occur with probability $.8 \times .3$, and outcome $a2$ could occur with probability $.1$. We then have that

$$Var(d1)$$

$$= (2000 - 1220)^2 P(2000) + (1000 - 1220)^2 P(1000) + (0 - 1220)^2 P(0)$$
$$= (2000 - 1220)^2 \times .56 + (1000 - 1220)^2 \times .1 + (0 - 1220)^2 \times .34$$
$$= 851{,}600$$

$$\sigma_{d1} = \sqrt{851{,}600} = 922.82.$$

It is left as an exercise to show that

$$Var(d2) = 160{,}000$$
$$\sigma_{d2} = 400.$$

So if we use the variance as our measure of risk, we deem $d1$ somewhat more risky, which means if Patricia is somewhat risk averse she might choose $d2$.

Using the variance alone as the measure of risk can sometimes be misleading. The next example illustrates this.

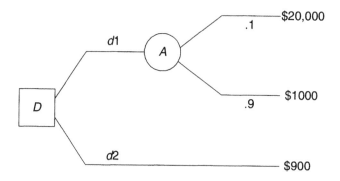

Figure 6.5: The decision tree discussed in Example 6.8.

Example 6.9 *Now suppose Patricia is going to make the decision modeled by the decision tree in Figure 6.5. It is left as an exercise to show that*

$$
\begin{aligned}
E(d1) &= \$2900 \\
Var(d1) &= 32{,}490{,}000 \\
\sigma_{d1} &= 5700
\end{aligned}
$$

and

$$
\begin{aligned}
E(d2) &= \$900 \\
Var(d2) &= 0 \\
\sigma_{d2} &= 0.
\end{aligned}
$$

If Patricia uses only the variance as her measure of risk, she might choose alternative d2 because d1 has such a large variance. Yet alternative d1 is sure to yield more return that alternative d2.

We see that the use of the variance alone as our measure of risk can be very misleading. This is a mistake some investors make. That is, they notice that one mutual fund has had a much higher expected return over the past 10 years than a second mutual fund, but they reject the first one because it also has had a much higher variance. Yet if they looked at each year, they would see that the first one always dominates the second one.

6.2.2 Risk Profiles

The expected value and variance are summary statistics, and, therefore, we lose information if all we report are these values. Alternatively, for each decision alternative, we could report the probability of all possible outcomes if the alternative is chosen. A graph that shows these probabilities is called a **risk profile**.

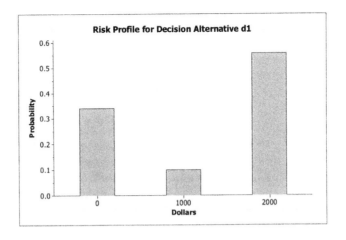

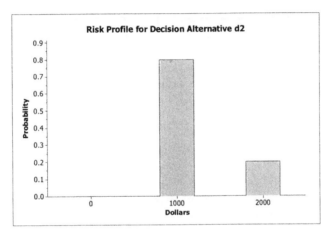

Figure 6.6: Risk profiles for the decision in Example 6.8.

Example 6.10 *Consider again Patricia's decision, which was discussed in Example 6.8. In that example, we computed the probability of all possible outcomes for each decision. We used those results to create the risk profiles in Figure 6.6. From these risk profiles, Patricia can see that there is a good chance she could end up with nothing if she chooses alternative d1, but she also has a good chance of obtaining $2000. On the other hand, the least she could end up with is $1000 if she chooses alternative d2, but probably this is all she will obtain.*

A **cumulative risk profile** shows for each amount x the probability that the payoff will be less than or equal to x if the decision alternative is chosen. A cumulative risk profile is a cumulative distribution function. Figure 6.7 shows the cumulative risk profiles for the decision in Example 6.8.

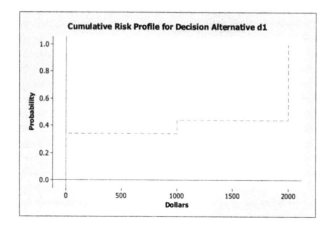

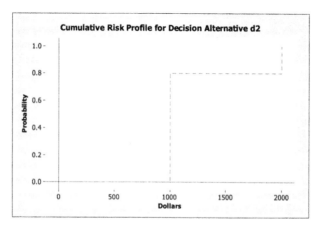

Figure 6.7: Cumulative risk profiles for the decision in Example 6.8.

6.3   Dominance

Some decisions do not require the use of utility functions or risk profiles because one decision alternative dominates the other for all decision makers. We discuss dominance next.

6.3.1   Deterministic Dominance

Suppose we have a decision that can be modeled using the decision tree in Figure 6.8. If we choose alternative $d1$, the least amount of money we will realize is $4, whereas if we choose alternative $d2$, the most amount of money we will realize is $3. Assuming that maximizing wealth is the only consideration in this decision, there is then no reasonable argument one can offer for choosing $d2$ over $d1$, and we say $d1$ deterministically dominates $d2$. In general, decision alternative $d1$ **deterministically dominates** decision alternative $d2$ if the utility obtained

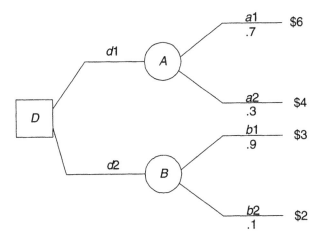

Figure 6.8: Decision alternative $d1$ deterministically dominates decision alternative $d2$.

from choosing $d1$ is greater than the utility obtained from choosing $d2$ regardless of the outcomes of chance nodes. When we observe deterministic dominance, there is no need to compute expected utility or develop a risk profile.

6.3.2 Stochastic Dominance

Suppose we have a decision that can be modeled using the decision tree in Figure 6.9. If the outcomes are $a1$ and $b2$, we will realize more money if we choose $d1$, while if the outcomes are $a2$ and $b1$, we will realize more money if we choose $d2$. So there is no deterministic dominance. However, the outcomes are the same for both decisions, namely \$6 and \$4, and, if we choose $d2$, the probability is higher that we will receive \$6. So again, assuming that maximizing wealth is the only consideration in this decision, there is no reasonable argument for choosing $d1$ over $d2$, and we say alternative $d2$ stochastically dominates alternative $d1$.

A different case of stochastic dominance is illustrated by the decision tree in Figure 6.10. In that tree, the probabilities are the same for both chance nodes, but the utilities for the outcomes of B are higher. That is, if $b1$ occurs we realize \$7, while if $a1$ occurs we realize only \$6, and if $b2$ occurs we realize \$5, while if $a2$ occurs we realize only \$4. So again, assuming that maximizing wealth is the only consideration in this decision, there is no reasonable argument for choosing $d1$ over $d2$, and we say alternative $d2$ stochastically dominates alternative $d1$.

Although it often is not hard to recognize stochastic dominance, it is a bit tricky to define the concept. We do so next in terms of cumulative risk profiles. We say that alternative $d2$ **stochastically dominates** alternative $d1$ if the cumulative risk profile $F_2(x)$ for $d2$ lies under the cumulative risk profile $F_1(x)$ for $d1$ for at least one value of x and does not lie over it for any values of x.

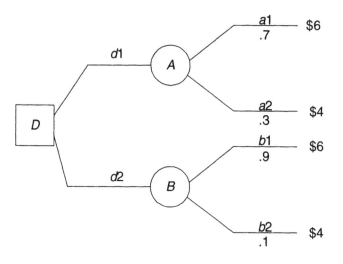

Figure 6.9: Decision alternative $d2$ stochastically dominates decision alternative $d1$.

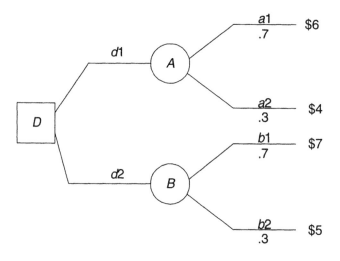

Figure 6.10: Decision alternative $d2$ stochastically dominates decision alternative $d1$.

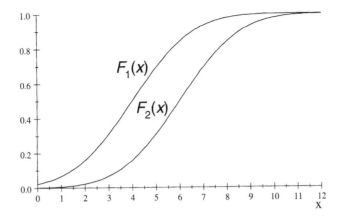

Figure 6.11: If $F_1(x)$ is the cumulative risk profile for $d1$ and $F_2(x)$ is the cumulative risk profile for $d2$, then $d2$ stochastically dominates $d1$.

That is, for at least one value of x

$$F_2(x) < F_1(x),$$

and for all values of x

$$F_2(x) \leq F_1(x).$$

This is illustrated in Figure 6.11. Why should this be the definition of stochastic dominance? Look again at Figure 6.11. There is no value of x such that the probability of realizing $\$x$ or less is smaller if we choose $d1$ than if we choose $d2$. So there is no amount of money that we may want or require that would make $d1$ the better choice.

Figure 6.12 shows two cumulative risk profiles that cross, which means we do not have stochastic dominance. Now the decision alternative chosen can depend on an individual's preference. For example, if the amounts are in units of $\$100$, and Mary needs at least $\$400$ to pay her rent or else be evicted, she may choose alternative $d1$. On the other hand, if Sam needs at least $\$800$ to pay his rent or else be evicted, he may choose alternative $d2$.

6.3.3 Good Decision versus Good Outcome

Suppose Scott and Sue are each about to make the decision modeled by the decision tree in Figure 6.10, Scott chooses alternative $d1$, and Sue chooses alternative $d2$. Suppose further that outcomes $a1$ and $b2$ occur. So Scott ends up with $\$6$, and Sue ends up with $\$5$. Did Scott make a better decision than Sue? We just claimed that there is no reasonable argument for choosing $d1$ over $d2$. If we accept that claim, we cannot now conclude that Scott made the better decision. Rather, Scott made a **bad decision with a good outcome**, while Sue made a **good decision with a bad outcome**. The quality of a decision

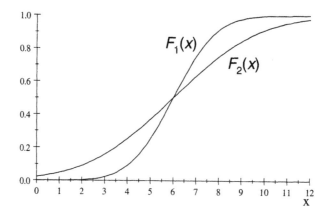

Figure 6.12: There is no stochastic dominance.

must be judged based on the information available when the decision is made, not on outcomes realized after the decision is made. One of the authors (Rich Neapolitan) amusingly remembers the following story from his youth. When his uncle Hershell got out of the army, he used his savings to buy a farm in Texas next to his parents' farm. The ostensible reason was that he wanted to live near his parents and resume his life as a farmer. Somewhat later, oil was discovered on his farm, and Hershell became wealthy as a result. After that Rich's dad used to say, "Everyone thought Hershell was not too bright when he wasted money on a farm with such poor soil, but it turns out he was shrewd like a fox."

6.4   Sensitivity Analysis

Both influence diagrams and decision trees require that we assess probabilities and outcomes. Sometimes assessing these values precisely can be a difficult and laborious task. For example, it would be difficult and time consuming to determine whether the probability that the S&P 500 will be above 1500 in January 2008 is .3 or .35. Sometimes further refinement of these values would not affect our decision anyway. Next, we discuss **sensitivity analysis**, which is an analysis of how the values of outcomes and probabilities can affect our decision. After introducing the concept with simple models, we show a more detailed model.

6.4.1   Simple Models

We show a sequence of examples.

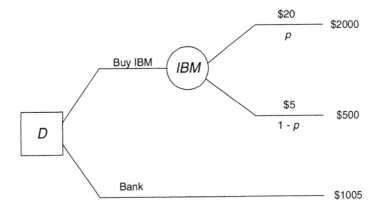

Figure 6.13: As long as p is greater than .337, buying IBM maximizes expected value.

Example 6.11 *Suppose that currently IBM is at $10 a share, and you feel there is a .5 probability it will be go down to $5 by the end of the month and a .5 probability it will go up to $20. You have $1000 to invest, and you will either buy 100 shares of IBM or put the money in the bank and earn a monthly interest rate of .005. Although you are fairly confident of your assessment of the outcomes, you are not very confident of your assessment of the probabilities. In this case, you can represent your decision using the decision tree in Figure 6.13. Notice in that tree that we represented the probability of IBM going up by a variable p. We then have*

$$E(\text{Buy IBM}) = p(2000) + (1 - p)(500)$$

$$E(\text{Bank}) = \$1005.$$

We will buy IBM if $E(\text{Buy IBM}) > E(\text{Bank})$, which is the case if

$$p(2000) + (1 - p)(500) > 1005.$$

Solving this inequality for p, we have

$$p > .337.$$

We have determined how sensitive our decision is to the value of p. As long as we feel that the probability of IBM going up is at least equal to .337, we will buy IBM. We need not refine our probabilistic assessment further.

Example 6.12 *Suppose you are in the same situation as in the previous example, except that you feel that the value of IBM will be affected by the overall value of the Dow Jones Industrial Average in one month. Currently, the Dow is at 10,500 and you assess that it will either be at 10,000 or 11,000 at the end of the month. You feel confident assessing the probabilities of your stock going*

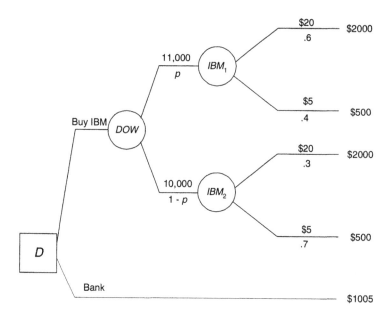

Figure 6.14: As long as p is greater than .122, buying IBM maximizes expected value.

up dependent on whether the Dow goes up or down, but you are not confident assessing the probability of the Dow going up or down. Specifically, you model your decision using the decision tree in Figure 6.14. We then have

$$E(Buy\ IBM) = p(.6 \times 2000 + .4 \times 500) + (1 - p)(.3 \times 2000 + .7 \times 500)$$

$$E(Bank) = 1005.$$

We will buy IBM if $E(Buy\ IBM) > E(Bank)$, which is the case if

$$p(.6 \times 2000 + .4 \times 500) + (1 - p)(.3 \times 2000 + .7 \times 500) > 1005.$$

Solving this inequality for p, we have

$$p > .122.$$

As long as we feel that the probability of the Dow going up is at least equal to .122, we will buy IBM.

In a **two-way sensitivity analysis** we simultaneously analyze the sensitivity of our decision to two quantities. The next example shows such an analysis.

Example 6.13 *Suppose you are in the same situation as in the previous example, except you are confident in your assessment of the probability of the*

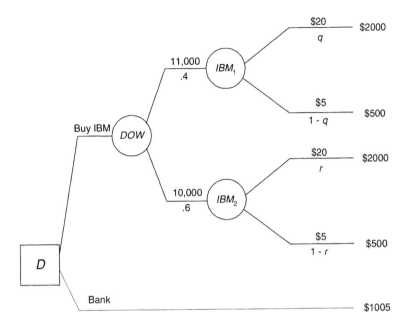

Figure 6.15: For this decision we need to do a two-way sensitivity analysis.

Dow going up, but you are not confident in your assessment of the probabilities of your stock going up dependent on whether the Dow goes up or down. Specifically, you model your decision using the decision tree in Figure 6.15. We then have

$$E(\text{Buy IBM}) = .4(q \times 2000 + (1-q) \times 500) + .6(r \times 2000 + (1-r) \times 500)$$

$$E(\text{Bank}) = 1005.$$

We will buy IBM if $E(\text{Buy IBM}) > E(\text{Bank})$, which is the case if

$$.4(q \times 2000 + (1-q) \times 500) + .6(r \times 2000 + (1-r) \times 500) > 1005.$$

Simplifying this inequality, we obtain

$$q > \frac{101}{120} - \frac{3r}{2}.$$

The line $q = 101/120 - 3r/2$ is plotted in Figure 6.16. Owing to the previous inequality, the decision which maximizes expected value is to buy IBM as long as the point (r, q) lies above that line. For example, if $r = .6$ and $q = .1$ or $r = .3$ and $q = .8$, this would be our decision. However, if $r = .3$ and $q = .1$, it would not.

Example 6.14 *Suppose you are in the same situation as in the previous example, except you are not comfortable assessing any probabilities in the model.*

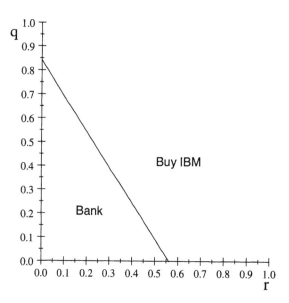

Figure 6.16: The line $q = 101/120 - 3r/2$. As long as (r, q) is above this line, the decision that mazimizes expected value in Example 6.13 is to buy IBM.

However, you do feel that the probability of IBM going up if the Dow goes up is twice the probability of IBM going up if the Dow goes down. Specifically, you model your decision using the decision tree in Figure 6.17. We then have

$$E(Buy\ IBM) = p(q \times 2000 + (1-q) \times 500) + (1-p)((q/2) \times 2000 + (1-q/2) \times 500)$$

$$E(Bank) = 1005.$$

We will buy IBM if $E(Buy\ IBM) > E(Bank)$, which is the case if

$$p(q \times 2000 + (1 - q) \times 500) + (1 - p)((q/2) \times 2000 + (1 - q/2) \times 500) > 1005.$$

Simplifying this inequality yields

$$q > \frac{101}{150 + 150p}.$$

Figure 6.18 plots the curve $q = 101/(150+150p)$. The decision which maximizes expected value is to buy IBM as long as the point (p, q) lies above that curve.

We can also investigate how sensitive our decision is to the values of the outcomes. The next example illustrates this sensitivity.

Example 6.15 *Suppose you are in the situation discussed in Example 6.11. However, you are not confident as to your assessment of how high or low IBM*

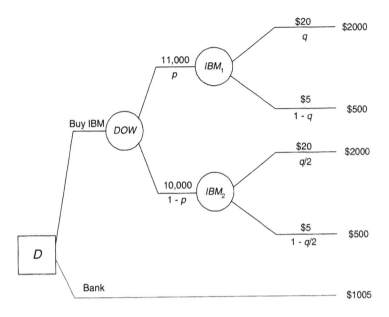

Figure 6.17: For this decision we need to do a two-way sensitivity analysis.

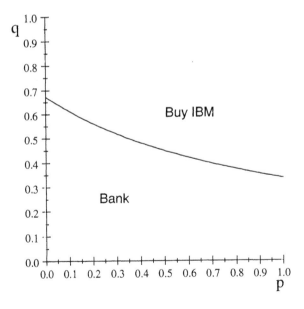

Figure 6.18: The curve $q = 101/(150 + 150p)$. As long as (p, q) is above this curve, the decision that mazimizes expected value in Example 6.14 is to buy IBM.

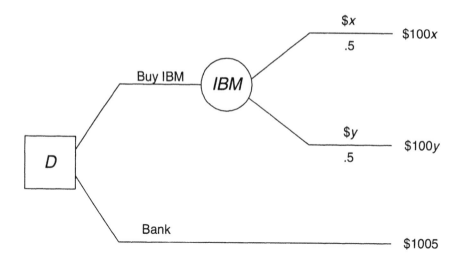

Figure 6.19: This decision requires a two-way sensitivity analysis of the outcomes.

*will go. That is, currently IBM is at $10 a share, and you feel there is a .5
probability it will go up by the end of the month and a .5 probability it will go
down, but you do not assess how high or low it will be. As before, you have
$1000 to invest, and you will either buy 100 shares of IBM or put the money
in the bank and earn a monthly interest rate of .005. In this case, you can
represent your decision using the decision tree in Figure 6.19. We then have*

$$E(Buy\ IBM) = .5(100x) + .5(100y)$$

$$E(Bank) = \$1005.$$

We will buy IBM if $E(Buy\ IBM) > E(Bank)$, which is the case if

$$.5(100x) + .5(100y) > 1005.$$

Simplifying this inequality yields

$$y > \frac{201}{10} - x.$$

*Figure 6.20 plots the curve $y = 201/10 - x$. The decision which maximizes
expected value is to buy IBM as long as the point (x, y) lies above that curve.*

6.4.2   A More Detailed Model

The examples given so far have been oversimplified and therefore do not represent decision models one would ordinarily use in practice. We did this to
illustrate the concepts without burdening you with too many details. Next, we
show an example of a more detailed model, which an investor might actually
use to model a decision.

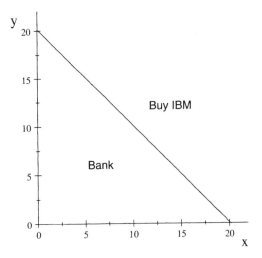

Figure 6.20: The line $y = 201/10 - x$. As long as (x, y) is above this line, the decision that maximizes expected value in Example 6.15 is to buy IBM.

Example 6.16 *Some financial analysts argue that the small investor is best served by investing in mutual funds rather than in individual stocks. The investor has a choice of many different types of mutual funds. Here we consider two that are quite different. "Aggressive" or "growth" mutual funds invest in companies that have high growth potential and ordinarily do quite well when the market in general performs well, but do quite poorly when the market does poorly. "Allocation" funds distribute their investments among cash, bonds, and stocks. Furthermore, the stocks are often in companies that are considered "value" investments because for some reason they are considered currently undervalued. Such funds are ordinarily more stable. That is, they do not do as well when the market does well or as poorly when the market performs poorly. Assume that Christina plans to invest $1000 for the coming year, and she decides she will either invest the money in a growth fund or in an allocation fund or will put it in a one-year CD which pays 5%. To make her decision, she must model how these funds will do based on what the market does in the coming year, and she must assess the probability of the market performance in the coming year. Based on how these funds have done during market performances in previous years, she is comfortable assessing how the funds will do based on market performance. She feels that the market probably will not do well in the coming year because economists are predicting a recession, but she is not comfortable assessing precise probabilities. So she develops the decision tree in Figure 6.21. Given that tree, the growth fund should be preferred to the bank if*

$$1600p + 1000q + 300(1 - p - q) > 1050.$$

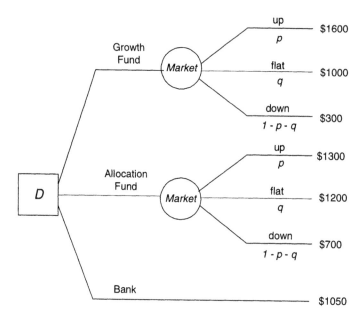

Figure 6.21: A decision tree modeling the decision whose alternatives are a growth mutual fund, an allocation mutual fund, and the bank.

Simplifying this expression yields

$$q > (15 - 26p)/14.$$

Based on this inequality, Figure 6.22 (a) shows the region in which she should choose the growth fund and the region in which she should choose the bank if these were her only choices. Notice that the regions are bounded above by the line $q = 1 - p$. The reason is that we must have $q + p \leq 1$.

The allocation fund should be preferred to the bank if

$$1300p + 1200q + 900(1 - p - q) > 1050.$$

Simplifying this expression yields

$$q > (3 - 8p)/6.$$

Based on this inequality, Figure 6.22 (b) shows the region in which she should choose the allocation fund and the region in which she should choose the bank if these were her only choices.

The growth fund should be preferred to the allocation fund if

$$1600p + 1000q + 300(1 - p - q) > 1300p + 1200q + 900(1 - p - q).$$

Simplifying this expression yields

$$q > (6 - 9p)/4.$$

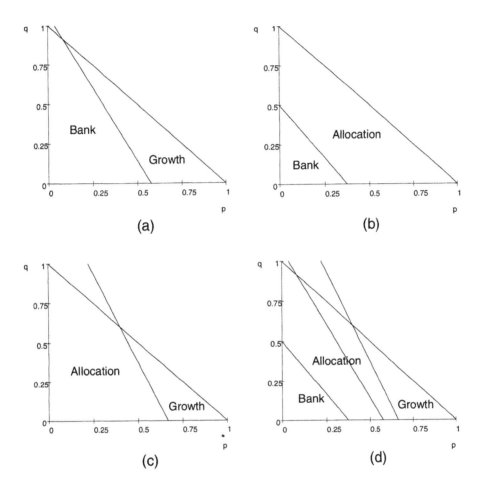

Figure 6.22: If the only choice is between the growth fund and the bank, Christina should use (a); if it is between the allocation fund and the bank, she should use (b); if it is between the growth fund and the allocation fund, she should use (c); and if she is choosing from all three alternatives, she should use (d).

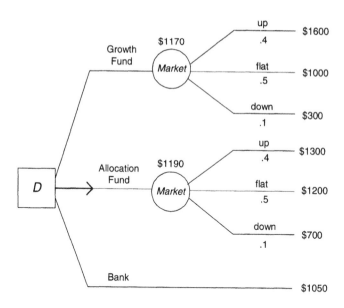

Figure 6.23: Buying the allocation fund maximizes expected value.

Based on this inequality, Figure 6.22 (c) shows the region in which she should choose the growth fund and the region in which she should choose the allocation fund if these were her only choices.

Finally, the lines for all of the previous three comparisons are plotted in Figure 6.22 (d). This diagram shows the regions in which she should make each of the three choices when all three are being considered.

The values in the previous examples were exaggerated from what we (the authors) actually believe. For example, we do not believe growth funds will go up about 60% in good market conditions. We exaggerated the values so that it would be easy to see the regions in the diagram in Figure 6.22 (d). Other than that, this example illustrates how one of us makes personal investment decisions. It is left as an exercise for you to assess your own values and determine the resultant region corresponding to each investment choice.

6.5   Value of Information

Figure 6.23 shows the decision tree in Figure 6.21, but with values assigned to the probabilities. It is left as an exercise to show that, given these values, the decision that maximizes expected value is to buy the allocation fund, and

$$E(D) = E(\text{allocation fund}) = \$1190.$$

This is shown in Figure 6.23. Before making a decision, we often have the chance to consult with an expert in the domain which the decision concerns.

Suppose in the current decision we can consult with an expert financial analyst who is perfect at predicting the market. That is, if the market is going up, the analyst will say it is going up; if it will be flat, the analyst will say it will be flat; and if it is going down, the analyst will say it is going down. We should be willing to pay for this information, but not more than the information is worth. Next, we show how to compute the expected value (worth) of this information, which is called the expected value of perfect information.

6.5.1 Expected Value of Perfect Information

To compute the expected value of perfect information we add another decision alternative, which is to consult the perfect expert. Figure 6.24 shows the decision tree in Figure 6.23 with that alternative added. Next, we show how the probabilities for that tree were obtained. Since the expert is perfect, we have

$$P(\text{Expert} = \text{says up} \mid \text{Market} = \text{up}) = 1$$

$$P(\text{Expert} = \text{says flat} \mid \text{Market} = \text{flat}) = 1$$

$$P(\text{Expert} = \text{says down} \mid \text{Market} = \text{down}) = 1.$$

We therefore have

$$P(\text{up} \mid \text{says up})$$

$$= \frac{P(\text{says up} \mid \text{up})P(\text{up})}{P(\text{says up} \mid \text{up})P(\text{up}) + P(\text{says flat} \mid \text{up})P(\text{flat}) + P(\text{says down} \mid \text{down})P(\text{down})}$$

$$= \frac{1 \times .4}{1 \times .4 + 0 \times .5 + 0 \times .1} = 1.$$

It is not surprising that this value is 1, since the expert is perfect. This value is the far right and uppermost probability in the decision tree in Figure 6.24. It is left as an exercise to compute the other probabilities and solve the tree. We see that

$$E(\text{Consult Perfect Analyst}) = \$1345.$$

Recall that without consulting this analyst the decision alternative that maximizes expected utility is to buy the allocation fund, and

$$E(D) = E(\text{allocation fund}) = \$1190.$$

The difference between these two expected values is the **expected value of perfect information (EVPI)**. That is,

$$
\begin{aligned}
EVPI &= E(\text{Consult Perfect Analyst}) - E(D) \\
&= \$1345 - \$1190 = \$155.
\end{aligned}
$$

This is the most we should be willing to pay for the information. If we pay less than this amount, we will have increased our expected value by consulting the expert, while if we pay more, we will have decreased our expected value.

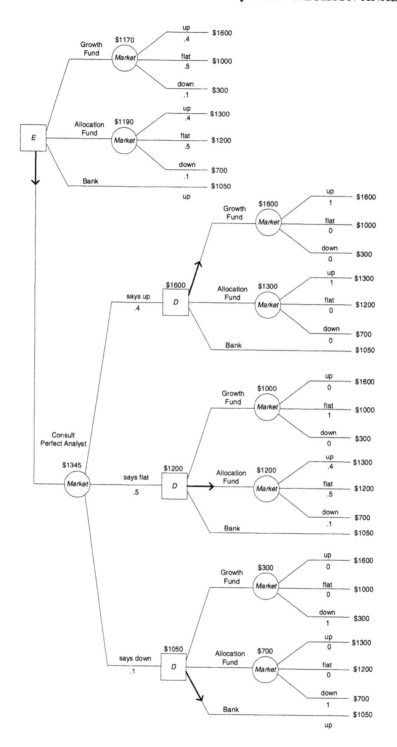

Figure 6.24: The maximum expected value without consulting the perfect expert is $1190, while the expected value of consulting that expert is $1345.

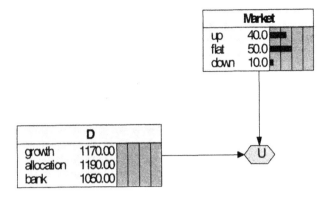

Figure 6.25: The decision tree in Figure 6.23 represented as an influence diagram and solved using Netica.

We showed decision trees in Figures 6.23 and 6.24 so that you could see how the expected value of perfect information is computed. However, as is usually the case, it is much easier to represent the decisions using influence diagrams. Figure 6.25 shows the decision tree in Figure 6.23 represented as an influence diagram and solved using Netica. Figure 6.26 shows the decision tree in Figure 6.24 represented as an influence diagram and solved using Netica. We have added the conditional probabilities of the Expert node to that diagram. (Recall that Netica does not show conditional probabilities.) Notice that we can obtain the EVPI directly from the values listed at decision node E in the influence diagram in Figure 6.26. That is,

$$
\begin{aligned}
EVPI &= E(\text{consult}) - E(\text{do not consult}) \\
&= \$1345 - \$1190 = \$155.
\end{aligned}
$$

6.5.2 Expected Value of Imperfect Information

Real experts and tests ordinarily are not perfect. Rather, they are only able to give estimates which are often correct. Let's say we have a financial analyst who has been predicting market activity for 30 years and has had the following results:

1. When the market went up, the analyst said it would go up 80% of the time, would be flat 10% of the time, and would go down 10% of the time.

2. When the market was flat, the analyst said it would go up 20% of the time, would be flat 70% of the time, and would go down 10% of the time.

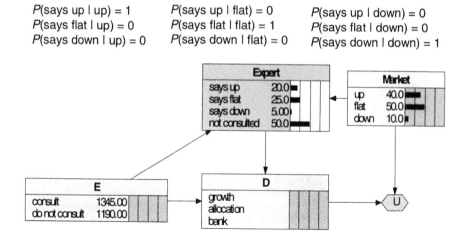

P(says up | up) = 1 P(says up | flat) = 0 P(says up | down) = 0
P(says flat | up) = 0 P(says flat | flat) = 1 P(says flat | down) = 0
P(says down | up) = 0 P(says down | flat) = 0 P(says down | down) = 1

Figure 6.26: The decision tree in Figure 6.24 represented as an influence diagram and solved using Netica.

3. When the market went down, the analyst said it would go up 20% of the time, would be flat 20% of the time, and would go down 60% of the time.

We therefore estimate the following conditional probabilities for this expert:

$$P(\text{Expert} = \text{says up} \mid \text{Market} = \text{up}) = .8$$

$$P(\text{Expert} = \text{says flat} \mid \text{Market} = \text{up}) = .1$$

$$P(\text{Expert} = \text{says down} \mid \text{Market} = \text{up}) = .1$$

$$P(\text{Expert} = \text{says up} \mid \text{Market} = \text{flat}) = .2$$

$$P(\text{Expert} = \text{says flat} \mid \text{Market} = \text{flat}) = .7$$

$$P(\text{Expert} = \text{says down} \mid \text{Market} = \text{flat}) = .1$$

$$P(\text{Expert} = \text{says up} \mid \text{Market} = \text{down}) = .2$$

$$P(\text{Expert} = \text{says flat} \mid \text{Market} = \text{down}) = .2$$

$$P(\text{Expert} = \text{says down} \mid \text{Market} = \text{down}) = .6.$$

Figure 6.27 shows the influence diagram in Figure 6.25 with the additional decision alternative that we can consult this imperfect expert. We also show the conditional probabilities of the Expert node in that diagram. The increased expected value we realize by consulting such an expert is called the **expected value of imperfect information (EVII)**. It is given by

$$
\begin{aligned}
EVII &= E(\text{consult}) - E(\text{do not consult}) \\
&= \$1261.50 - \$1190 = \$71.50.
\end{aligned}
$$

This is the most we should pay for this expert's information.

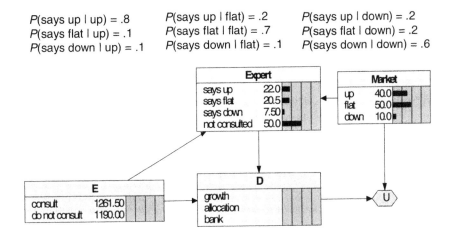

Figure 6.27: An influence diagram that enables us to compute the expected value of imperfect information.

6.6 Normative Decision Analysis

The analysis methodology presented in this and the previous chapter for recommending decisions is called **normative decision analysis** because the methodology prescribes how people should make decisions rather than describes how people do make decisions. In 1954 L. Jimmie Savage developed axioms concerning an individual's preferences and beliefs. If an individual accepts these axioms, Savage showed that the individual must prefer the decisions obtained using decision analysis. Tversky and Kahneman [1981] conducted a number of studies showing that individuals do not make decisions consistent with the methodology of decision analysis. That is, their studies indicate that decision analysis is not a descriptive theory. Kahneman and Tversky [1979] developed **prospect theory** to describe how people actually make decisions when they are not guided by decision analysis. In 2002 Dan Kahneman won the Nobel Prize in economics for this effort. An alternative descriptive theory of decision making is **regret theory** [Bell, 1982].

EXERCISES

Section 6.1

Exercise 6.1 Using the technique illustrated in Example 6.3, assess your personal risk tolerance r.

Exercise 6.2 Using the value of r assessed in the previous exercise, determine the decision that maximizes expected utility for the decision in Example 5.1.

Exercise 6.3 Suppose Mary has an investment opportunity that entails a .9 probability of gaining $8000 and a .1 probability of losing $20,000. Let $d1$ be the decision alternative to take the investment opportunity and $d2$ be the decision alternative to reject it. Suppose Mary's risk preferences can be modeled using $\ln(x)$.

1. Determine the decision alternative that maximizes expected utility when Mary has a wealth of $21,000.

2. Determine the decision alternative that maximizes expected utility when Mary has a wealth of $25,000.

Section 6.2

Exercise 6.4 Compute the variance of the decision alternatives for the decision in Example 5.2. Plot risk profiles and cumulative risk profiles for the decision alternatives. Discuss whether you find the variance or the risk profiles more helpful in determining the risk inherent in each alternative.

Exercise 6.5 Compute the variance of the decision alternatives for the decision in Example 5.3. Plot risk profiles and cumulative risk profiles for the decision alternatives. Discuss whether you find the variance or the risk profiles more helpful in determining the risk inherent in each alternative.

Exercise 6.6 Compute the variance of the decision alternatives for the decision in Example 5.5. Plot risk profiles and cumulative risk profiles for the decision alternatives. Discuss whether you find the variance or the risk profiles more helpful in determining the risk inherent in each alternative.

Section 6.3

Exercise 6.7 Does one of the decision alternatives in the decision tree in Figure 6.28 deterministically dominate? If so, which one?

Exercise 6.8 Does one of the decision alternatives in the decision tree in Figure 6.29 stochastically dominate? If so, which one? Create cumulative risk profiles for the decision alternatives.

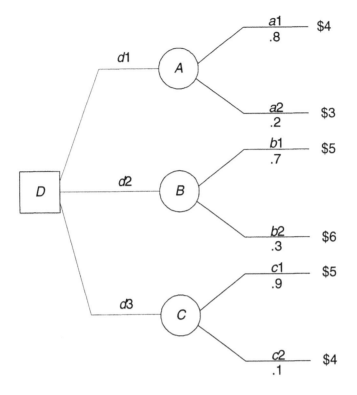

Figure 6.28: A decision tree.

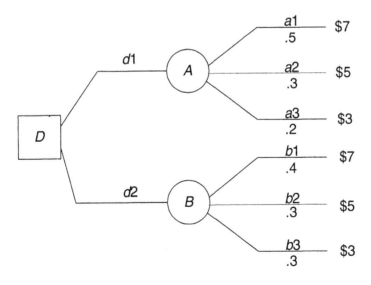

Figure 6.29: A decision tree.

Section 6.4

Exercise 6.9 Suppose that currently Lucent is at $3 a share, and you feel there is a .6 probability it will be go down to $2 by the end of the month and a .4 probability it will go up to $5. You have $3000 to invest, and you will either buy 1000 shares of Lucent or put the money in the bank and earn a monthly interest rate of .004. Although you are fairly confident of your assessment of the outcomes, you are not very confident of your assessment of the probabilities. Let p be the probability Lucent will go down. Determine the largest value of p for which you would decide to buy Lucent.

Exercise 6.10 Suppose you are in the same situation as in the previous exercise, except you feel that the value of Lucent will be affected by the overall value of the NASDAQ in one month. Currently, the NASDAQ is at 2300, and you assess that it will either be at 2000 or 2500 at the end of the month. You feel confident assessing the probabilities of your stock going up dependent on whether the NASDAQ goes up or down, but you are not confident assessing the probability of the NASDAQ going up or down. Specifically, you feel the probability of Lucent going up if the NASDAQ goes up is .8 and the probability of Lucent going up given the NASDAQ goes down is .3. Let p be the probability the NASDAQ will go up. Determine the smallest value of p for which you would decide to buy Lucent.

Exercise 6.11 Suppose you are in the same situation as in the previous exercise, except you are confident in your assessment of the probability of the NASDAQ going up, but you are not confident in your assessment of the probabilities of your stock going up dependent on whether the NASDAQ goes up or down. Specifically, you feel the probability that the NASDAQ will go up is .7. Let p be the probability of Lucent going up given the NASDAQ goes up, and let q be the probability of Lucent going up given the NASDAQ goes down. Do a two-way sensitivity analysis on p and q.

Exercise 6.12 Suppose you are in the same situation as in the previous exercise, except you are not comfortable assessing any probabilities in the model. However, you do feel that the probability of Lucent going up if the NASDAQ goes up is three times the probability of Lucent going up if the NASDAQ goes down. Let p be the probability of the NASDAQ going up, and let q be the probability of Lucent going up given the NASDAQ goes down. Do a two-way sensitivity analysis on p and q.

Exercise 6.13 Suppose you are in the situation discussed in Exercise 6.9. However, you are not confident of your assessment of how high or low Lucent will go. That is, currently Lucent is at $3 a share, and you feel there is a .5 probability it will go up by the end of the month and a .5 probability it will go down, but you do not assess how high or low it will be. As before, you have $3000 to invest, and you will either buy 1000 shares of IBM or put the money in the bank and earn a monthly interest rate of .004. Let x be the value of Lucent if it goes up and y be the value of Lucent if it goes down. Do a two-way sensitivity analysis on x and y.

Exercise 6.14 Suppose we are in the situation in Example 6.16, except the growth fund will be at $1800, $1100, and $200 if the market is, respectively, up, flat, or down, while the allocation fund will be at $1400, $1000, or $400. Perform the analysis in Example 6.16.

Exercise 6.15 Investigate some actual mutual funds, and articulate where you feel they will be if the market is up, flat, or down. Furthermore, determine the actual fixed rate you could get from a bank. Then perform the analysis in Example 6.16.

Exercise 6.16 Suppose you can invest $18,000 in a multi-year project. Annual cash flow will be $5000 for the next five years. That is, you will receive $5000 in one year, $5000 in two years,..., and $5000 in five years. Alternatively, you could put the $10,000 in a five-year CD paying 8% annually.

1. Should you invest in the project or buy the CD? Note that you need to consider the net present value (NPV) of the project.

2. Suppose you are not really certain the project will pay $5000 annually. Rather, you assess there is a .3 probability it will pay $4000, a .5 probability it will pay $5000, and a .2 probability it will pay $6000. Determine the decision alternative that maximizes expected value.

3. Suppose you are not only uncertain the project will pay $5000 annually, but you cannot readily access the probabilities of it paying $4000, $5000, or $6000. Let p be the probability it will pay $4000, q be the probability it will pay $5000, and $1 - p - q$ be the probability it will pay $6000. Do a two-way sensitivity analysis on p and q.

Exercise 6.17 Assess your own values for the decision tree in Figure 6.21, and determine the resultant region corresponding to each investment choice.

Section 6.5

Exercise 6.18 Suppose we have the decision tree in Figure 6.24, except the growth fund will be at $1800, $1100, and $200 if the market is, respectively, up, flat, or down, while the allocation fund will be at $1400, $1000, or $400.

1. Compute the expected value of perfect information by hand.

2. Model the problem instance as an influence diagram using Netica, and determine the expected value of perfect information using that influence diagram.

Exercise 6.19 Suppose we have the same decision as in the previous example, except we can consult an expert who is not perfect. Specifically, the expert's accuracy is as follows:

$$P(\text{Expert} = \text{says up} \mid \text{Market} = \text{up}) = .7$$

$$P(\text{Expert} = \text{says flat} \mid \text{Market} = \text{up}) = .2$$

$$P(\text{Expert} = \text{says down} \mid \text{Market} = \text{up}) = .1$$

$$P(\text{Expert} = \text{says up} \mid \text{Market} = \text{flat}) = .1$$

$$P(\text{Expert} = \text{says flat} \mid \text{Market} = \text{flat}) = .8$$

$$P(\text{Expert} = \text{says down} \mid \text{Market} = \text{flat}) = .1.$$

Model the problem instance as an influence diagram using Netica, and determine the expected value of consulting the expert using that influence diagram.

Exercise 6.20 Consider the decision problem discussed in Chapter 5, Exercise 5.10. Represent the problem with an influence diagram using Netica, and, using that influence diagram, determine the EVPI concerning Texaco's reaction to a $5 billion counteroffer.

Exercise 6.21 Recall Chapter 2, Exercise 2.18, in which Professor Neapolitan has the opportunity to drill for oil on his farm in Texas. It costs $25,000 to drill. Suppose that if he drills and oil is present, he will receive $100,000 from the sale of the oil. If only natural gas is present, he will receive $30,000 from the sale of the natural gas. If neither are present, he will receive nothing. The alternative to drilling is to do nothing, which would definitely result in no profit, but he will not have spent the $25,000.

1. Represent the decision problem with an influence diagram, and solve the influence diagram.

2. Now include a node for the test discussed in Exercise 6.28, and determine the expected value of running the test.

Part II

Financial Applications

Chapter 7

Investment Science

An important problem in investment science is modeling the risk of a portfolio. In this chapter we present a Bayesian network for portfolio risk analysis. However, first we need to review standard topics in investment science because knowledge of this material is necessary to an understanding of portfolio risk analysis.

7.1 Basics of Investment Science

This section discusses the basics of the stock market and investing in stocks. Before that, however, we review the concept of interest.

7.1.1 Interest

We start with the following example:

Example 7.1 *Suppose you currently have $1000, and you wish to save it for the future. You could put the money in a jar or a safety deposit box. However, if you did this in one year you would still have only $1000. Alternatively, you could purchase a one-year* **certificate of deposit**, *or CD, from a bank that pays a fixed annual interest rate equal to .06. The bank pays the interest only at the end of the fixed term, which in this case is one year. So at the end of one year the bank would pay you*

$$\$1000 \times .06 = \$60$$

in return for holding your money. This is your profit from the investment. Furthermore, at the end of the year the value of your investment is

$$\$1000(1 + .06) = \$1060.$$

CDs are treated as bank deposits by the Federal Deposit Insurance Corporation, which means they are insured for amounts up to $100,000 if the bank suffers insolvency. Therefore, we treat an investment such as this as having no risk, realizing, however, that everything has some risk. For example, the country itself could fold. However, risks that are extremely small relative to other risks under consideration are treated as if they do not exist.

In general, the fraction of money earned on an investment is called the **rate of return** r on the investment. So in the case of CDs, the rate of return is the interest rate. The amount of money invested is called the **principal** P, and the time it takes to realize the interest is called the **holding period**. In the previous example, the principal was $1000, and the holding period was one year. Unless we state otherwise, the holding period is one year, and r represents an annual rate of return. For CDs the interest is paid at the end of the holding period. This is called **simple interest**. In this case, the profit received from the investment is given by

$$Profit = P \times r, \tag{7.1}$$

and the value V of the investment at the end of the holding period is given by

$$V = P(1 + r).$$

In the case of CDs, one ordinarily is required to leave the money deposited for the entire holding period. Banks also offer ordinary savings accounts which do not have any required holding period. That is, you may take your money back at any time and realize the interest accumulated up to that point. If P is the principal, r is the annual interest rate, and t is the time in years that your money has been in the savings account, then at time t the value of the investment is

$$V = P(1 + rt). \tag{7.2}$$

Example 7.2 *Suppose you have an ordinary savings account, you deposit $1000, and $r = .06$. Then after six months (.5 years) the value of your investment is*

$$V = P(1 + rt) = \$1000(1 + .06 \times .5) = \$1030.$$

In the previous example, the time was less than one year. If the time is greater than one year, ordinarily Formula 7.2 is not used. Rather after one year, interest for the following year is also paid on the interest earned during the first year. That is, the principal for the second year is $P(1 + r)$, which means at the end of two years the value of the investment is equal to $P(1 + r)(1 + r) = P(1 + r)^2$. This procedure is repeated in each following year so that at the end of t years the value of the investment is

$$V = P(1 + r)^t.$$

This is called **compound interest**, and we say the interest is compounded yearly. We say "compounded" because the principal for the next year is the sum (combination) of the previous principle and the interest earned.

Example 7.3 *Suppose you deposit $1000, the annual interest rate $r = .06$, and the interest is compounded yearly. Then after $t = 20$ years the value of the investment is*

$$V = P(1 + r)^t = \$1000(1 + .06)^{20} = \$3207.14.$$

Notice that if the interest were not compounded, we would have

$$V = P(1 + rt) = \$1000(1 + .06 \times 20) = \$2200,$$

which is substantially less.

Often interest is compounded more frequently than once a year. For example, it may be compounded monthly. The following example shows how this is beneficial to the investor.

Example 7.4 *Suppose the annual interest rate is .06, the interest is compounded monthly, and the principal is $1000. Then the monthly interest rate is $.06/12 = 0.005$, which means at the end of 12 months (1 year) we have*

$$V = \$1000(1 + .005)^{12} = \$1061.68.$$

Notice that this value is $1.68 more than if the interest were compounded annually.

In general, if the annual interest rate is r, we compound monthly, and the principal is P, then at the end of one year we have

$$V = P\left(1 + \frac{r}{12}\right)^{12}.$$

Instead of compounding monthly, we could compound weekly, daily, hourly, or even continuously. Continuous compounding is obtained by taking the limit, as the compounding frequency approaches infinity, of the final value of the investment. If we compound n times per year, at the end of one year

$$V = P\left(1 + \frac{r}{n}\right)^n.$$

We obtain continuous compounding by taking the limit of this expression as n approaches ∞. To that end, we have

$$\lim_{n \to \infty} P\left(1 + \frac{r}{n}\right)^n = Pe^r.$$

The preceding limit is a standard result in calculus. We summarize these results in Theorem 7.1.

Theorem 7.1 *Suppose P is the principal, r is the annual interest rate, and t is the time in years the money is invested. If interest is compounded n times per year, then the value of the investment after t years is*

$$V = P\left(1 + \frac{r}{n}\right)^{nt}.$$

If interest is compounded continuously, then the value of the investment after n years is

$$V = Pe^{rt}.$$

Example 7.5 *Suppose the annual interest rate is .06, the interest is compounded continuously, and the principal is $1000. Then after two years*

$$V = Pe^{rt} = \$1000e^{.06 \times 2} = \$1127.50.$$

7.1.2   Net Present Value

We present the notion of net present value via a few examples.

Example 7.6 *Suppose George decides to sell his car to his friend Pete for $3000. However, Pete tells George that he is a bit strapped for cash and asks if he can give him the money one year from today. Is Pete really offering George $3000? If George received the money today, he could put it in the bank at whatever current interest rate is available to him and end up with more than $3000 in one year. Suppose the interest rate is .06 and is compounded annually. If George received $3000 today and he put the money in the bank, he would end up with*

$$\$3000(1 + .06) = \$3180.$$

So Pete should actually pay George $3180 in one year.

Often, it is useful to look at this matter in reverse. That is, we determine how much $3000 received in one year is worth today. This is called the net present value (NPV) of the money. To do this we simply divide by 1.06 instead of multiplying. We then have

$$NPV = \frac{\$3000}{1 + .06} = \$2830.19.$$

So Pete is really offering George $2830.19.

Example 7.7 *When George points out to Pete that he is shortchanging him, Pete says "Look, I can come up with $1000 today, but then I can only give you $1000 in one year, and will have to give you the final $1000 in 2 years." How much is Pete really offering George now? The $1000 received today is actually worth $1000. To determine the present value of the $1000 received in one year, we have to divide by 1.06 as before. To determine the present value of the $1000 received in two years, we must divide by 1.06 to determine its present value in one year, and then we must divide that result by 1.06 to determine its present value today. That is equivalent to dividing $1000 by $(1.06)^2$. We therefore have*

$$NPV = \$1000 + \frac{\$1000}{1 + .06} + \frac{\$1000}{(1 + .06)^2} = \$2833.40.$$

Example 7.8 *Finally, George tells Pete that he can pay in three equal installments as Pete indicated, but he wants the total to be equivalent to $3000 received today. To determine the amount of the installments we must solve*

$$\$3000 = x + \frac{x}{1 + .06} + \frac{x}{(1 + .06)^2}$$

for x. The answer is $x = \$1058.80$.

In general, if we have a cash flow stream $(x_0, x_1, x_2, \ldots, x_n)$ such that x_0 is received today, x_1 is received in one time period, x_2 is received in two time periods,..., x_n is received in n time periods, and the interest rate for each time period is r, then the **net present value (NPV)** of the cash flow stream is given by

$$NPV = x_0 + \frac{x_1}{(1+r)} + \frac{x_2}{(1+r)^2} \cdots + \frac{x_n}{(1+r)^n}.$$

7.1.3 Stocks

Suppose you start a small company that manufactures shoes which provide a battery-driven massage, it does very well, and you realize that with more capital you could expand it into a larger company. In order to raise this capital you could hire an investment banking firm to sell shares of your company to the public. Once you do this, your company is called a corporation, and the totality of these shares is called the **stock** in the company. Each share of stock represents one vote on matters of corporate governance concerning the company. The first issuance of stock in a formerly privately owned company is called an **initial public offering (IPO)**. At this time a document, called a **prospectus**, describing the prospects of the company is approved by the Securities and Exchange Commission (SEC), and the price at which the shares will be offered to the public is determined.

After the IPO, the stock can be bought and sold on a **stock exchange** such as the New York Stock Exchange (NYSE). A stock exchange provides facilities for individuals to trade shares of stocks, and only members who own seats on the exchange can trade on it. Most seats are owned by large-scale brokerage firms.

An individual investor can trade stocks by going through a broker. However, the individual must pay a commission for the trade. Today, investors can easily trade through on-line brokerage firms such as Ameritrade. The term asset is more general than the term stock. An **asset** is any instrument that can be traded easily. Therefore, stocks are assets, but there are other assets such as bonds, which will not be discussed here.

As with any commodity, the price of a share of a stock goes up when there are more buyers than there are sellers. At the close of today, Microsoft might sell for $28 a share, while tomorrow it could go up to $30 a share in the morning and back down to $29 a share by evening. There are many factors affecting the demand for a stock, including the current prospects of the company, macro-economic variables such as inflation and productivity, and perhaps ill-founded expectations of exuberant investors.

Return on Stocks

Many companies pay dividends to their shareholders. A **dividend** is a share of a company's net profits distributed by the company to its stockholders. The dividend is paid in a fixed amount for each share of stock held. Dividends are usually paid quarterly (four times a year). When determining the rate of return on an investment in a stock we must take into account any dividends paid. For stocks, we often call the rate of return, for the period the stock is held, the **holding period return rate (HPR)**.

Example 7.9 *Suppose one share of company XYZ is bought today for $10, sold six months later for $12, and no dividends are paid. Then the holding period is six months, and*

$$HPR = \frac{\$12 - \$10}{\$10} = .2.$$

Example 7.10 *Suppose we have the same situation as in the previous example except a dividend D_1 of $1 per share is paid at the end of the holding period. Then*

$$HPR = \frac{\$1 + \$12 - \$10}{\$10} = .3.$$

In general, if P_0 is the price today of one share of the stock, P_1 is the price at the end of the holding period of one share of the stock, D_1 is the dividend per share paid at the end of the holding period, and no other dividends are paid during the holding period, then

$$HPR = \frac{D_1 + P_1 - P_0}{P_0}. \tag{7.3}$$

When we purchase shares of a stock, we do not know what the return rate will be. The stock could go up or down in value. So there is risk involved in stock purchases. Although we cannot know the return, we can try to estimate it based on the return rates in previous time periods. Usually, we take the time period to be one year. Consider the next example.

Example 7.11 *Suppose stock XYZ had the annual return rates shown in the following table:*

Year	Annual Return r
2000	$-.05$
2001	$-.08$
2002	$-.12$
2003	$.25$
2004	$.20$

Then the expected value $\bar{r}$ and standard deviation σ_r of r are given by

$$\bar{r} = \frac{(-.05) + (-.08) + (-.12) + .25 + .20)}{5} = .04.$$

$$\sigma_r = \sqrt{\frac{(-.05-.04)^2+(-.08-.04)^2+(-.12-.04)^2+(.25-.04)^2+(.20-.04)^2}{5}}$$
$$= .153.$$

Do the expected return rate and standard deviation computed in the previous example tell us what to expect from XYZ's stock for the coming year? Not necessarily. In order to give investors an idea as to what to expect for the coming year, 3-year, 5-year, 10-year, and life-of-stock average annual return rates and corresponding standard deviations are ordinarily computed for companies and made available to the public. However, these values are relative to the probability distribution obtained from the given time period. They are not relative to our subjective probability distribution obtained from what we expect to happen in the coming year. Therefore, they are merely one piece of information that helps form our probability distribution. There are many other factors concerning the current prospects of the company and the state of the economy which are needed to formulate an informed subjective probability distribution for the coming year. We discuss this matter more in the sections that follow. Presently, we only offer the following related example to clarify the matter. Suppose a friend of yours received mostly C grades all through high school and during her first two years of college. However, right before her third year of college, she told you that she was going to get serious about her future and study hard. The result was that she achieved an A− average during her third year. You would not base your expectations concerning her performance in her fourth year of college solely on her grade distribution since she started high school. Rather, you would also consider your current information about her, namely that she said she was going to study hard and did receive good grades the past year. In the same way, we evaluate the prospects for the rate of return of a company's stock using all current information about the economy and the company.

Once we have an informed subjective distribution for the return rate, we can make a decision based on our preference for risk. The next example illustrates this.

Example 7.12 *Suppose we are considering two stocks, XYZ and ABC, and, based on our careful analysis, we believe their return rates for the coming year have the following expected values $\bar{r}$ and standard deviations σ_r:*

	$\bar{r}$	σ_r
ABC	.1	.02
XYZ	.1	.04

Suppose further that the return rates are normally distributed. Figure 7.1 shows the probability density function for the two stocks. Although both density functions have means equal to .1, the density function for ABC's return rate is more concentrated around .1, which means we are more confident the return rate will be close to .1. Recall that for the normal density function 95% of the mass falls within 1.96 standard deviations of the mean. So for ABC the probability is .95 that the return rate will be in the interval

$$(.1 - 1.96 \times .02, .1 + 1.96 \times .02) = (.061, .139),$$

whereas for XYZ the probability is .95 that the return rate will be in the interval

$$(.1 - 1.96 \times .04, .1 + 1.96 \times .04) = (.022, .178).$$

Most investors would choose ABC because when expected return rates are the same, they would choose the less risky investment. However, there is nothing irrational about choosing XYZ. A less risk-averse individual might be willing to suffer the greater chance of having less return or even losing money in order to have the chance at a larger return.

Short Sale of Stocks

Suppose you expect company XYZ's stock to fall in price during the next year. You make money if this happens by selling the stock short. A **short sale** of a stock consists of borrowing shares of the stock from a broker and selling the stock today. In the future the investor must purchase the stock in the market and replace the borrowed shares. This is called covering the short position. In practice, the brokerage firm ordinarily borrows the shares from another investor. The short seller's account must typically have 50% of the amount of the short sale in other assets as collateral for the short sale. This required percentage is called the required **margin**. Furthermore, the short seller typically cannot use the money obtained from the short sale to generate income.

Example 7.13 *Suppose a share of XYZ is currently selling for $100, and you sell 50 shares short. Suppose further that the required margin is 50%. Then the amount of your short sale is $100 \times 50 = \$5000$, which means you must have $2500 in other assets in your account. Let's say that this $2500 is in a U.S. Treasury Bill. After the short sale, your account will be as follows:*

Assets		Liabilities		Equity
Cash:	$5000	Short position in XYZ:	$5000	$2500
T-Bill:	$2500	(100 shares owed)		

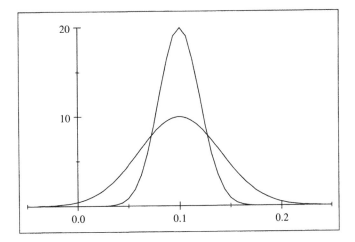

Figure 7.1: The density functions for the return rates discussed in Example 7.12. The wider curve is the normal density function with $\bar{r} = .1$ and $\sigma_r = .04$, while the thinner curve is the normal density function with $\bar{r} = .1$ and $\sigma_r = .02$.

Notice that your equity is exactly the same as it was before the short sale (as it should be). That is, although you have an additional $5000 in cash, you also have the additional liability of $5000 worth of stock.

Suppose the company pays no dividends, and you cover your short position one year later when one share of the stock is worth $80. Then your profit is

$$\$100 \times 50 - \$80 \times 50 = \$1000,$$

and your holding period return is given by

$$HPR = \frac{P_2 - P_1}{P_1} = \frac{-\$4000 - (-\$5000)}{-\$5000} = -.2.$$

It may seem odd that your return rate is negative while you made a profit. However, your investment was negative, namely $-\$5000$. So using Equality 7.1,

$$Profit = P_1 \times HPR = (-\$5000)(-.2) = \$1000.$$

Although a margin of 50% is ordinarily required to place a short sale, a lesser amount is often required to maintain it. That percentage is called the **maintenance margin**.

Example 7.14 *Suppose you place the short sale in Example 7.13, and your maintenance margin is 35%. Suppose further that the stock goes up in value to $150 per share. Then your liability becomes*

$$50 \times \$150 = \$7500,$$

and your margin is now

$$\frac{\$2500}{\$7500} \approx .333.$$

Since 33.3% < 35%, *you get a* **margin call**, *meaning you must deposit sufficient additional cash to bring your margin up to* 35%.

7.1.4   Portfolios

There are thousands of stocks, both in the United States and internationally, which are available to an investor. A **portfolio** is a group of investments held by an investor. For example, your portfolio might consist of 100 shares of XYZ and 200 shares of ABC. Briefly, one reason for buying more than one stock is to lessen risk. Stock ABC could go up even if stock XYZ goes down.

The fraction of the money value of your portfolio which is invested in a particular stock x_i is called the **weight** w_i of that stock in your portfolio. Clearly, if there are n stocks in a portfolio, then

$$\sum_{i=1}^{n} w_i = 1.$$

Example 7.15 *Suppose one share of IBM is worth* $25, *one share of Lucent is worth* $5, *and one share of Disney is worth* $30. *Suppose further that your portfolio consists of* 200 *shares of IBM,* 400 *shares of Lucent, and* 100 *shares of Disney. Then we have the following:*

Stock	# Shares in Portfolio	Price/Share	Total Value	Weight in Portfolio
IBM	200	$25	$5000	.5
Lucent	400	$5	$2000	.2
Disney	100	$30	$3000	.3
Portfolio total			$10,000	1

The weight, for example, of IBM in the portfolio was obtained as follows:

$$w_{IBM} = \frac{\$5000}{\$10,000} = .5.$$

7.1.5   The Market Portfolio

One particular portfolio which will interest us is the market portfolio. The **market portfolio** consists of all traded stocks. If an investor wants to duplicate the market portfolio in the investor's own portfolio, then the percentage representation of each stock in that portfolio must be according to the stock's total capitalization in the market. The **total capitalization** of a stock is the total value of all its shares. The next example illustrates this notion.

Example 7.16 *Suppose there are only four stocks, ABC, XYZ, DEF, and, GHK in the market, and their capitalization in the market is as follows:*

Stock	# Shares in Market	Price/Share	Total Capitalization	Weight in Market
ABC	1000	$20	$20,000	.1
XYZ	10,000	$3	$30,000	.15
DEF	5000	$10	$50,000	.25
GHK	20,000	$5	$100,000	.5
Total in market			$200,000	1

If Mary wanted to mimic the market portfolio, Mary would invest 10% of her money in ABC, 15% in XYZ, 25% in DEF, and 50% in GHK.

You may ask why Mary may want to mimic the market portfolio. In Section 7.2.2 we discuss why one may want to do this.

We supposed that there are only four stocks in the market portfolio for the sake of a simple illustration. In actuality, there are thousands of traded stocks. So, in practice, it would be difficult, if not impossible, for an individual investor to mimic the market portfolio. However, there are mutual funds and electronic traded funds (ETFs), which trade like stocks and which enable an investor to invest in large segments of the market by making a single purchase. We discuss these next.

7.1.6 Market Indices

A **stock market index** is an indicator that keeps track of the performance of some subset of stocks. Perhaps the most well-known index is the Dow Jones Industrial Average which is based on a list of "blue-chip" companies. The list was first developed in 1896 by Charles Dow and Edward Jones, and at that time only 20 companies were included in it. However, in 1928 it was expanded to include 30 companies. The total stock of the 30 companies is worth 25% of the value of all stocks on the NYSE, and the editors of *The Wall Street Journal* decide which companies to list and when to make changes. Since 1928 it has been updated more than 20 times, and only General Electric remains from the original 1986 list. Current companies on the list include AT&T, IBM, McDonald's, and Walt Disney. The average itself is based on a portfolio that owns one share of stock of each of the 30 companies. It is not based on the companies' total capitalization. This is called a **price-weighted index**.

Another well-known index is the S&P 500, which is an index of 500 common stocks of large U.S. companies chosen for market size, liquidity, and industry group representation. Its purpose is to represent U.S. equity performance, and it is often used as the market portfolio when by the market we mean the U.S. market. Each company enters into the value of this index according to its total capitalization. This is the same method used to illustrate mimicking the market portfolio in Example 7.16. An index of this type is called a **market-value-weighted index**.

The MSCI U.S. Small-Capitalization 1750 Index consists of small U.S. companies. Perhaps the most representative U.S. index is the Wilshire 5000 Index, which includes all NYSE and American Stock Exchange (Amex) stocks plus many NASDAQ stocks.

There are a number of non-U.S. indices, including the TSX (Canada), Nikkei (Japan), FTSE (United Kingdom), DAX (Germany), and Hang Seng (Hong Kong). The Morgan Stanley Capital International Europe, Australia and Far East (MSCI EAFE) Index is designed to represent the performance of developed stock markets outside the United States and Canada.

There are mutual funds and Electronic Trading Funds (ETFs) that represent many of the indices. Without going into detail, this means that one share of these funds invests in all the companies included in the index. For example, Vanguard sells mutual funds that represent the Wilshire 5000 Index and the MSCI U.S. Small-Capitalization Index. So if an investor wanted to invest in the market portfolio represented by an index, that investor could simply purchase the corresponding mutual fund or ETF. If by the market portfolio we mean all stocks throughout the world, the investor would have to invest in several indices to obtain a portfolio that closely corresponded to the market portfolio.

7.2 Advanced Topics in Investment Science★

We reviewed the basics of investing in the stock market in Section 7.1. This section discusses more advanced topics in investment science. First, we present mean-variance portfolio theory, then the Capital Asset Pricing Model (CAPM), next factor models and Arbitrage Pricing Theory, and, finally, equity valuation models. We can only cover the essentials of these topics here. You are referred to an investment science text such as [Bodie et al., 2004] for a thorough treatment.

7.2.1 Mean-Variance Portfolio Theory

Previously, we discussed investing in a portfolio that mimics the market portfolio. In Section 7.2.2 we show why an investor might want to do this. Before that, we need to discuss mean-variance portfolio theory.

Return and Variance of a Portfolio

We start with Theorem 7.2.

Theorem 7.2 *Suppose portfolio A consists of the n assets $x_1, x_2, \ldots, x_n$, each with weight $w_1, w_2, \ldots, w_n$. Furthermore, let r_i be the rate of return of x_i, $\bar{r}_i$ be the expected value of r_i, and σ_{ij} be $Cov(r_i, r_j)$. Then the rate of return r_A of the portfolio is given by*

$$r_A = \sum_{i=1}^{n} r_i.$$

Furthermore, the expected value and variance of r_A are given by

$$\bar{r}_A = \sum_{i=1}^{n} w_i \bar{r}_i$$

$$Var(r_A) = \sum_{i=1}^{n} \sum_{j=1}^{n} w_i w_j \sigma_{ij}.$$

Proof. The proof is left as an exercise. ∎

Example 7.17 *Let portfolio A consist of two assets such that*

$$w_1 = .25 \qquad w_2 = .75$$

$$\bar{r}_1 = .12 \qquad \bar{r}_2 = .15$$

$$\sigma_1 = .20 \qquad \sigma_2 = .18$$

$$\sigma_{12} = .01 \ (this \ value \ is \ typical \ for \ two \ stocks),$$

where σ_i is the standard deviation of r_i. Recall that the correlation coefficient of a random variable with itself is its variance. So $\sigma_{ii} = \sigma_i^2$. Owing to the previous theorem, we then have

$$\begin{aligned} \bar{r}_A &= w_1 \bar{r}_1 + w_2 \bar{r}_2 \\ &= (.25)(.12) + (.75)(.15) = .1425 \end{aligned}$$

and

$$Var(r_A)$$

$$\begin{aligned} &= w_1 w_1 \sigma_{11} + w_1 w_2 \sigma_{12} + w_2 w_1 \sigma_{21} + w_2 w_2 \sigma_{22} \\ &= w_1 w_1 (\sigma_1)^2 + w_1 w_2 \sigma_{12} + w_2 w_1 \sigma_{21} + w_2 w_2 (\sigma_2)^2 \\ &= (.25)(.25)(.20)^2 + (.25)(.75)(.01) + (.75)(.25)(.01) + (.75)(.75)(.18)^2 \\ &= .024. \end{aligned}$$

Therefore, the standard deviation of r_A is

$$\sigma_A = \sqrt{.024} = .155.$$

Note that this standard deviation is smaller than the standard deviations of r_1 and r_2. By diversifying, we have reduced our risk while maintaining an expected return close to that of the higher expected return.

Example 7.18 *Let portfolio A consist of 20 assets $x_1, x_2, \ldots, x_{20}$ such that for all i*

$$w_i = \frac{1}{20} \qquad \bar{r}_i = .10 \qquad \sigma_i = .15$$

and for all $i \neq j$

$$\sigma_{ij} = 0.$$

Then

$$\begin{aligned} \bar{r}_A &= \sum_{i=1}^{20} w_i \bar{r}_i \\ &= \sum_{i=1}^{20} \frac{1}{20} \times .10 = .10 \end{aligned}$$

and

$$
\begin{aligned}
Var(r_A) &= \sum_{i-1}^{20}\sum_{j-1}^{20} w_i w_j \sigma_{ij} \\
&= \sum_{i=1}^{20} w_i w_i \sigma_{ii} + \sum_{i\neq j} w_i w_j \sigma_{ij} \\
&= \sum_{i=1}^{20} \frac{1}{20} \times \frac{1}{20} \times (.15)^2 + \sum_{i\neq j} \frac{1}{20} \times \frac{1}{20} \times 0 \\
&= .00005625.
\end{aligned}
$$

Therefore, the standard deviation of r_A is

$$
\sigma_A = \sqrt{.00005625} = .0075.
$$

Notice that each individual asset has a standard deviation (risk) equal to .15, while the portfolio has the same expected value of each individual asset but a standard deviation of only .0075. We have significantly decreased our risk by diversifying in uncorrelated assets.

The previous example shows the value of diversification. If assets are uncorrelated, the limit, as the number of assets approaches infinity, of the standard deviation (risk) is equal to 0. So by investing in assets that are uncorrelated or have low correlation, we can significantly reduce our risk. An example of a well-diversified portfolio would be one containing domestic and foreign stocks and bonds, domestic and foreign real estate, and commodities such as gold, precious metals, and oil.

Mean-Standard Deviation Diagrams

We can compare two securities by plotting the expected values and standard deviations of their return rates on a graph that plots standard deviations on the x-axis and expected return rates on the y-axis. Such a graph, called a **mean-standard deviation diagram**, is shown in Figure 7.2. In this particular diagram, we plotted the points $(\sigma_1, \bar{r}_1)$ and $(\sigma_2, \bar{r}_2)$, which are the expected return rates and standard deviations for two securities x_1 and x_2. Notice that x_2 has a greater expected return rate, but also a greater risk (larger standard deviation). Suppose we want a portfolio that has an expected return rate and standard deviation in between that of the two securities. Then we can include precisely these two securities in the portfolio according to some weights. Let's investigate the possible portfolios we can obtain by varying these weights. Let w be the weight given to x_1 and $1 - w$ be the weight given to x_2 in a given portfolio. Assume for the present that we will not sell either asset short. In this case $0 \leq w \leq 1$. Therefore, owing to Theorem 7.2, the expected value and variance of the return rate of the portfolio are given by

$$
\bar{r}_w = w\bar{r}_1 + (1 - w)\bar{r}_2 \tag{7.4}
$$

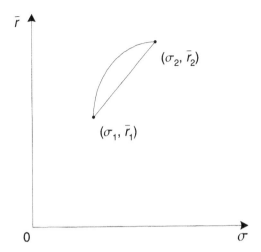

Figure 7.2: If $\rho_{12} = 1$, the set of all portfolios consisting of x_1 and x_2 determines a line segment connecting $(\sigma_1, \bar{r}_1)$ and $(\sigma_2, \bar{r}_2)$. If $\rho_{12} < 1$, the set of all portfolios consisting of x_1 and x_2 determines a curve connecting $(\sigma_1, \bar{r}_1)$ and $(\sigma_2, \bar{r}_2)$, which is convex to the left. It is assumed that short selling is not allowed.

$$Var(r_w) = w^2 (\sigma_1)^2 + w(1-w)\sigma_{12} + (1-w)w\sigma_{21} + (1-w)^2 (\sigma_2)^2. \quad (7.5)$$

Assume initially that the correlation coefficient ρ_{12} of the two return rates is 1. Then

$$\rho_{12} = \frac{\sigma_{12}}{\sigma_1 \sigma_2} = 1,$$

which means $\sigma_{12} = \sigma_1 \sigma_2$. Substituting this equality in Equality 7.5 and rearranging terms yield

$$Var(r_w) = [w\sigma_1 + (1-w)\sigma_2]^2,$$

which means

$$\sigma_w = w\sigma_1 + (1-w)\sigma_2. \quad (7.6)$$

Together, Equalities 7.4 and 7.6 are the parametric expression for a line through the points $(\sigma_1, \bar{r}_1)$ and $(\sigma_2, \bar{r}_2)$. When $w = 0$, we are at the point $(\sigma_2, \bar{r}_2)$, which means we are fully invested in x_2, while when $w = 1$, we are at the point $(\sigma_1, \bar{r}_1)$, which means we are fully invested in x_1. For $0 < w < 1$, we are somewhere on the line segment connecting these two points. We conclude that if short selling is not allowed and $\rho_{12} = 1$, then the set of all portfolios consisting of two assets determines a line segment in the mean-standard deviation diagram. If short selling is not allowed and $\rho_{12} < 1$, it is possible to show that the set of all portfolios consisting of x_1 and x_2 lies on a curve connecting $(\sigma_1, \bar{r}_1)$ and $(\sigma_2, \bar{r}_2)$, which is convex to the left. This is illustrated in Figure 7.2.

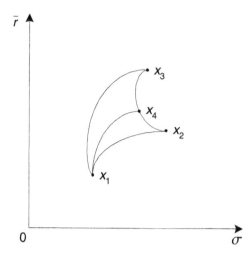

Figure 7.3: The set of all portfolios consisting of x_1, x_2, and, x_3 determines a solid region which is convex to the left. It is assumed that short selling is not allowed.

Consider now a portfolio consisting of three securities: x_1, x_2, and x_3. Assume again that short selling is not allowed. All portfolios consisting only of x_2 and x_3 must lie on a curve connecting x_2 and x_3, which is convex to the left. This is illustrated in Figure 7.3. Consider any portfolio x_4 on this curve. All portfolios consisting of x_4 and x_1 must lie on a curve connecting x_4 and x_1, which is convex to the left. Now as we vary the weight given to x_3 in x_4 from 0 to 1, x_4 ranges through all portfolios on the curve connecting x_3 and x_2, which means the curve connecting x_4 to x_1 sweeps through the entire interior of the region shown in Figure 7.3. We conclude that, if short selling is not allowed, the set of all portfolios consisting of three securities determines a solid region which is convex to the left.

This theory can be extended to the set of all portfolios consisting of n securities. If short selling is not allowed, this set would determine a sold region as shown in Figure 7.4 (a). If short selling is allowed, this set would determine an unbounded region such as the one to the right of the left-most curve shown in Figure 7.4 (b) (including the curve itself). Note that this region includes the region determined when short selling is not allowed. Unlike the depiction in Figure 7.4 (b), the left-most edges of the two regions can coincide.

Next, let the n securities be the market portfolio, which consists of all securities which an investor can trade. Based on what we just demonstrated, every investor's portfolio in the market falls somewhere in the unbounded region to the right of the curve shown in Figure 7.5 (or on the curve). Portfolio x_0 is called the **minimum variance portfolio** because it is the portfolio with the smallest variance. Consider the three portfolios x_1, x_2, and x_3 shown in Figure

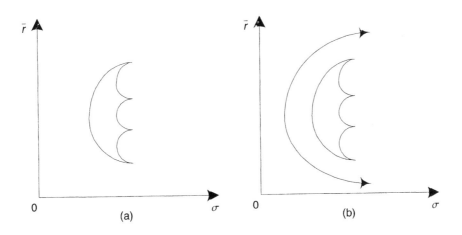

Figure 7.4: If short selling is not allowed, the set of all porfolios consisting of n securities determines a solid region as shown in (a). If short selling is allowed, this set would determine an unbounded region such as the one to the right of the left-most curve shown in (b).

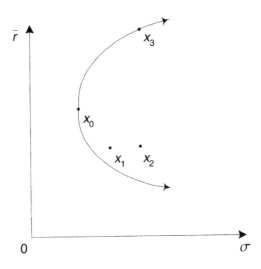

Figure 7.5: Portfolio x_0 is the minimum variance portfolio. Portfolio x_3 would be preferred to x_2.

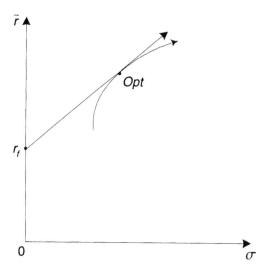

Figure 7.6: Portfolio *Opt* is optimal.

7.5. We have that $\bar{r}_1 = \bar{r}_2$ and $\sigma_1 < \sigma_2$. As discussed in Example 7.12, many investors would prefer x_1 to x_2 because they have the same expected returns, but x_1 is less risky. However, as also discussed in that example, there is nothing irrational about preferring x_2 to x_1. On the other hand, if we assume investors prefer more return over less return, there is no rational argument for preferring x_2 to x_3 because $\sigma_2 = \sigma_3$ and $\bar{r}_3 > \bar{r}_2$. For the same amount of risk, x_3 offers more expected return. We see then that any portfolio x_2 that is interior to the curve or on the bottom part of the curve has a corresponding portfolio x_3 on the top part of the curve such that x_3 is preferred to x_2. We call the top part of the curve the **efficient frontier** because all desired portfolios lie there. Portfolios on this frontier are called efficient. The graph in Figure 7.6 shows the efficient frontier. Notice that the efficient frontier begins on the left with the minimum variance portfolio. Then, by proceeding to the right on the efficient frontier, we improve our expected return, but at the expense of incurring more risk.

Notice the point $(0, r_f)$ in the graph in Figure 7.6. This point corresponds to the investment consisting of the risk-free asset. The risk-free asset is the best interest-bearing investment available that carries no risk. That is, $\sigma_f = 0$. The CD discussed in example 7.1 is an example of an asset that carries essentially no risk. Usually, r_f is taken to be the return on U.S. Treasury Bills. If we draw a ray emanating from $(0, r_f)$ that is tangent to the efficient frontier, we obtain a unique point where the line touches the efficient frontier. Let *Opt* (for optimal) be a portfolio corresponding to this point, and, for simplicity, let's also call this point *Opt*. Consider now the set of all portfolios consisting of *Opt* and the risk-free asset. Since $\sigma_f = 0$, it is not hard to see that this set consists of the ray emanating from $(0, r_f)$ and passing through the point *Opt*. The points on this ray to the left of *Opt* correspond to portfolios which have positive weights in the risk-free asset. That is, some money is invested in the risk-free asset. The points to the right of *Opt* correspond to portfolios which have negative weights in the risk-free asset. That is, money is borrowed to purchase more of *Opt*. Notice that this ray lies above the efficient frontier, which means that a portfolio on it is preferred to the one lying beneath it on the efficient frontier. So when we include the risk-free asset in our portfolio, we obtain a new efficient set of portfolios. This is the One-Fund Theorem.

Theorem 7.3 One-Fund Theorem. *There is a fund Opt of risky securities such that any efficient portfolio consists only of that fund and the risk-free asset.*

This theorem is quite remarkable. One can obtain any efficient portfolio simply by including a unique portfolio *Opt* and the risk-free asset according to some weights in one's portfolio. To determine *Opt* one must first determine subjective probability distributions of the rates of return of all securities in the market and then mathematically find the weights that determine the tangent point to the efficient frontier. Although this is possible, it would be very laborious and difficult for an individual investor. In the next section we discuss one argument for finding *Opt* readily without even determining a probability distribution.

Before closing, we note that the situation is slightly more complicated than we depicted since the rates at which an investor can borrow and invest (lend) are ordinarily not the same. Bodie et al. [2004] discuss the modifications necessary to accommodate this nuance.

7.2.2   Market Efficiency and CAPM

After discussing market equilibrium and the capital asset pricing model (CAPM), we briefly investigate the model's accuracy and industry's use of the model.

Market Equilibrium

Suppose all major investors have access to the same information and possess the same tools for developing subjective probability distributions of the rates of return of all securities in the market. Then these investors should arrive at the same subjective probability distributions and determine the same optimal portfolio *Opt*. They therefore would purchase securities to obtain this portfolio, and *Opt* would become the market portfolio. Suppose some new information becomes available which changes the probability distributions. Temporarily, *Opt* would not be the market portfolio, but investors would quickly purchase securities to obtain *Opt*, and rapidly the market portfolio would once again become *Opt*. So although, temporarily, *Opt* may deviate from the market portfolio, supply and demand for assets quickly drive *Opt* and the market portfolio back to being the same, and we say equilibrium is restored. We have offered a market equilibrium argument that *Opt* is the market portfolio. If this argument is correct, we say the **market is efficient**. In this case, all the small investor

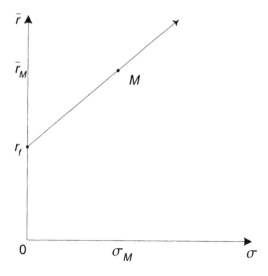

Figure 7.7: The capital market line.

needs to do in order to invest in Opt is to purchase an index fund that represents the market portfolio.

The Capital Market Line

If the market is efficient, then all efficient portfolios lie on the line shown in Figure 7.7, where M denotes the market portfolio. That line is called the **capital market line**. Since the line passes through the points $(0, r_f)$ and $(\sigma_M, \bar{r}_M)$, it has a slope equal to $(\bar{r}_M - r_f)/(\sigma_M - 0)$. Therefore, since the y-intercept is r_f, we have the equation for the line as

$$\bar{r} = r_f + \frac{\bar{r}_M - r_f}{\sigma_M}\sigma. \tag{7.7}$$

Notice that the expected return rate increases with the standard deviation according to the constant $K = (\bar{r}_M - r_f)/\sigma_M$. This constant determines how much additional expected return we get for a unit increase in standard deviation.

Example 7.19 *Suppose $r_f = .04$, $\bar{r}_M = .12$, and $\sigma_M = .15$. Then*

$$K = \frac{\bar{r}_M - r_f}{\sigma_M} = \frac{.12 - .04}{.15} = .533.$$

So if we are willing to suffer a standard deviation of $\sigma = .1$, our expected annual rate of return will be

$$\begin{aligned} \bar{r} &= r_f + K\sigma \\ &= .04 + .533(.1) = .093. \end{aligned}$$

Note that we have less risk than we would have if we invested all our money in the market, but we also have less expected return.

Example 7.20 *In theory, an investor can obtain any large expected return rate desired at the expense of suffering more risk. For example, suppose we want an expected rate of return equal to 1, which means the expected value of the profit is 100% of the investment. Assuming again that $K = .533$, we set*

$$
\begin{aligned}
1 &= r_f + K\sigma \\
&= .04 + .533\sigma
\end{aligned}
$$

and solve for σ to yield $\sigma = 1.80$. This is a very large standard deviation. If we assume a normal distribution, the probability is .025 that the rate of return will fall more than 1.96 standard deviations to the left of the mean. So if we make this assumption, the probability is .025 that the rate of return will be less than

$$
1 - 1.96 \times 1.8 = -2.53.
$$

This means there is a .025 probability that the investor will lose 253% or more of the investment amount. The reason the investor can lose more than the amount invested is that money must be borrowed in order to go to the right of M on the capital market line. This borrowed money would also be lost if the market performed poorly.

CAPM

Next, we discuss an important consequence of the efficient market hypothesis, namely the capital asset pricing model (CAPM). First, we develop some preliminary concepts.

Preliminary Concepts   For a given asset x_i, let its return rate r_i be given by

$$
r_i = r_f + \beta_i(r_M - r_f) + \epsilon_i, \tag{7.8}
$$

where r_f is the rate of return of the risk-free asset, r_M is the rate of return of the market,

$$
\beta_i = \frac{Cov(r_i, r_M)}{\sigma_M^2},
$$

and ϵ_i is a random variable which is defined to make Equality 7.8 hold (that is, $\epsilon_i \equiv r_i - [r_f + \beta_i(r_M - r_f)]$). The term β_i is called the **beta** of asset x_i. The following theorem separates the variance of r_i into two components.

Theorem 7.4 *We have that*

$$
Var(r_i) = \beta_i^2 Var(r_M) + Var(\epsilon_i) \tag{7.9}
$$

and

$$
Cov(\epsilon_i, r_M) = 0.
$$

Proof. It is left as an exercise to show that $Cov(\epsilon_i, r_M) = 0$ can be obtained by computing $Cov(r_i, r_M)$ using the expression for r_i in Equality 7.8.

As to the first equality in the theorem, using Equality 7.8 we have

$$
\begin{aligned}
Var(r_i) &= Var(r_f + \beta_i(r_M - r_f) + \epsilon_i) \\
&= Var(\beta_i r_M + \epsilon_i) \\
&= \beta_i^2 Var(r_M) + Var(\epsilon_i) + Cov(\epsilon_i, r_M).
\end{aligned}
$$

The proof now follows since we have already concluded $Cov(\epsilon_i, r_M) = 0$. ∎

We see from Equality 7.9 that the risk of an asset consists of two components, namely the risk due to the market as a whole and the risk due to a random variable ϵ_i that is uncorrelated with the market. Every asset in the market is correlated with the market as a whole and therefore has a risk component related to the market. This component, namely $\beta_i^2 Var(r_M)$, is termed **market risk** or **systematic risk**. However, there is another component of risk that is unrelated to the market. This component, namely $Var(\epsilon_i)$, is termed **firm-specific** or **nonsystematic risk**.

Example 7.21 Suppose $Var(r_M) = .0225$, and for stock XYZ we have that $Cov(r_{XYZ}, r_M) = .045$. Then

$$
\beta_{XYZ} = \frac{Cov(r_{XYZ}, r_M)}{\sigma_M^2} = \frac{.045}{.0225} = 2,
$$

and, owing to Theorem 7.4,

$$
\begin{aligned}
Var(r_{XYZ}) &= \beta_{XYZ}^2 Var(r_M) + Var(\epsilon_{XYZ}) \\
&= 2^2(.0225) + Var(\epsilon_{XYZ}) \\
&= .09 + Var(\epsilon_{XYZ}).
\end{aligned}
$$

This means the asset has at least four times the variance of the market.

The beta (β_i) of asset x_i can tell us much about the riskiness of the asset. If $\beta_i > 1$ the asset is more risky than the market, while if $0 < \beta_i < 1$, the systematic component of the risk is less than the market risk. If $\beta_i < 0$, the asset is negatively correlated with the market.

Let's determine the expected value of r_i using Equality 7.8. To that end, we have

$$
\begin{aligned}
\bar{r}_i &= E(r_f + \beta_i(r_M - r_f) + \epsilon_i) \\
&= r_f + \beta_i(\bar{r}_M - r_f) + \bar{\epsilon}_i. \tag{7.10}
\end{aligned}
$$

We see then that the expected rate of return of an asset also increases with its beta. This is a reasonable result since we'd expect to receive increased expected return for increased risk.

Example 7.22 *Suppose we have same values as in Example 7.21, which means*
$\beta_{XYZ} = 2$. *Suppose further that* $r_f = .04$ *and* $\bar{r}_M = .10$. *Then*

$$
\begin{aligned}
\bar{r}_{XYZ} &= r_f + \beta_{XYZ}(\bar{r}_M - r_f) + \bar{\epsilon}_{XYZ} \\
&= .04 + 2(.10 - .04) + \bar{\epsilon}_{XYZ} \\
&= .16 + \bar{\epsilon}_{XYZ}.
\end{aligned}
$$

In the previous example, we cannot even say that $\bar{r}_{XYZ}$ must be at least .16 because $\bar{\epsilon}_{XYZ}$ could be negative. So, in general, without knowing $\bar{\epsilon}_i$ the theory just developed gives us little help in determining the risk/reward trade-off in investing in an asset. However, it is a remarkable result of the efficient market hypothesis that $\bar{\epsilon}_i = 0$ for all securities x_i in the market. We show that result next.

The Pricing Model The following theorem, whose result is called the **capital asset pricing model (CAPM)**, is a consequence of the efficient market hypothesis. Its proof can be found in [Luenberger, 1998].

Theorem 7.5 *If the market is efficient, and M is the market portfolio, then for any asset x_i in the market, its expected return rate $\bar{r}_i$ satisfies*

$$
\bar{r}_i = r_f + \beta_i(\bar{r}_M - r_f), \tag{7.11}
$$

where

$$
\beta_i = \frac{Cov(r_i, r_M)}{\sigma_M^2}.
$$

This result is not surprising. Recall that Equality 7.10 says that

$$
\bar{r}_i = r_f + \beta_i(\bar{r}_M - r_f) + \bar{\epsilon}_i.
$$

So if the market is efficient, Theorem 7.5 entails that $\bar{\epsilon}_i = 0$. In an efficient market, it seems that factors other than the market should, on the average, have no effect on return rates.

Example 7.23 *Suppose we have the situation in Example 7.22, and the market is efficient. Then, owing to Theorem 7.5,*

$$
\begin{aligned}
\bar{r}_{XYZ} &= r_f + \beta_{XYZ}(\bar{r}_M - r_f) \\
&= .04 + 2(.10 - .04) \\
&= .16.
\end{aligned}
$$

If the market is efficient, Theorem 7.5 says that

$$
\bar{r}_i = r_f + \beta_i(\bar{r}_M - r_f),
$$

which means $\bar{r}_i$ is a linear function of β_i with slope $\bar{r}_M - r_f$ and y-intercept equal to r_f. This function, shown in Figure 7.8, is called the **security market**

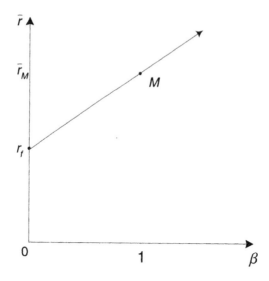

Figure 7.8: The security market line.

line. Notice how $\bar{r}_i$ increases with β_i. Recall that $Var(r_i)$ also increases with β_i. For this reason, β_i is called the **risk exposure** of asset x_i to the market, and $(\bar{r}_M - r_f)$ the **risk premium** for the market. The risk premium tells us how much additional expected rate of return (relative to the risk-free rate) we obtain by investing in the market, and the risk exposure of asset x_i tells us the degree to which this premium affects the expected return of x_i.

Example 7.24 *In Example 7.23, the market risk premium is .06, the risk exposure is 2, and the additional expected return rate we obtain for this exposure to the market is .12. Of course, we pay for this additional expected return rate by incurring more risk. Recall from Example 7.21 that $Var(r_{XYZ}) \geq .09$, which mean $\sigma_{XYZ} \geq (.09)^{1/2} = 0.3$. So there is a good chance we could lose money.*

Beta of a Portfolio   A portfolio also has a beta, and this beta can readily be calculated from the betas of the assets in the portfolio. We have the following theorem:

Theorem 7.6 *Suppose portfolio A consists of the n securities $x_1, x_2, \ldots, x_n$ with weight $w_1, w_2, \ldots, w_n$. Then the beta of portfolio A is given by*

$$\beta_F = \sum_{i=1}^{n} w_i \beta_i.$$

Proof. *The proof is left as an exercise.* ∎

Since portfolios have betas, the theory just developed applies to portfolios as well as individual assets. Practically speaking, we can simply treat a portfolio as an asset.

Nonsystematic Risk of an Efficient Portfolio   Consider a portfolio on the capital market line (see Figure 7.7). Such a portfolio consists of the market portfolio and the risk-free asset according to some weights. We show next that such a portfolio has no nonsystematic risk.

Theorem 7.7 *Any portfolio A on the capital market line has no nonsystematic risk. That is, $Var(\epsilon_A) = 0$.*
Proof. *If portfolio A is on that line, then for some weight w*

$$r_A = (1 - w)r_f + wr_M.$$

Therefore,

$$
\begin{aligned}
\beta_A &= \frac{Cov(r_A, r_M)}{\sigma_M^2} \\
&= \frac{Cov((1-w)r_f + wr_M, r_M)}{\sigma_M^2} \\
&= \frac{w\sigma_M^2}{\sigma_M^2} = w.
\end{aligned}
\tag{7.12}
$$

Furthermore,

$$
\begin{aligned}
Var(r_A) &= Var((1-w)r_f + wr_M) \\
&= w^2\sigma_M^2.
\end{aligned}
\tag{7.13}
$$

Owing to Theorem 7.4,

$$Var(r_A) = \beta_A^2\sigma_M^2 + Var(\epsilon_A) \tag{7.14}$$

Combining Equalities 7.12, 7.13, and 7.14, we have

$$w^2\sigma_M^2 = w^2\sigma_M^2 + Var(\epsilon_A),$$

which means $Var(\epsilon_A) = 0$. ■

Notice that the result in the previous theorem holds regardless of whether we assume that the market is efficient. However, if we do not assume that the market is efficient, we cannot conclude that all efficient portfolios lie on the capital market line. If, on the other hand, we assume that the market is efficient, all efficient portfolios lie on the capital market line, and all other portfolios lie beneath it and are riskier because they have both market risk and nonsystematic risk. A portfolio on the capital market line is diversified into the entire market and therefore is left with only market risk.

Is the Market Efficient?

The efficient market hypothesis is reasonable and leads eloquently to CAPM. But is the market really efficient? Since CAPM is a consequence of this hypothesis, we could see if historical data contradict CAPM, which would thereby contradict the efficient market hypothesis. However, the efficient market hypothesis says that major investors have access to the same information and form the same subjective probability distributions based on this information. Each year this distribution could change, and, furthermore, this distribution might not represent what ends up happening in reality. So to test CAPM based on historical data, we need to make the following assumptions:

1. There is an actual probability distribution which is generating the returns in each of the time slices, and the return in each time slice is one item sampled according to this distribution.

2. The consensus subjective probability distribution is the same as the distribution mentioned in (1) above.

For example, suppose we sample data from each of the past 60 months. We must assume that the return in each month is one item sampled from an actual probability distribution of the rates of return. Furthermore, we must assume that the consensus subjective probability distribution is this distribution.

Once we make these assumptions, we can test CAPM as follows. First, recall from Equality 7.10 that for each asset x_i,

$$\bar{r}_i = r_f + \beta_i(\bar{r}_M - r_f) + \bar{\epsilon}_i.$$

CAPM says that $\bar{\epsilon}_i = 0$ for all i. To test whether this is the case, for some period of time we collect data on the annual rate of return r_i of each asset x_i, the annual rate of return r_M of the market, and the annual risk-free rate of return r_f. We collect one data item for each time slice in our time period. Let's assume that the time period is five years and the time slice is one month. Then we have 60 sample points. Let's further assume that there are 500 assets in our market portfolio. Then for each of these 60 sample points, we have data consisting of 500 values of r_i, 1 value of r_M, and 1 value of r_f.

Using these data, we then do a first-pass regression to estimate the values of β_i for all i. That is, for $1 \leq i \leq 500$, we do the regression

$$r_{it} - r_{ft} = a_i + b_i(r_{Mt} - r_{ft}) + \epsilon_{it}. \tag{7.15}$$

In this equality, r_{it} is the rate of return for asset x_i in time slice t, r_{Mt} is the rate of return of the market in time slice t, and r_{ft} is the risk-free rate in time splice t. Note that we do 500 regressions, and in each regression we have 60 sample points.

For $1 \leq i \leq 500$, the value of b_i obtained from the regression is our estimate of β_i.

Next, for $1 \leq i \leq 500$, we compute the sample averages of $r_{it} - r_{ft}$ and $r_{Mt} - r_{ft}$ over the 60 holding periods. Denote these averages as

$$\overline{r_i - r_f} \qquad \text{and} \qquad \overline{r_M - r_f}.$$

We then do a second-pass regression using the 500 values of b_i obtained from the first-pass regression as the values of our independent variable and the values of $\overline{r_i - r_f}$ as the values of our dependent variable. The regression is as follows:

$$\overline{r_i - r_f} = c + db_i + \epsilon_i.$$

Now according to CAPM, the result of this regression should be

$$c = 0 \quad \text{and} \quad d = \overline{r_M - r_f}.$$

Improvements to this method are discussed in [Bodie et al., 2004]. A number of studies have been done using the method with and without the improvements. One problem in many of these studies is that some index, such as all stocks in the S&P 500, is used as a proxy for the entire market, and this proxy may not be efficient while the market is, and vice versa. A study in [Fama and MacBeth, 1973] overcame this difficulty. This study, which is somewhat more complex than what was outlined here, substantiated that $\bar{r}_i - r_f$ is a linear function of $\bar{r}_M - r_f$ as CAPM entails, but it did not show that $c = 0$. However, this latter result was not conclusive. The study also indicated that the expected return rate does not increase with nonsystematic risk.

Industry Version of CAPM

Industry finds it useful to evaluate and report betas of assets as indications of their risks and expected return rates. Ordinarily, they do an analysis over the past 60 months similar to the analysis just described. Besides reporting the beta estimates, they also report estimates of alpha. We describe alpha next.

Suppose we do the regression in Equality 7.15 for asset x_i. We obtain values of both a_i and b_i. As already noted, b_i is our estimate of β_i. The intercept a_i is our estimate of α_i, which is the **alpha** of the asset. Alpha shows how much the expected return rate deviates from the CAPM predictions. That is, CAPM says that $\alpha_i = 0$.

Industry also reports the **coefficient of determination**, denoted R^2, which is the square of the correlation coefficient of the return rate of an asset and the market return rate. That is, for asset x_i

$$R_i^2 = [\rho(r_i, r_M)]^2 = \left[\frac{Cov(r_i, r_M)}{\sigma_i \sigma_M} \right]^2.$$

It is not hard to see that this means

$$\beta_i^2 \sigma_M^2 = R_i^2 \sigma_i^2. \tag{7.16}$$

Recall that Equality 7.9 says

$$\sigma_i^2 = \beta_i^2 \sigma_M^2 + Var(\epsilon_i),$$

where $\beta_i^2 \sigma_M^2$ is the market risk, and $Var(\epsilon_i)$ is the nonsystematic risk. So R_i^2 is a measure of what fraction of the total risk is market risk.

Example 7.25 *Suppose for asset ABC we obtain the following estimates:*

$$\beta_{ABC} = 1.2 \qquad and \qquad R^2_{ABC} = 1.$$

Furthermore, assume $\sigma^2_M = .04$. *Then, owing to Equality 7.16,*

$$\beta^2_{ABC}\sigma^2_M = R^2_{ABC}\sigma^2_{ABC},$$

which means

$$\sigma^2_{ABC} = \frac{\beta^2_{ABC}\sigma^2_M}{R^2_{ABC}} = \frac{(1.2)^2 \times .04}{1} = .058.$$

All the risk is market risk since $R^2_{ABC} = 1$.

Example 7.26 *Suppose for asset XYZ we obtain the following estimates:*

$$\beta_{XYZ} = 1.2 \qquad and \qquad R^2_{XYZ} = .8.$$

Again, assume $\sigma^2_M = .04$. *Since Equality 7.16 says*

$$\beta^2_{XYZ}\sigma^2_M = R^2_{XYZ}\sigma^2_{XYZ},$$

which means

$$\sigma^2_{XYZ} = \frac{\beta^2_{XYZ}\sigma^2_M}{R^2_{XYZ}} = \frac{(1.2)^2 \times .04}{.8} = .072.$$

The total risk is .072, while the market risk is 80% of that or .058. The remaining 20% of the risk (.014) is nonsystematic risk.

It is insightful to rewrite Equality 7.16 as follows:

$$\sigma^2_i = \left(\frac{\beta^2_i}{R^2_i}\right)\sigma^2_M. \tag{7.17}$$

This equality shows that if $\beta^2_i > R^2_i$, the asset is more risky than the market, while if $\beta^2_i < R^2_i$, the asset is less risky than the market.

Merrill Lynch publishes the book *Security Risk Evaluation*, which is commonly called the "Beta Book," and which shows the betas, alphas, and R^2 of most publicly traded U.S. stocks. The S&P 500 Index is used as a proxy for the market portfolio, the data consist of the monthly rate of returns of each of the most recent 60 months, and straight regression is used to determine the values. The alphas they report are slightly different than what we described, but they are usually close. Bodie et al. [2004] discuss the difference. Table 7.1 shows some listings from the January 2005 edition, which includes prices as of December 31, 2005.

The alphas shown in Table 7.1 are percents and monthly returns. For example, if $\bar{r}_M = .01$ and $r_f = .0025$, for stock AVII we have the following:

$$\begin{aligned}
\bar{r}_{AVII} &= \alpha_{AVII} + \beta_{AVII}(\bar{r}_M - r_f) \\
&= .0287 + .79(.01 - .0025) \\
&= .0346.
\end{aligned}$$

Ticker Symbol	Security Name	Beta	Alpha	R^2	σ Resid.	σ Beta	σ Alpha	Adj. Beta
DOWI	Dow Ind.	0.93	0.20	0.86	1.78	0.05	0.23	0.95
SPALNS	S&P 500	1.00	0.00	1.00	0.00	0.00	0.00	1.00
ASBH	ASB Hold.	1.22	−1.30	0.20	5.27	0.57	1.56	1.14
AVII	AVI Biophm.	0.79	2.87	0.00	32.40	0.90	4.19	0.86
AVX	AVX Corp.	2.18	0.70	0.39	12.79	0.35	1.45	1.78

Table 7.1: Some entries from the January 2005 edition of Merrill Lynch's Security Risk Evaluation.

This result is only approximate because, as mentioned earlier, Merrill Lynch's alphas are slightly different than what we call alpha. AVII may look attractive because its alpha is .0287 greater than the 0 predicted by CAPM. However, there are a couple of problems with drawing this conclusion. First, the computed values of $\bar{r}_i$, α_i, and β_i are themselves random variables. That is, the values we obtain from these 60 data points are only estimates of the "true" values. If we had a different set of 60 data points, we would obtain different estimates. So, for example, if we call the computed value of the expected return rate $\hat{r}_i$, we will have $E(\hat{r}_i) = \bar{r}_i$, and there will be a $Var(\hat{r}_i)$. The same holds for alphas and betas. As shown in [Luenberger, 1998], the variances in the estimates of return rates and alphas are very large unless we have a lot of data points, whereas the variances in estimates of betas are not. In Table 7.1, look at the columns that show the estimates of the standard deviations of alpha and beta. Standard deviations of alphas are typically larger than the estimates of alphas themselves. So we cannot have a lot of confidence in these estimates. If we try to use sufficient years of data to obtain a reliable estimate, we encounter a parameter shift. That is, the underlying nature of the market could not be assumed to be the same in the years considered; so the parameters would not maintain the same values.

A second problem is that if these alphas and betas are to be useful, we must assume that there is no parameter shift in the time span including the past five years and the coming year. Our goal, after all, is to determine the expected return for the coming year. If there is a parameter shift for the coming year, these computed values are not very useful.

Note the column labeled "Std. Dev. Resid." in Table 7.1. This is the standard deviation of the error term ϵ_{it} (called the residual) in the regression shown in Equality 7.15. Finally, note the column labeled "Adj. Beta" in Table 7.1. It is assumed that betas do change over time and drift towards 1. New companies often produce a specific, innovative product. However, as time goes on, they ordinarily survive only if that product is accepted and/or they diversify. So eventually they become less risky and have betas about equal to 1. Merrill Lynch uses a formula to adjust the beta obtained from the regression toward 1. This adjusted value appears in the column "Adj. Beta."

7.2.3 Factor Models and APT

You may have noticed that when the monthly report of U.S. non-farm payroll is greater than expected, the U.S. stock market often goes up, and when inflation is greater than expected, the market often goes down. Factor models try to model the relationship between expected return rates of assets and these economic variables, which are called **macroeconomic risk factors**. The theory does not say what the macroeconomic factors need to be. After discussing the assumptions in the factor models, we present arbitrage pricing theory (APT), which yields an important result concerning factor models. Then we discuss the relationship between APT and CAPM. Finally, we present a commonly used factor model.

The Assumptions in Factor Models

A **factor model** assumes that we have k risk factors F_1, F_2, ..., F_k such that the following holds:

1. For every asset x_i in the market, we have

    $$r_i(t) - \bar{r}_i(t) = b_{i1} f_1(t) + b_{i2} f_2(t) + \ldots + b_{ik} f_k(t) + \epsilon_i(t), \qquad (7.18)$$

    where

    (a) $r_i(t)$ is the total rate of return of asset x_i (capital gains and dividends) realized at the end of time period t.

    (b) $\bar{r}_i(t)$ is the expected rate of return of asset x_i at the beginning of time period t.

    (c) b_{ij} is the **risk exposure** of asset x_i to the jth risk factor.

    (d) $f_j(t)$ is the value of risk factor F_i at the end of time period t.

    (e) $\epsilon_i(t)$ is the value of the **nonsystematic shock** of asset x_i at the end of time period t.

2. $\bar{f}_j(t) = 0$ for all j.

3. $\bar{\epsilon}_i(t) = 0$ for all i.

4. $Cov(\epsilon_i(t), f_j(t)) = 0$ for all i and all j.

5. $Cov(\epsilon_i(t), e_i(t')) = 0$ for all i and for all $t \neq t'$.

6. $Cov(f_j(t), f_j(t')) = 0$ for all j and for all $t \neq t'$.

The above assumptions entail that the excess rate return (the amount that the actual rate exceeds the expected rate) of every asset x_i in the market is a function of that asset's exposure to macroeconomic risk factors plus a component ϵ_i, which is not due to the risk factors. Assumption (2) says the value of each risk factor is adjusted so that its expected value is 0. The nonsystematic shock (ϵ_i) is the component of the return rate that is not due to any of the factors. Assumption (3) says the expected value of the nonsystematic shock of each asset x_i is 0. Assumption (4) says the nonsystematic shock is uncorrelated with the value of each of the risk factors. Assumption (5) says the nonsystematic shock at time t for a given asset is uncorrelated with its nonsystematic shock at another point in time t', while Assumption (6) says the value of a given risk factor at time t is uncorrelated with its value at another point in time t'.

Note that $Cov(\epsilon_i(t), e_k(t))$ need not be 0. Oddly, many authors do make this a requirement. For example, the energy price shock in late summer 2005 stemming from Hurricane Katrina caused correlation among the asset-specific shocks of firms in economic sectors that use a lot of energy.[1] Also, $Cov(f_j(t), f_m(t))$ need not be 0. For example, if F_1 is inflation and F_2 is interest rates, their values should be correlated.

Note that Equality 7.18 does not give us what we wish to know, namely the expected rate of return of a given asset. Rather, it tells us the actual return rate if we know the expected return rate, the values of the risk factors, and the nonsystematic shock. The factor model assumptions become more useful when we also assume no arbitrage. We discuss this next.

Arbitrage Pricing Theory (APT)

Suppose Stock XYZ is currently selling on the NYSE for \$30 a share and on the NASDAQ for \$40 a share. You could simultaneously purchase one share for \$30 and sell one share for \$40, thereby realizing an immediate profit of \$10 a share without incurring any risk. This is an example of **arbitrage**, which is defined to be an action which enables one to realize sure instantaneous profit or sure future profit with zero investment. The **no arbitrage principle** entails that if two assets are equivalent in all economically relevant ways, they should have the same price. It is based on the assumption that if a price discrepancy arises, investors will quickly rush to buy where the asset is cheap, thereby driving the price up, and sell where it is expensive, thereby driving the price down. Arbitrage pricing theory assumes that markets move in this manner, and therefore the no arbitrage principle holds (other than for anomalies which quickly disappear).

Specifically, the assumptions of **arbitrage pricing theory (APT)** are as follows:

1. We have identified k risk factors F_1, F_2, ..., and F_k such that the factor model assumptions hold.

2. The no arbitrage principle holds.

3. There are a large number of assets in the market under consideration.

Given these assumptions, it is possible to prove (e.g., [Ingersoll, 1987]) that there are k number $P_1, P_2, \ldots, P_k$ such that the expected return rate of asset x_i is given by

$$\bar{r}_i(t) = r_f(t) + b_{i1}P_1 + b_{i2}P_2 + \ldots + b_{ik}P_k, \tag{7.19}$$

---

[1] I thank Edwin Burmeister for this example.

where $r_f(t)$ is the return rate of the risk-free asset. $P_j(t)$ is called the **risk premium** for the jth risk factor. When multiplied by a particular asset's risk exposure (b_{ij}) to the factor, it gives the premium in expected return that is realized in return for being exposed to the factor's risk. The risk premiums are not actually constant over time. However, researchers (e.g., [Campbell, 1999]) have argued that if the time frame is relatively small, the variation in risk premium is of secondary importance. Researchers have therefore assumed them to be constant when developing the theory.

Example 7.27 *Suppose we have identified that the following two factors satisfy the factor model assumptions for the U.S. Stock Market:*

1. F_1: *Business Cycle*

   The time slice is one month. The value $f_1(t)$ of F_1 is determined as follows: The expected value of a business index is computed at both the beginning and the end of month t. Then $f_1(t)$ is the difference between the end-of-month value and the beginning-of-month value.

2. F_2: *Inflation*

   The value $f_2(t)$ of F_2 is the difference between the actual inflation observed at the end of month t and the inflation predicted at the beginning of month t.

Suppose further that we make the APT assumptions. Finally, for two companies XYZ and ABC assume the following:

Company	b_{i1}	b_{i2}
XYZ (x_1)	4.5	$-.60$
ABC (x_2)	3.5	.40

Notice that XYZ's risk exposure to inflation is negative, while ABC's risk exposure is positive. Inflation erodes real income, which means people have less money for superfluous activities. So industries like retailers, services, and restaurants do poorly in inflationary times. On the other hand, companies with large asset holdings such as real estate or oil often do well in such times.

As is often done, even though the time period is one month, we report the return rates as yearly rates. Suppose for time slice t (say, April 2005) the risk-free rate $r_f(t) = .03$, and we determine the following:

$$P_1 = .043$$
$$P_2 = .015.$$

Then

$$\bar{r}_1(t) = r_f + b_{11}P_1 + b_{12}P_2$$
$$= .03 + (4.5)(.043) + (-.60)(.015) = .2145$$

$$\bar{r}_2(t) = r_f + b_{21}P_1 + b_{22}P_2$$
$$= .03 + (3.5)(.043) + (.40)(.015) = .1865.$$

You may wonder how we obtain the values of b_{ij} and $P_i(t)$. They are estimated from historical data using nonlinear seemingly unrelated regression (NLSUR). This is similar to the way alphas and betas are obtained for CAPM (see Section 7.2.2), but is somewhat more complex. McElroy and Burmeister [1988] show the technique.

APT and CAPM

You probably noticed similarities between APT and CAPM. Next, we make the connection explicit.

Theorem 7.8 *Suppose the factor model assumptions are satisfied for the single factor $r_M - \bar{r}_M$, where r_M is the rate of return of the market. Then, if we make the APT assumptions, for every asset x_i,*

$$\bar{r}_i = r_f + b_i(\bar{r}_M - r_f)$$

and

$$b_i = \beta_i.$$

That is, the CAPM formula is satisfied.
Proof. *For the sake of consistency with the notation used when we discussed CAPM, we do not explicitly show the dependency of return rates, etc. on t. Suppose we have one factor F_1, which is the market, and its value is given by*

$$f_1 = r_M - \bar{r}_M.$$

Then, due to Equality 7.18, for every asset x_i,

$$r_i = \bar{r}_i + b_{i1}(r_M - \bar{r}_M) + \epsilon_i, \tag{7.20}$$

where $Cov(r_M - \bar{r}_M, \epsilon_i) = 0$. It is left as an exercise to show that if we compute $Cov(r_i, r_M)$ using Equality 7.20, we can conclude that $b_{i1} = Cov(r_i, r_M)/\sigma_M^2 = \beta_i$.

Therefore, owing to Equality 7.19, there exists a constant $P_1(t)$ such that for every asset x_i

$$\bar{r}_i = r_f + \beta_i P_1.$$

If we apply this equality to the market itself, we obtain

$$\bar{r}_M = r_f + \beta_M P_1.$$

Since $\beta_M = 1$, we conclude that

$$P_1 = \bar{r}_M - r_f.$$

This completes the proof. ■

We have deduced the CAPM formula from the APT assumptions for a single factor which is the market. We stress that this does not mean the two theories are equivalent. The APT assumptions say nothing about an efficient market. Rather, they entail assumptions about the covariance matrix. Previously, we deduced the CAPM formula by assuming an efficient market without saying anything about the nature of the covariance matrix. So the efficient market hypothesis and the CAPM formula could be correct even if the single factor model is not. Indeed, it is widely accepted that we need more than one factor.

Intuition for the Risk Premiums

Notice that for the single factor model just discussed, we found that the risk premium is $\bar{r}_M - r_f$. This risk premium is how much additional expected rate of return (relative to the risk-free rate) we obtain by investing in an asset that has a risk exposure of 1 to the market. Such an asset would be the market portfolio itself. In general, in a multi-factor index model, we can think of the risk premium P_k as being the additional expected return we obtain (relative to the risk-free rate) by investing in an asset that has a risk exposure of 1 to factor F_k and risk exposures of 0 to all other factors. For example, if some asset had a risk exposure of 1 to factor Business Cycle and was not affected by other factors, the additional expected return we would obtain by investing in that asset would be the risk premium for Business Cycle.

A Popular Factor Model

Berry et al. [1988] have identified five risk factors that they argue are robust over time and explain as much as possible of the variation in stock returns. First, we describe the factors, and then we show how one might use them.

The Risk Factors   The risk factors are as follows (notice that the first two are the ones used in Example 7.27):

1. F_1: Business Cycle

   The time slice is one month. The value $f_1(t)$ of F_1 is computed as follows: The expected value of a business index is computed at both the beginning and end of month t. Then $f_1(t)$ is the difference between the end-of-month value and the beginning-of-month value. A positive value of $f_1(t)$ means that the expected growth rate of the economy has increased. Firms such as retail stores that do well when business activity increases would have relatively high risk exposures to this risk, whereas utility companies, for example, would have little exposure to business cycle risk.

2. F_2: Inflation

   The value $f_2(t)$ of F_2 is the difference between the actual inflation observed at the end of month t and the inflation predicted at the beginning of month t. Most stocks have a negative risk premium for inflation. That is, if inflation surprise is positive ($f_2(t) > 0$), then most stocks' returns are less. Industries whose products are luxuries, such as hotel chains and toy manufacturers, are most sensitive to inflation risk, whereas companies whose products are necessities, such as foods, tires, and shoes, are less sensitive to inflation risk. Companies that have a large amount of assets, such as real estate or oil, may have positive risk exposures to inflation.

3. F_3: Investor Confidence

The value $f_2(t)$ of F_2 is the difference between the end-of-month rate of return on relatively risky corporate bonds and the end-of-month rate of return on government bonds, adjusted so that the mean of the difference is 0 over a long historical sample period. When this difference is relatively high, it indicates that investor confidence has increased because it shows that investors have preferred buying the riskier corporate bonds more than they do on the average. Most stocks do better when confidence is higher. So most stocks have positive risk exposures to "investor confidence."

4. F_4: Time Horizon

The value $f_4(t)$ of F_4 is the difference between the end-of-month return on 20-year government bonds and 30-day Treasury Bills, adjusted so that the mean of the difference is 0 over a long historical sample period. When this difference is relatively high, it means that the price of long-term bonds has increased relative to the price of the short-term Treasury Bill. This means that investors require lower compensation for holding an investment with a longer time to payout. **Growth stocks** are stocks in companies that may not be producing well now, but the expectation is that they will produce more (grow) in the future. **Income stocks** are stocks in companies that pay good dividends, but are not expected to grow much in the future. Growth stocks tend to have higher risk exposures to "time horizon."

5. F_5: Market Timing

Market timing risk is calculated to be that part of the S&P 500 return rate that is not explained by the first four risk factors and an error term. That is, if $r_M(t)$ is the rate of return of the S&P 500 at the end of month t, then $f_5(t)$ satisfies the following equality:

$$r_M(t) - \bar{r}_M(t) = b_{M1}f_1(t) + b_{M2}f_2(t) + b_{M3}f_3(t) + b_{M4}f_4(t) + 1 \times f_5(t).$$

Note that $b_{M5} = 1$ since the S&P 500 has a risk exposure of 1 to itself.

Burmeister and colleagues have developed a commercially available software package called BIRR® Risks and Return Analyzer® . They use this analyzer to reestimate every month the values of the five risk premiums and each stock's exposures to these premiums. The following table shows some values from the April 1992 analysis, which is based on monthly data through the end of March 1992:

Risk Factor	Risk Premium P_i (per year)	S&P 500 Risk Exposure	Reebok Risk Exposure
Business Cycle	1.49	1.71	4.59
Inflation	−4.32	−.37	−.48
Confidence	2.59	.27	.73
Time Horizon	−.66	.56	.77
Market Timing	3.61	1.00	1.50

The third column shows the risk exposure for the S&P 500, while the fourth column shows the risk exposure for Reebok International Ltd., a company involved in manufacturing athletic apparel. Notice how much greater Reebok's exposure to business cycle risk is compared to that of the S&P 500. Notice further that both have negative exposure to inflation.

Let's compute the expected annual return rates using Equality 7.19. In April 1992, the annual return rate for 30-day Treasury Bills (which is used as the risk-free rate) was .05. If we let $\bar{r}_M(t)$ be the expected annual return rate of the S&P 500 and $\bar{r}_R(t)$ be the expected annual return rate of Reebok, we then have

$$
\begin{aligned}
\bar{r}_M(t) &= r_f(t) + b_{M1}P_1(t) + b_{M2}P_2(t) + b_{M3}P_3(t) + b_{M4}P_4(t) + b_{M5}P_5(t) \\
&= .05 + (1.71)(1.49) + (-.37)(-4.32) + (.27)(2.59) \\
&\quad + (.56)(-.66) + (1.00)(3.61) \\
&= .1309,
\end{aligned}
$$

and

$$
\begin{aligned}
\bar{r}_R(t) &= r_f(t) + b_{R1}P_1(t) + b_{R2}P_2(t) + b_{R3}P_3(t) + b_{R4}P_4(t) + b_{R5}P_5(t) \\
&= .05 + (4.59)(1.49) + (-.48)(-4.32) + (.73)(2.59) \\
&\quad + (.77)(-.66) + (1.50)(3.61) \\
&= .2071.
\end{aligned}
$$

Using APT Risk Analysis   Burmeister et al. [1994] suggests several ways an investment analyst can use the results of the risk analysis just described. We mention only a couple of them here.

1. **Attribution of Return:** It is important for a mutual fund manager to compare the risk exposure profile of the managed fund to that of an appropriate benchmark. For example, suppose Joe manages a small-capitalization mutual fund. Such a fund invests in small companies. Joe can determine the overall risk exposure for his fund to that of an index of small-capitalization firms. If the risk exposures are the same, and Joe's fund outperforms the index, then we can attribute his superior performance to his stock selections. That is, he successfully chose stocks that returned more than would be expected based on the risk undertaken. On

the other hand, if the risk exposures of his fund were different than that of the index, then the superior performance could have been due to the fact that some risk factor had a greater than expected value.

2. **Betting on a Factor:** A manager may possess special knowledge about the economy. For example, suppose Nancy has knowledge which leads her to believe the economy will recover from a recession faster than is commonly believed. If she is correct, the value of business cycle risk will be positive. That is, $f_1(t)$ will be greater than 0. This means stocks which have greater exposure to business cycle risk (i.e., b_{i1} is large) will perform better than expected. So Nancy can make a factor bet on business cycle and change her portfolio to include more stocks with large values of b_{i1}.

7.2.4 Equity Valuation Models

The efficient market hypothesis entails that information about assets is rapidly disseminated until equilibrium is restored. Arbitrage pricing theory entails that if a price discrepancy arises, investors will quickly rush to buy or sell, thereby eliminating that discrepancy. Both these models assume that investors are constantly analyzing the market and modify their portfolios accordingly. That is, it is the ongoing search for mispriced assets that restores equilibrium or destroys arbitrage opportunities. Next, we discuss valuation models that analysts use to discover mispriced assets. A **valuation model** represents the current and prospective profitability of a company. Using such a model, an analyst attempts to determine the fair market value of the stock in the company. This is called a **fundamental analysis**. We discuss valuation models and fundamental analysis next. First, we introduce a new notion of return, namely the required rate of return of a stock.

Required Rate of Return

Recall that the beta of an asset was defined in Section 7.2.2. One common use of beta is to determine the required rate of return of a stock. We develop this concept next. Let β_i be the beta of a given stock. Consider portfolio A consisting only of the market portfolio and the risk-free asset, whose weight w in the market portfolio is β_i. Then

$$
\begin{aligned}
\bar{r}_A &= (1-w)r_f + w\bar{r}_M \\
&= r_f + w(\bar{r}_M - r_f) \\
&= r_f + \beta_i(\bar{r}_M - r_f)
\end{aligned}
$$

and

$$
\begin{aligned}
\sigma_A^2 &= Var((1-w)r_f + wr_M) \\
&= w^2 \sigma_M^2 \\
&= \beta_i^2 \sigma_M^2.
\end{aligned}
$$

Now, due to Equality 7.10,

$$\begin{aligned} \bar{r}_i &= r_f + \beta_i\left(\bar{r}_M - r_f\right) + \bar{\epsilon}_i \\ &= \bar{r}_A + \bar{\epsilon}_i \end{aligned}$$

and due to Equality 7.9,

$$\begin{aligned} \sigma_i^2 &= \beta_i^2 \sigma_M^2 + Var(\epsilon_i) \\ &= \sigma_A^2 + Var(\epsilon_i). \end{aligned}$$

Notice that the return of x_i has at least as large a variance as the return of A, and its expected value differs by $\bar{\epsilon}_i$. Now CAPM entails that $\bar{\epsilon}_i = 0$. However, regardless of whether CAPM is correct, if $\bar{\epsilon}_i \leq 0$ most investors would not invest in x_i ahead of A.[2] For this reason,

$$k_i \equiv r_f + \beta_i\left(\bar{r}_M - r_f\right) \tag{7.21}$$

is called the **required rate of return** of asset x_i. Often, we omit the subscript and just write k.

Example 7.28 *Suppose stock XYZ has a beta equal to 1.5, $\bar{r}_M = .10$ annually, and $r_f = .03$ annually. Then according to Equality 7.21, the required rate of return for XYZ is*

$$\begin{aligned} k_{XYZ} &= r_f + \beta_{XYZ}(\bar{r}_M - r_f) \\ &= .03 + 1.5(.10 - .03) \\ &= .135. \end{aligned}$$

Let's review the three notions of return that we have introduced. Recall that Equality 7.3 states that if P_1 is the amount invested today in an asset, P_2 is the amount received in the future, and D_1 is the total of the dividends paid during the holding period, then the holding period return (HPR) is given by

$$HPR = \frac{D_1 + P_1 - P_0}{P_0}.$$

If the holding period is one year, we denote the HPR by r. The holding period return is the actual return realized by the asset at the end of the holding period. We know precisely what it is when the holding period is over. For example, in Example 7.9 it was .2. When the holding period begins, we have an expectation concerning the HPR realized by the end of the holding period. This is the expected value of the return and is denoted $\bar{r}$ if the holding period is one year. We have that

$$\bar{r} = \frac{\bar{D}_1 + \bar{P}_1 - P_0}{P_0}.$$

---

[2]However, it is not absolutely cogent that an investor should prefer A. A given investor might want to give up some expected return in exchange for a greater variance in order to have a chance at a larger return. For example, if the fixed rate is .04, and an investment has an expected return of .03 and standard deviation of .02, a given investor might prefer the investment owing to the chance at returns greater than .04.

The required rate of return k of an asset is the expected return rate computed using the CAPM formula. This is the expected return rate the stock should have based on its risk. If we simply use CAPM to determine the expected return, then $\bar{r} = k$. However, we may use some other analysis to compute $\bar{r}$. We show such an analysis in the next subsection. If according to this other analysis $\bar{r}$ is less than k, we would not consider the stock a good buy. On the other hand, if it is greater than k, we might consider it a good buy.

Example 7.29 *In Example 7.28 we showed that the required rate of return k_{XYZ} for company XYZ is .135. Now suppose that currently the stock's price P_0 is \$40 a share, our expected price $\bar{P}_1$ in one year is \$44 a share, and our expected dividend $\bar{D}_1$ to be paid at the end of the coming year is \$2 per share. Then*

$$
\begin{aligned}
\bar{r} &= \frac{\bar{D}_1 + \bar{P}_1 - P_0}{P_0} \\
&= \frac{\$2 + \$44 - \$40}{\$40} \\
&= .15.
\end{aligned}
$$

Based on this analysis, XYZ might be a good buy since $\bar{r} > k_{XYZ}$.

Intrinsic Value

In Example 7.29 you may wonder where we obtained $\bar{P}_1$ and $\bar{D}_1$. We discuss this shortly. First, let's look at the problem a bit differently. Recall from Section 7.1.2 that we can compute the net present value (NPV) of an amount of money received in one year by dividing that amount by $1 + r_f$, where r_f is the one-year risk-free rate. For example, if some deal will definitely pay \$105 one year from today and $r_f = .05$, then the NPV of the deal is given by

$$
NPV = \frac{\$105}{1 + .05} = \$100.
$$

We consider \$100 the fair amount to pay for the deal because we could also realize \$105 in one year by investing \$100 at the risk-free rate. This notion can be extended to an asset. If we expect to realize $\bar{D}_1 + \bar{P}_1$ in one year by purchasing one share of the asset, then we can think of the present value of this realization as being the number obtained by dividing $\bar{D}_1 + \bar{P}_1$ by $1 + k$, where k is the required rate of return of the asset. We call this present value the **intrinsic value** V_0 of a share of the asset. That is,

$$
V_0 = \frac{\bar{D}_1 + \bar{P}_1}{1 + k}.
$$

Example 7.30 *Suppose again that $\bar{P}_1 = \$44$, $\bar{D}_1 = \$2$, and $k = .135$ for company XYZ. Then*

$$
V_0 = \frac{\bar{D}_1 + \bar{P}_1}{1 + k} = \frac{\$2 + \$44}{1 + .135} = \$40.53.
$$

The idea with intrinsic value is that we could invest $40.53 today in a portfolio A, which is a combination of the market portfolio and the risk-free asset, such that A has the same beta as XYZ and therefore has expected return k (see Section 7.2.4). We can expect that investment to be worth $40.53(1 + .135) = \$46$ in one year. Since one share of XYZ and portfolio A have the same expected return, and owning one share of XYZ has at least as much risk as owning A, most investors would not prefer one share of XYZ. Therefore, the value today of one share of XYZ should be no greater than the value today of portfolio A. So technically the intrinsic value of one share of XYZ is no more than $40.63. However, this value is still called the intrinsic value rather than an upper bound on the intrinsic value. If one share of XYZ is currently selling for $42, then XYZ is not a good buy because it is selling for more than its intrinsic value. However, if it is selling for $40, it may be considered a good buy.

Dividend Discount Models

In the following we assume for simplicity that all dividends are paid at the end of the year.

General Dividend Discount Model   We saw that the intrinsic value of one share of an asset is given by

$$V_0 = \frac{\bar{D}_1 + \bar{P}_1}{1 + k}.$$

Using the same formula, the intrinsic value V_1 in one year is given by

$$V_1 = \frac{\bar{D}_2 + \bar{P}_2}{1 + k},$$

where $\bar{P}_2$ is the expected price of the stock in two years and $\bar{D}_2$ is the expected dividend paid in two years. Now assume that we obtain our value of $\bar{P}_1$ from what we compute to be the intrinsic value in one year. That is, assume

$$\bar{P}_1 = V_1.$$

We then have

$$
\begin{aligned}
V_0 &= \frac{\bar{D}_1 + V_1}{1 + k} \\
&= \frac{\bar{D}_1 + \dfrac{\bar{D}_2 + \bar{P}_2}{1 + k}}{1 + k} \\
&= \frac{\bar{D}_1}{1 + k} + \frac{\bar{D}_2 + \bar{P}_2}{(1 + k)^2}.
\end{aligned}
$$

If we repeat this process indefinitely, we obtain

$$V_0 = \frac{\bar{D}_1}{1 + k} + \frac{\bar{D}_2}{(1 + k)^2} + \frac{\bar{D}_3}{(1 + k)^3} + \cdots + . \tag{7.22}$$

This formula is called the **Dividend Discount Model (DDM)** for asset value. It may seem odd to you that expected price does not enter the formula. Price is implicitly included in the formula because it entails that the price the asset is worth in the future depends on the expected dividends at that time in the future.

Constant Growth Dividend Discount Model A problem with the general DDM is that we must somehow know the expected dividend return indefinitely into the future. A simplifying assumption, which is often used, is that the expected yearly growth rate of the expected value of the dividend is a constant g. (Note that g is the expected value of the growth. However, it is standard to show only the dividend as an expected value.) That is, $\bar{D}_2 = \bar{D}_1(1 + g)$, $\bar{D}_3 = \bar{D}_2(1 + g) = \bar{D}_1(1 + g)^2$, etc. With this assumption, Equality 7.22 is as follows:

$$
\begin{aligned}
V_0 &= \frac{\bar{D}_1}{1 + k} + \frac{\bar{D}_1(1 + g)}{(1 + k)^2} + \frac{\bar{D}_1(1 + g)^2}{(1 + k)^3} + \cdots + \\
&= \frac{\bar{D}_1}{1 + g}\left[\left(\frac{1 + g}{1 + k}\right) + \left(\frac{1 + g}{1 + k}\right)^2 + \left(\frac{1 + g}{1 + k}\right)^3 + \cdots +\right] \\
&= \frac{\bar{D}_1}{1 + g}\left[\frac{1}{1 - (1 + g)/(1 + k)} - 1\right] \\
&= \frac{\bar{D}_1}{1 + g}\left[\frac{1 + g}{k - g}\right] \\
&= \frac{\bar{D}_1}{k - g}.
\end{aligned}
$$

The third equality above is a standard result in calculus. It holds only if $g < k$. So we have established that if we have a constant **expected dividend growth rate** g and $g < k$, then

$$
V_0 = \frac{\bar{D}_1}{k - g}. \tag{7.23}
$$

Equality 7.23 is called the **Constant-Growth Dividend Discount Model**.

Example 7.31 *As in Example 7.30, suppose $\bar{D}_1 = \$2$ and $k = .135$ for company XYZ. However, now suppose the constant expected dividend growth rate $g = .09$ instead of assuming $\bar{P}_1 = \$44$. Then owing to Equality 7.23,*

$$
V_0 = \frac{\bar{D}_1}{k - g} = \frac{2}{.135 - .09} = \$44.44.
$$

Note that in the previous example we needed to know the constant-growth rate g to determine the intrinsic value, whereas in Example 7.30 we needed to know the expected value $\bar{P}_1$ of the price of a share in one year. Based on previous performance, it may be more accurate to assume constant growth and estimate g than to try to estimate $\bar{P}_1$.

The constant-growth DDM is valid only if $g < k$. If g were equal to or greater than k, it would say the intrinsic value of a share was infinite. The explanation is that a growth rate this high is not sustainable, which means the model is not appropriate. Instead, we should use a multi-stage DDM, which is discussed in [Bodie et al., 2004].

Investment Opportunities

Next, we discuss how companies can increase their value, and thereby the value of their stock, by investing some of their earnings in growth opportunities rather than paying them all out as dividends.

First, we need a number of definitions. The **earnings** of a company are the net income or profit available to the shareholders. The **retained earnings** are any earnings that are not paid out in dividends. The **equity** of a company is the value of the company minus any mortgages or other debts it may have. The **share capital** in a company is the portion of a company's equity that is obtained from issuing shares in the company. The **treasury shares** in the company are any shares the company bought back from the investors. The **shareholder's equity** in a company is the share capital plus retained earnings minus the value of the treasury shares. That is, shareholder equity is the current value of the money the company obtained by going public. **Return on equity (ROE)** is the ratio of the earnings for the trailing (past) 12 months to the shareholder's equity. That is, ROE is the rate of return of the shareholder's equity. The ratio of the total annual earnings of a company to the number of outstanding shares of stock in the company is called the **earnings per share (EPS)** of the company.

Companies realize earnings continually throughout the year, and they usually pay dividends quarterly. However, for the purpose of simplifying calculations in the following examples, we assume earnings are realized at the end of the year and, at that same time, some of the earnings are paid as dividends and some are reinvested.

Example 7.32 *Suppose the current shareholder's equity S_0 in company XYZ is \$100,000,000, and the earnings E_0 in the trailing 12 months were \$15,000,000. Then*

$$ROE = \frac{E_0}{S_0} = \frac{\$15,000,000}{\$100,000,000} = .15.$$

Furthermore, if there are 3,000,000 shares outstanding,

$$EPS_0 = \frac{\$15,000,000}{3,000,000} = \$5.$$

Suppose further that company XYZ pays its entire earnings out as dividends. Then the dividend paid this year would be

$$D_0 = EPS_0 = \$5.$$

Since none of the earnings are reinvested, there is no growth in the coming year. Assume that we expect the earnings for the coming year to be the same and that the entire earnings will again be paid out as dividends. Then $\bar{D}_1 = D_0 = EPS_0 = \$5$, and if the required rate of return k for XYZ is .125, according to Equality 7.23

$$V_0 = \frac{\bar{D}_1}{k-g} = \frac{EPS_0}{k-g} = \frac{\$5(1+0)}{.125-0} = \$40.$$

If this procedure continues indefinitely (i.e., the same earnings are realized and all earnings are paid out in dividends) the intrinsic value will stay fixed at $40, which means the stock should never increase in value.

Instead of paying out all its earnings as dividends, company XYZ could reinvest some of its earnings. The fraction of earnings that are reinvested is called the **plowback ratio** b. The following example shows what could happen if XYZ reinvests some of its earnings.

Example 7.33 *Assume the same situation as in the previous example. Suppose, however, that company XYZ decides to use a plowback ratio $b = .6$ on the past year's earnings and that any reinvested money can be expected to have the same return on equity as realized in the past year. Then the shareholder's equity S_1 for the coming year is given by*

$$
\begin{aligned}
S_1 &= S_0 + b \times E_0 \\
&= \$100,000,000 + .6 \times \$15,000,000 \\
&= \$109,000,000.
\end{aligned}
$$

If we expect the same ROE in the coming year as in the previous year, then

$$
\begin{aligned}
\bar{E}_1 &= \overline{ROE} \times S_1 \\
&= .15 \times \$109,000,000 \\
& \$16,350,000.
\end{aligned}
$$

We then have an expected growth g in earnings of

$$g = \frac{\bar{E}_1 - E_0}{E_0} = \frac{\$16,350,000 - \$15,000,000}{\$15,000,000} = .09.$$

We have used g to represent the growth in earnings, because we can use this growth in earnings to enable dividends to grow at the same rate. Since we plowed back b fraction of our earnings, the fraction of EPS_0 available for dividends is only $1 - b$. So the dividend D_0 paid this year is only $EPS_0(1 - b)$. We therefore have

$$
\begin{aligned}
V_0 = \frac{\bar{D}_1}{k-g} &= \frac{D_0(1+g)}{k-g} \\
&= \frac{EPS_0(1-b)(1+g)}{k-g} \\
&= \frac{\$5(1-.6)(1+.09)}{.125-.09} = \$62.29.
\end{aligned}
$$

By reinvesting some of their earnings, XYZ has significantly increased the intrinsic value of its stock.

We can generalize the results in the previous example to obtain a simple expression for g in terms of ROE and b. Assuming $\overline{ROE} = ROE$, we have

$$
\begin{aligned}
g = \frac{\bar{E}_1 - E_0}{E_0} &= \frac{\overline{ROE} \times S_1 - E_0}{E_0} \\
&= \frac{ROE\,(S_0 + b \times E_0) - ROE \times S_0}{E_0} \\
&= \frac{ROE \times b \times E_0}{E_0} = ROE \times b. \tag{7.24}
\end{aligned}
$$

Company XYZ did well to reinvest its earnings because its expected value of ROE exceeds its required rate of return k. This means XYZ could realize a greater expected return by reinvesting than its risk requires. This is often true for growth companies such as those in the computer industry in the 1990s. At that time there were breakthroughs in computer hardware, which enabled new and varied applications of computer software. As a result, there were many opportunities for companies in this industry to grow. This was especially true for newer companies. On the other hand, established companies in older industries do not have as many growth opportunities. An example would be the automotive industry today. Such companies have progressed beyond their growth phase and, therefore, no longer benefit sufficiently by reinvesting their earnings. Consider the next example.

Example 7.34 *Suppose the current shareholder's equity S_0 in company ABC is $\$100,000,000$, and the earnings E_0 in the trailing 12 months were $\$10,000,000$. Then*

$$
ROE = \frac{E_0}{S_0} = \frac{\$10,000,000}{\$100,000,000} = .10.
$$

Suppose further that company ABC pays its entire earnings out as dividends and that there are 2,000,000 shares outstanding. Then the dividend paid this year would be

$$
D_0 = EPS_0 = \frac{\$10,000,000}{2,000,000} = \$5.
$$

If we assume that we expect the earnings for the coming year to be the same and that the entire earnings will again be paid out as dividends, then if the required rate of return k for ABC is .125, we have

$$
\begin{aligned}
V_0 = \frac{\bar{D}_1}{k - g} &= \frac{D_0(1 + g)}{k - g} \\
&= \frac{EPS_0(1 + g)}{k - g} \\
&= \frac{\$5}{.125 - 0} = \$40.
\end{aligned}
$$

Now if ABC instead uses a plowback ratio $b = .6$, the dividend paid this year will only be $EPS_0(1 - b)$. Furthermore, owing to Equality 7.24,

$$
g = ROE \times b = .10 \times .6 = .06.
$$

We then have

$$V_0 = \frac{\bar{D}_1}{k-g} = \frac{EPS_0(1-b)(1+g)}{k-g}$$

$$= \frac{\$5(1-.6)(1+.06)}{.125-.06} = \$32.6.$$

ABC has reduced the intrinsic value of its stock by reinvesting. The problem is that by reinvesting ABC is realizing less expected return for its shareholders than its risk requires. That is, ROE is too small compared to k.

In general, a company can only increase the intrinsic value of its stock by reinvesting if ROE is not too much smaller than k. Stated precisely, the intrinsic value can be increased by reinvesting if and only if $ROE > k/(k+1)$. You are asked to show this in the exercises.

Consider again Examples 7.32 and 7.33. When company XYZ decided to distribute all its earning for the coming year as dividends, we had

$$V_0 = \frac{EPS_0}{k} = \frac{\$5}{.125} = \$40.$$

On the other hand, when it decided to reinvest part of its earnings with a plowback ratio of $b = .6$, we had

$$V_0 = \frac{EPS_0(1-b)(1+ROE \times b)}{k-ROE \times b}$$

$$= \frac{\$5(1-.6)(1+.15 \times .6)}{.125-.15 \times .6} = \$62.29.$$

The difference between these two values is called the present value of growth opportunities (PVGO). That is,

$$PVGO = \frac{EPS_0(1-b)(1+ROE \times b)}{k-ROE \times b} - \frac{EPS_0}{k}. \tag{7.25}$$

In this example, $PVGO = \$62.29 - \$40 = \$22.29$. PVGO is the amount by which a company can increase the value of its stock by reinvesting.

P/E Ratio

The ratio of the price per share P of a stock to the EPS of the stock is called the **P/E ratio** for the stock. In practice, we compute the trailing P/E ratio and the forward P/E ratio. The **trailing P/E ratio** is the ratio of the stock's latest closing price to the EPS based on the company's last reported 12 months of earnings. We simply denote this ratio as P/E. The **forward P/E ratio** is the ratio of the stock's latest closing price to the EPS as provided by Zach's Investment Research. In the first three quarters of the fiscal year they use earnings estimates for the current year. In the fourth quarter they use earnings estimates for the next fiscal year. The estimate is the mean estimate derived from all polled estimates of Wall Street analysts. We denote the forward P/E ratio as $(P/E)_F$.

Example 7.35 *Suppose the last reported 12-month earnings of company XYZ is $10,000,000, there are 2,000,000 shares of stock outstanding, and today's closing price per share is $50. Then*

$$EPS = \frac{\$10,000,000}{2,000,000} = \$5,$$

and the trailing P/E ratio is given by

$$P/E = \frac{P}{EPS} = \frac{\$50}{\$5} = 10.$$

Suppose it is August, which is in the third quarter, and the average estimate of earnings for the current fiscal year is $12,000,000. Then

$$EPS_F = \frac{\$12,000,000}{2,000,000} = \$6,$$

and the forward P/E ratio is given by

$$(P/E)_F = \frac{P}{EPS_F} = \frac{\$50}{\$6} = 8.33.$$

Equality 7.25 says that

$$PVGO = V_0 - \frac{EPS_0}{k}, \tag{7.26}$$

where

$$V_0 = \frac{EPS_0(1-b)(1+ROE \times b)}{k - ROE \times b} \tag{7.27}$$

is the intrinsic value if the company takes advantage of its growth opportunities. Assuming it does this, and further assuming the price P_0 is equal to the intrinsic value V_0, a rearrangement of Equality 7.26 yields

$$\frac{P_0}{EPS_0} = \frac{1}{k}\left(1 + \frac{PVGO}{EPS_0/k}\right).$$

Now P_0/EPS_0 is approximately equal to the trailing P/E ratio. So we have

$$P/E \approx \frac{1}{k}\left(1 + \frac{PVGO}{EPS_0/k}\right). \tag{7.28}$$

Furthermore, since $\overline{EPS}_1 = EPS_0(1 + ROE \times b)$, a rearrangement of Equality 7.27 yields (assuming again P_0 is equal to V_0)

$$\frac{P_0}{\overline{EPS}_1} = \frac{(1-b)}{k - ROE \times b}.$$

Since P_0/EPS_1 is approximately equal to the forward P/E ratio, we have

$$(P/E)_F \approx \frac{(1-b)}{k - ROE \times b} \tag{7.29}$$
$$= \frac{(1-b)}{k - g}.$$

The last equality is due to Equality 7.24.

Equalities 7.28 and 7.29 show what the trailing and forward P/E ratios should be based on intrinsic value, not their values computed according to the definitions given at the beginning of this subsection. Therefore, they can be used to estimate whether a security is mispriced relative to the values obtained using those definitions. The next example illustrates this.

Example 7.36 *Suppose again that company XYZ has $k = .125$, $ROE = .15$, and $b = .6$. Then according to Equality 7.29,*

$$(P/E)_F \approx \frac{(1-b)}{k - ROE \times b} = \frac{(1 - .6)}{.125 - .15 \times .6} = 11.429.$$

This is the forward P/E ratio according to intrinsic value. If the forward P/E ratio obtained by taking the ratio of the stock's latest closing price to the EPS as provided by Zach's Investment Research is only 8, based on this analysis we would consider XYZ a good buy.

Notice in the previous example that we can make the P/E ratio arbitrarily large by taking b arbitrarily close to $.125/.15 \approx .833$. (A value of $b = .125/.15$ makes the denominator 0.) You may ask why company XYZ would not do this. Recall that Equalities 7.28 and 7.29 were developed starting with Equality 7.23, which assumes the constant growth dividend model. In practice this model would always be only an approximation, as no company can continue to grow at a constant rate. As mentioned earlier, newer companies in new areas can often achieve significant growth (e.g., Google at the time of this writing), but established companies in older industries do not have as many growth opportunities. So we cannot assume that we can choose a plowback ratio which yields an arbitrarily large P/E ratio. If a company plows back too much, it might be wasted since it may not have that capacity for growth. So we must assume that the company will only plow back as much as can reasonably be used towards growth.

Notice that Equality 7.29 says that the P/E ratio should increase with growth $g = ROE \times b$. This has led analysts to develop rules of thumb for when a stock is mispriced. For example, Peter Lynch [Lynch and Rothchild, 2000] defined the PEG ratio by

$$PEG = \frac{(P/E)_F}{g \times 100}.$$

He says "The P/E ratio of any company that's fairly priced will equal its growth rate.....if the P/E ratio is less than the growth rate, you have found yourself a bargain." That is, if the $PEG < 1$ the stock may be a good buy.

Example 7.37 *Suppose we have the situation described in Example 7.36, and XYZ's actual current P/E ratio is 8. Then*

$$PEG = \frac{(P/E)_F}{g \times 100} = \frac{8}{.15 \times .6 \times 100} = .89.$$

Based on Lynch's rule of thumb, XYZ may be a good buy.

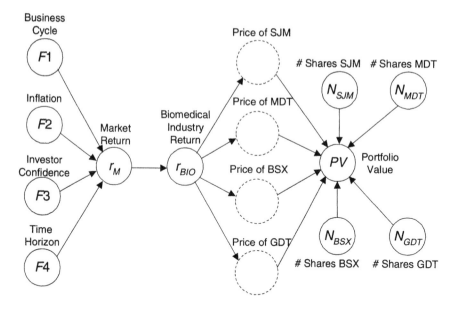

Figure 7.9: A Bayesian network for portfolio risk analysis. Each dotted circle represents a set of nodes.

Benjamin Graham [Graham and Zweig, 2003] has developed seven rules for discovering undervalued stocks.

7.3   A Bayesian Network Portfolio Risk Analyzer★

Recall that arbitrage pricing theory (APT) entails that if a price discrepancy arises, investors will quickly rush to buy or sell, thereby eliminating that discrepancy. This model assumes that investors are constantly analyzing the market and modifying their portfolios accordingly, and they use valuation models to discover mispriced assets. Ordinarily, application of the models proceed independently. That is, we use either APT (e.g., the factor model described in [Berry et al., 1988]) to determine expected returns of securities or a valuation model to determine the intrinsic value of a security. Demirer et al. [2007] developed a portfolio risk analyzer using a Bayesian network that combines the two approaches. Specifically, factors are used to determine a probability distribution of the return of the market, and then that information is used in a valuation model to determine a probability distribution of the intrinsic value of the security. We discuss this analyzer next. After showing the structure of the Bayesian network in the analyzer, we discuss the parameters.

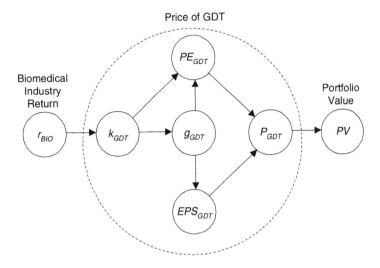

Figure 7.10: The set of nodes called "Price of GDT" in Figure 7.9 is enclosed in the dotted circle.

7.3.1 Network Structure

The structure of the Bayesian network for portfolio risk analysis appears in Figure 7.9. In general, the network could contain any stocks. We show it for four particular stocks in the biomedical industry. Those stocks are Guidant Corporation (GDT), St. John's Medical (STJ), Boston Scientific (BSX), and Medtronic, Inc. (MDT). The first four risk factors, developed in [Berry et al., 1988] (see Section 7.2.3), appear on the far left side of the network. The values of these nodes are the actual values of the risk factors, not the risk premiums. For example, the value $f_2(t)$ of F_2 is the difference between the actual inflation observed at the end of time period t and the inflation predicted at the beginning of time period t. It is assumed that the risk factors are independent and that the rate of return of the market realized at the end of time period t depends on the risk factors' values at the end of the time period. So there is an edge from each risk factor to the market return. Next, there is an edge from the market return to the return of the biomedical industry. It is assumed that the return of a particular industry is independent of the risk factors given the return of the market. There is an edge from the return of the biomedical industry to the price of each stock. This price is actually the intrinsic value of the stock at the beginning of the time period, and it depends on the return of the biomedical industry for the time period. The node for the price of a stock in Figure 7.9 represents the set of nodes appearing in Figure 7.10.

Figure 7.10 shows the nodes for GDT. There is a corresponding set of nodes for each of the four stocks. Our discussion concerning GDT pertains to each of them. We discuss, in turn, each node and why the node depends on its parents.

1. k_{GDT}: This is the required rate of return of the stock. An innovative feature of this model is that k is a random variable rather than an expected value. Recall from Section 7.2.4 that in classical valuation models

$$k_i = r_f + \beta_i \left(\bar{r}_M - r_f \right).$$

However, the entire theory could have been developed using r_M instead of $\bar{r}_M$. The current model does something that is not ordinarily done in valuation models; namely, it obtains a probability distribution of the market return rather than just an expected value and variance. Owing to this feature, it also obtains a probability distribution of the required rate of return. It is assumed that this required rate is independent of the market return given the biomedical industry return. So there is an edge from r_{BIO} to k_{GDT}.

2. g_{GDT}: This is the annual growth in earnings (see Equality 7.24). In general, companies with more growth have higher required rates of returns. So there is an edge from k_{GDT} to g_{GDT}.

3. EPS_{GDT}: This is the expected earnings per share the next time it is reported. If we let EPS be the expected earnings per share for a company, EPS_0 be the most recent reported value of earnings per share, and g be the expected annual growth rate, then

$$EPS = EPS_0(1 + g).$$

So there is an edge from g_{GDT} to EPS_{GDT}.

4. PE_{GDT}: This is the forward P/E ratio, where P represents intrinsic value rather than actual price. Recall that Equality 7.29 says the forward P/E ratio $(P/E)_F$ is given by

$$(P/E)_F \approx \frac{(1 - b)}{k - g}.$$

So there are edges from k_{GDT} and g_{GDT} to PE_{GDT}.

5. P_{GDT}: This is the stock's intrinsic value. It depends deterministically on EPS_{GDT} and PE_{GDT} as follows:

$$P_{GDT} = EPS_{GDT} \times PE_{GDT}.$$

Since the price in the forward P/E ratio is the current price, P_{GDT} represents current intrinsic value. On any given day the stock will have an actual price. So there will not be a probability distribution of actual price on a given day. The variable P_{GDT} represents the possible values the price could have based on nonsystematic shock and fluctuations in risk factors. So the probability distribution of this node is the distribution of the price on each future day until new information is added to the network. We will discuss this more when describing the parameters in the network.

We return now to Figure 7.9. The node labeled PV, which represents the total value of the portfolio, depends on the prices of the stocks and how many shares of each stock we own. It is simply the sum over all the stocks of the price of the stock times the number of shares of the stock. Like the prices of the individual stocks, the portfolio will have an actual value on a given day. This variable represents the values the portfolio could have until new information is added to the network. Note the importance of this information to the investor. The probability distribution of this node informs the investor of the risk inherent in the portfolio.

7.3.2 Network Parameters

Next, we discuss how the parameter values for each node in the network were determined.

1. $F1$, $F2$, $F3$, $F4$: The values of each of the risk factors were obtained for each month from 1971 to 1999. These values were discretized into three ranges: *low*, *medium*, and *high*. The probability of each of these states (ranges) for a given factor was then set equal to the fraction of months the factor's value fell in the range.

2. r_M: Using again monthly data from 1971 to 1999, the values of the market return were also discretized into the three ranges: *low*, *medium*, and *high*. The conditional probability of each of these states (ranges) given a combination of states of the four factors was then computed to be the fraction of months the return fell in the range taken over all months the four factors had this combination of states.

3. r_{BIO}: Again using monthly data from 1971 to 1999, the values of the biomedical industry return were also discretized into the three ranges: *low*, *medium*, and *high*. The conditional probability of each of these states (ranges) given a state of the market return was then computed to be the fraction of months the biomedical return fell in the range taken over all months the market return was in the state.

4. k_{GDT}, k_{BSX}, k_{MDT}, k_{SJM}: Using monthly data from 1992 to 1999, each stock's return was discretized into three ranges representing low, medium, and high returns. However, now the midpoint of each range was used as the state of the variable. For example, the set of states of k_{GDT} turned out to be

$$\{low = .08, \ medium = .10, \ high = .12\}.$$

Numeric values were used because numbers were needed in the calculation of the P/E ratios. The conditional probability of each of these values given a state of the biomedical industry return was then computed to be the fraction of months the stock's return fell in the range taken over all months the biomedical industry return was in the state.

5. g_{GDT}, g_{BSX}, g_{MDT}, g_{SJM}:   For each stock, a number of analysts' first call estimates of next year's earnings per share were obtained. Based on these estimates and the most recently reported earnings per share (EPS_0), low, medium, and high growth rates for the stock were determined. For example, the set of states of g_{GDT} was determined to be

$$\{low = -.01, \ medium = .04, \ high = .07\}.$$

Conditional probability distributions given values of required return rates were then assigned. The same conditional distributions were used for all stocks. They are the following:

	g	low	medium	high
k			$P(g\|k)$	
low		.75	.25	.00
medium		.10	.80	.10
high		.00	.25	.75

6. EPS_{GDT}, EPS_{BSX}, EPS_{MDT}, EPS_{SJM}: These are deterministic nodes. For each stock, the node's value is deterministically computed from the growth rate and the most recently reported earnings per share. For example, if $EPS_{0,GDT}$ is the most recently reported earnings per share for GDT,

$$EPS_{GDT} = EPS_{0,GDT}(1 + g_{GDT}).$$

7. PE_{GDT}, PE_{BSX}, PE_{MDT}, PE_{SJM}: There are nine combinations of values of the parents of each of these nodes. Each node was assigned numeric low, medium, and high values based, respectively, on only the following three combinations of values of the parents: (k_high, g_low), (k_high, g_medium), and (k_low, g_medium). The numeric values were computed using the formula

$$(P/E)_F \approx \frac{1}{k - g}.$$

That is, they assume the plowback ratio b is sufficiently small so that it can be neglected. For example, we have these numeric values of PE_{GDT}:

$$low: \quad \frac{1}{k\_high - g\_low} = \frac{1}{.12 - .(-01)} = 8.33$$

$$medium: \quad \frac{1}{k\_high - g\_medium} = \frac{1}{.12 - (.04)} = 12.5$$

$$high: \quad \frac{1}{k\_low - g\_medium} = \frac{1}{.08 - (.04)} = 25.$$

The conditional probability distributions of each of these nodes given each combination of values of its parents were then assigned. The same conditional distributions were used for all stocks. They are the following:

k	g	PE	low	medium	high
				$P(PE|k,g)$	
low	low		.25	.75	.00
low	medium		0	.50	.50
low	high		0	0	1.00
medium	low		.75	.25	0
medium	medium		0	1.00	0
medium	high		0	.25	.75
high	low		100	0	0
high	medium		.50	.50	0
high	high		0	.75	.25

8. P_{GDT}, P_{BSX}, P_{MDT}, P_{SJM}: These are deterministic nodes. Each is the product of the values of its parents.

9. N_{GDT}, N_{BSX}, N_{MDT}, N_{SJM}: These are the number of shares of each stock in the portfolio. They are assigned values.

10. PV: This is a deterministic node. It is the sum over all the stocks of the price of the stock times the number of shares of the stock.

This completes the discussion of the parameter values in the network. Notice that a lot of the values in the network were hard-coded rather than being inputs to the network. For example, the possible values of EPS_{GDT} were computed from $EPS_{0,GDT}$ and the possible values of g. In the following year, when these latter values change, we would have to modify the parameter values in the network. The model just described is only a research model. A final production system would hard-code as little as possible. Indeed, a user should be able to add and delete different stocks from different industries without modifying the network.

7.3.3 The Portfolio Value and Adding Evidence

Figure 7.11 shows the probability distribution of the values of the portfolio when 100 shares of each stock are in the portfolio. This figure was produced using the software package Netica. The value 0 represents the range $0 to $1000, the value 1 represents the range $1 to $2000, and so on. The expected value and variance, which appear at the bottom, are $10,300 and $4900, respectively. Investors are concerned about the **value-at-risk (VaR)** at level α, which is the number VaR_α such that

$$P(\text{Portfolio Value} < VaR_\alpha) = \alpha.$$

The Bayesian network automatically provides this information. We see that $VaR_{.033}$ is somewhere between $0 and $1000. This means the probability that the portfolio will be worth less than $1000 is no more than .033.

We stress again that this is the probability distribution of the intrinsic value of the portfolio. On any given day the portfolio has an actual value. This

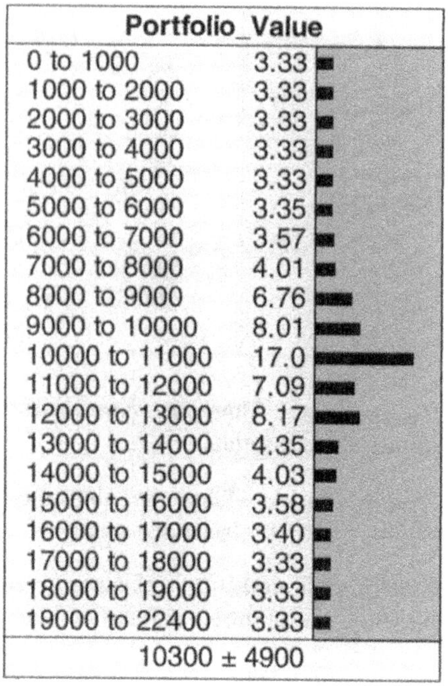

Portfolio_Value		
0 to 1000	3.33	
1000 to 2000	3.33	
2000 to 3000	3.33	
3000 to 4000	3.33	
4000 to 5000	3.33	
5000 to 6000	3.35	
6000 to 7000	3.57	
7000 to 8000	4.01	
8000 to 9000	6.76	
9000 to 10000	8.01	
10000 to 11000	17.0	
11000 to 12000	7.09	
12000 to 13000	8.17	
13000 to 14000	4.35	
14000 to 15000	4.03	
15000 to 16000	3.58	
16000 to 17000	3.40	
17000 to 18000	3.33	
18000 to 19000	3.33	
19000 to 22400	3.33	
10300 ± 4900		

Figure 7.11: Prior probability of the porfolio value.

distribution is based on years of historical data concerning the risk factors, the market return, the industry return, and the stock returns, and on the most recently reported stock earnings and analysts' estimates of the stock growth. It is therefore the distribution of the portfolio value on each future day until new information is added to the network. New information could come in the form of a parameter shift. For example, we could update the distributions of the risk factors based on additional data, new earnings per share could be announced $(EPS_{0,GDT})$, or analysts could estimate new values of the growth g. New information could also come in the form of instantiation of a node. Actually, in some cases the difference is only technical since, for example, a node could be included in the network for $EPS_{0,GDT}$. Next, we show examples of instantiating nodes in the current network, thereby changing the probability distribution.

The power of Bayesian networks is that we can add evidence to the network. In this way the portfolio manager can readily incorporate any special information that might affect the value of the portfolio. For example, if the manager feels inflation will be high, node $F2$ can be set to its high state. A more concrete example concerns a news report on July 10, 2001. Guidant Corporation (GDT) failed to win FDA panel backing for its pacemaker-like-heart-failure device. This should negatively affect earnings growth. Figure 7.12

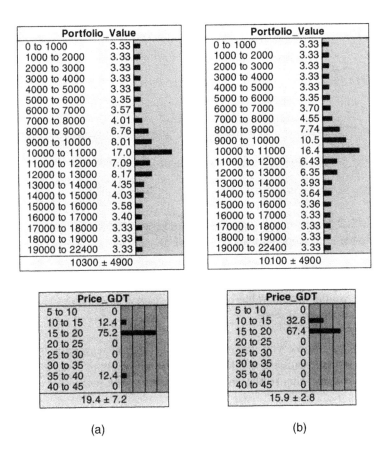

Figure 7.12: The prior probability distribution is in (a) while the posterior distribution after GDT's growth is instantiated to low appears in (b).

(a) shows the prior probability distributions of the portfolio value and the price of GDT (P_{GDT}), while Figure 7.12 (b) shows the posterior distributions when we instantiate GDT's growth (g_{GDT}) to low. Notice that the expected value of the price (intrinsic value) of GDT changed from \$19.4 to \$15.9. This is an 18.3 % drop in intrinsic value. The actual decline in the price of the stock was 13.75% by market close on July 10. However, the total decline from the time the announcement was made (12:37 PM on July 10) through the market close on July 11 was 22.98%. Given the expected decline of 18.3%, the portfolio manager could have sold GDT as soon as the announcement was made and avoided some of the loss.

EXERCISES

Section 7.1

Exercise 7.1 Suppose you deposit $2000, and the annual interest rate $r = .05$.

1. What is the value of your investment in one year?

2. What is the value of your investment in four months?

Exercise 7.2 Suppose you deposit $2000, the annual interest rate $r = .05$, and the interest is compounded yearly.

1. What is the value of your investment in 30 years?

2. What would be the value of your investment in 30 years if the interest were not compounded?

Exercise 7.3 Suppose you deposit $2000, the annual interest rate is .05, and the interest is compounded monthly. What is the value of your investment in 30 years? Compare this to the answers obtained in the previous exercise.

Exercise 7.4 Suppose you deposit $2000, the annual interest rate is .05, and the interest is compounded continuously. What is the value of your investment in 30 years? Compare this to the answers obtained in the previous two exercises.

Exercise 7.5 Suppose the current interest rate available to Clyde for the next four years is .05. Suppose further that Ralph owes Clyde $9500 and Ralph offers to pay Clyde back by giving him $2000 today and at the end of each of the next four years. Compute the net present value of this cash flow stream. Is Ralph offering Clyde a fair deal?

Exercise 7.6 Suppose the current interest rate available to Clyde for the next four years is .05, Ralph owes Clyde $9500, and Ralph offers to pay Clyde back by giving him five equal installments, one today and one at the end of each of the next four years. Compute the fair amount of the installments.

Exercise 7.7 Suppose one share of company ABC is bought today for $20, sold four months later for $25, and no dividends are paid. Compute the holding period return.

Exercise 7.8 Suppose the same situation as in the previous exercise, except a dividend of $2 per share is paid at the end of the holding period. Compute the holding period return.

Exercise 7.9 Suppose stock XYZ had the annual return rates shown in the following table:

Year	Annual Return r
2000	.05
2001	.08
2002	−.11
2003	.25
2004	−.02

Compute the expected value $\bar{r}$ of r and the standard deviation σ_r.

Exercise 7.10 Suppose stock DEF had the annual return rates shown in the following table:

Year	Annual Return r
2000	.05
2001	.06
2002	.04
2003	.07
2004	−.03

Compute the expected value $\bar{r}$ of r and the standard deviation σ_r.

Exercise 7.11 Assuming the return rates in the previous two examples are approximately normally distributed, plot the probability density functions for the returns of ABC and DEF. Compare the results.

Exercise 7.12 Suppose a share of XYZ is currently selling for $50, and you sell 200 shares short. Suppose further that the required margin is 50%.

1. Compute the following:

   (a) The amount of your short sale

   (b) The amount of money you must have in other assets in your account

2. Suppose the company pays no dividends and you cover your short position one month later when one share of the stock is worth $40. Compute the following:

   (a) Your profit

   (b) Your holding period return

Example 7.38 *Suppose one share of IBM is worth $30, one share of Lucent is worth $4, one share of Disney is worth $20, and one share of Microsoft is worth $25. Suppose further that your portfolio consists of 100 shares of IBM, 500 shares of Lucent, 200 shares of Disney, and 50 shares of Microsoft. Compute the weight each stock has in your portfolio.*

Section 7.2

Exercise 7.13 Let portfolio A consist of three assets such that

$$w_1 = .2 \qquad w_2 = .5 \qquad w_3 = .3$$

$$\bar{r}_1 = .10 \qquad \bar{r}_2 = .12 \qquad \bar{r}_3 = .15$$

$$\sigma_1 = .20 \qquad \sigma_2 = .18 \qquad \sigma_3 = .22$$

$$\sigma_{12} = .01 \qquad \sigma_{13} = .03 \qquad \sigma_{23} = .02.$$

Compute the expected value and standard deviation of the portfolio's return.

Exercise 7.14 Let portfolio A consist of 20 assets $x_1, x_2, \ldots, x_{20}$ such that for all i

$$w_i = \frac{1}{20} \qquad \bar{r}_i = .10 \qquad \sigma_i = .15$$

and for all $i \neq j$

$$\sigma_{ij} = .01.$$

Compute the expected value and standard deviation of the portfolio's return.

Exercise 7.15 Suppose the risk-free rate $r_f = .05$, the expected market return $\bar{r}_M = .10$, and the market return's standard deviation $\sigma_M = .15$.

1. If you want an expected return equal to .08 obtained from a portfolio consisting of the risk-free asset and an index fund that invests in the market, what weight should each have in your portfolio?

2. What is the standard deviation of your portfolio's return?

3. What is the nonsystematic risk of your portfolio?

Exercise 7.16 Suppose $Var(r_M) = .03$ and $Cov(r_{ABC}, r_M) = .05$ for stock ABC.

1. Compute the beta β_{ABC} for ABC. How does the variance of ABC's return compare to that of the market?

2. If $r_f = .05$, and $\bar{r}_M = .10$, determine the risk premium for the market.

3. If the market is efficient, $r_f = .05$, and $\bar{r}_M = .10$, determine the expected return for ABC.

Exercise 7.17 Suppose for asset ABC we obtain the following estimates:

$$\beta_{ABC} = 1.4 \qquad \text{and} \qquad R^2_{XYZ} = .6.$$

Assume $\sigma^2_M = .03$. Compute the total risk (variance) for ABC, the fraction of the risk which is market risk, and the fraction of the risk which is nonsystematic risk.

Exercise 7.18 Figure 7.13 shows a number of items of information that Fidelity (http://www.fidelity.com) makes available to the public about many mutual funds. The fund PREIX is an index fund for the S&P 500. So its behavior can be assumed to be the market behavior. We therefore have that $\sigma_M = .0772$.

1. Using Equality 7.17, compute the variance of the return of each of the mutual funds from its beta and R^2. Compare your results to the variances shown in Figure 7.13.

2. Notice that TWEIX has $\beta^2 < R^2$. Therefore, according to Equality 7.17, its variance should be less than that of the market (which it indeed is). Furthermore, since its beta is less than 1 if the efficient market hypothesis is correct, its expected return should be less than that of the market (see Equality 7.11). Yet its average return over the past 10 years is .1247, while that of the market is .0829. Discuss whether this contradicts the efficient market hypothesis. Consider the following in your discussion: (1) The efficient market hypothesis concerns the current probability distributions of the returns for the coming year. Are these the same as the distributions over the past 10 years? (2) At the beginning of Section 7.2.4 we noted that the efficient market hypothesis assumes that it is the ongoing search for mispriced assets that restores equilibrium, and analysts use valuation models to discover mispriced assets. Notice that TWEIX is a large value fund. This means that the managers of the fund look for mispriced assets.

3. Would you consider a fund like TWEIX a good buy? One way to investigate this would be to do an analysis such as the one suggested in Chapter 4, Exercise 4.22. Search for studies which have already been done that analyze this matter.

Exercise 7.19 Suppose the two factors in Example 7.27 satisfy the factor model assumptions for the U.S. Stock Market. Suppose further that we make the APT assumptions. For two companies CDE and FGH, assume the following:

Company	b_{i1}	b_{i2}
CDE (x_1)	3.1	2.6
FGH (x_2)	5.4	-1.4

Suppose that for time slice t the risk-free rate $r_f(t) = .05$, and we determine the following:

$$P_1 = .042$$
$$P_2 = .017.$$

Compute $\bar{r}_1(t)$ and $\bar{r}_2(t)$.

Exercise 7.20 Suppose stock ABC has a beta equal to 1.2, $\bar{r}_M = .10$ annually, and $r_f = .06$ annually. Compute the stock's required rate of return.

	PREIX	FCNTX	PRWCX	TWEIX	FESGX
Fund Name	T. Rowe Price Equity Index 500	Fidelity Contrafund	T. Rowe Price Capital Appreciation	American Cent Equity Income Inv CL	First Eagle Global CL C
General Information					
Investment Objective	Trade Growth and Income	Trade Growth	Trade Growth and Income	Trade Equity-Income	Trade Multi-Asset Global
Morningstar Category	Large Blend	Large Growth	Moderate Allocation	Large Value	World Allocation
Assets ($M)	6,676.80	65,030.30	8,528.30	1,994.40	4,786.40
Morningstar Overall Rating	★★★	★★★★★	★★★★★	★★★★	★★★★★
Overall # In Category	1517	1391	805	1115	65
Performance					
Non-Load Adj. Returns					
YTD	8.25	4.77	8.13	12.37	11.42
1 Yr	10.44	8.74	10.27	13.67	15.15
3 Yr	11.95	15.98	14.29	13.11	18.64
5 Yr	6.87	11.87	12.13	10.86	16.74
10 Year/Life	8.25	11.22	12.15	12.47	16.63†
Load Adj. Returns					
YTD	--	--	--	--	10.42
1 Yr	--	--	--	--	14.15
3 Yr	--	--	--	--	18.64
5 Yr	--	--	--	--	19.74
10 Year/Life	--	--	--	--	16.63†
Expenses					
Expense Ratio	0.35	0.91	0.76	0.98	1.95
Turnover Rate %	9	60	12	190	12
Transaction Fee	Yes	No	Yes	No	Yes
Volatility Measures					
Beta	1.00	1.01	0.94	0.70	1.02
R²	1.00	0.76	0.82	0.76	0.81
Standard Deviation	7.72	9.34	6.45	8.20	7.10

Figure 7.13: Information Fidelity supplies on several funds.

Exercise 7.21 Assume that the current price P_0 of stock ABC (from the previous exercise) is $30 a share, our expected price $\bar{P}_1$ in one year is $32 a share, and our expected dividend $\bar{D}_1$ to be paid at the end of the coming year is $1 per share. Compute the expected return $\bar{r}$ based on this analysis. Would you consider the stock a good buy?

Exercise 7.22 Suppose again for stock ABC that $\bar{P}_1 = \$32$, $\bar{D}_1 = \$1$, and the required rate of return is the value obtained in Exercise 7.20. Compute the intrinsic value of ABC. If ABC is currently selling for $28, would it be considered a good buy?

Exercise 7.23 Suppose the current shareholder's equity S_0 in company ABC is $200,000,000 and the earnings E_0 in the trailing 12 months were $40,000,000. Suppose further that there are 4,000,000 shares outstanding, and the required rate of return $k = .16$.

1. Compute the ROE and EPS_0.

2. Compute the intrinsic value of ABC if all the earnings are paid out as dividends.

3. Compute the intrinsic value of ABC if a plowback ratio of .7 is used.

Exercise 7.24 Suppose the current shareholder's equity S_0 in company DEF is $200,000,000 and the earnings E_0 in the trailing 12 months were $10,000,000.

Suppose further that there are 4,000,000 shares outstanding, and the required
rate of return $k = .16$.

1. Compute the ROE and EPS_0.

2. Compute the intrinsic value of DEF if all the earnings are paid out as
   dividends.

3. Compute the intrinsic value of DEF if a plowback ratio of .7 is used.

Exercise 7.25 Suppose the last reported 12 month earnings of company ABC
were $30,000,000, there are 1,000,000 shares of stock outstanding, and today's
closing price per share is $40. Suppose further that it is September, and the
average estimate of earnings for the current fiscal year are $29,000,000.

1. Compute the trailing P/E ratio.

2. Compute the forward P/E ratio.

Exercise 7.26 Suppose company ABC has $k = .16$, $ROE = .2$, and a plowback ratio of $b = .7$ is used.

1. Compute the forward P/E ratio according to intrinsic value.

2. If the forward P/E ratio obtained by taking the ratio of the stock's latest
   closing price to the EPS as provided by Zach's Investment Research is 17,
   based on this analysis would you consider ABC a good buy?

Exercise 7.27 Immediately following Example 7.33 we stated that company
XYZ did well to reinvest in growth opportunities because $ROE > k$. Show
that, in general, it is worthwhile to reinvest if and only if $ROE > k/(k + 1)$.
You can accomplish this by showing that there exist positive values of b such
that

$$\frac{EPS_0}{k} < \frac{EPS_0(1 - b)(1 + bROE)}{k - bROE}$$

if and only if $ROE > k/(k + 1)$.

Section 7.3

Exercise 7.28 Develop the Bayesian network discussed in this section using
Netica. Then answer the following questions using that network. For each
question start with a network with no instantiated nodes.

1. Instantiate inflation $(F2)$ to high. What effect does this have on the
   probability distribution of the value of the portfolio?

2. Instantiate r_{BIO} to high. What effect does this have on the probability distribution of the value of the portfolio? Now instantiate inflation ($F2$) to high. Is there any further effect on the probability distribution of the value of the portfolio?

3. Instantiate k_{GDT} to high. What effect does this have on the probability distributions of P_{GDT} and the value of the portfolio?

4. Instantiate PE_{GDT} to low and g_{GDT} to low. What effect does this have on the probability distributions of P_{GDT} and the value of the portfolio? Now instantiate k_{GDT} to high. Is there any further effect on the probability distributions of P_{GDT} or the value of the portfolio?

5. Instantiate business cycle ($F1$) to low. What effect does this have on the probability distribution of inflation ($F2$)?

6. Instantiate market return (r_m) to low. What effect does this have on the probability distribution of inflation ($F2$)? Now instantiate business cycle ($F1$) to low. Is there any further effect on the probability distribution of inflation ($F2$)?

Chapter 8

Modeling Real Options

When faced with the possibility of undertaking a new project, a company must make the decision of whether to pursue that project. A standard way to do this is to perform an analysis based on the expected cash that will be realized by the project. Recall from Chapter 7, Section 7.2.4, that a stock has a required rate of return k that is higher than the risk-free rate owing to its risk, and the expected net present value of the stock (intrinsic value) is computed using k. Similarly, a risky investment in a project has associated with it a rate, called the **risky discount rate**, which is used to compute the **net present expected value (NPEV)** of the project. The determination of this expected net present value is called a **discounted cash flow (DCF)** analysis. The next example illustrates such an analysis.

Example 8.1 *Suppose company Nutrastuff is considering developing a high-energy drink that does not contain caffeine. The cost of the project is $20* *million. Decision makers at Nutrastuff are uncertain of the demand for this product. However, they estimate the product will generate cash flow for the next three years. If the demand is low, they will realize a cash flow of $2 million in each of those years; if it is average, the cash flow will be $10 million each year; and if it is high the cash flow will be $15 million each year. Finally, they estimate that the probabilities of the demand being low, average, and high are .25, .50, and .25, respectively, and that the risky discount rate for this project is .12. This information is summarized in the following table:*

Demand	Annual Cash Flow	Probability
low	$2 million	.25
average	$10 million	.50
high	$15 million	.25

We then have the following for this project(all values are in millions):

$$
\begin{aligned}
NPEV \;=\; & -\$20 + .25\left(\frac{\$2}{1.12} + \frac{\$2}{(1.12)^2} + \frac{\$2}{(1.12)^3}\right) + \\
& .5\left(\frac{\$10}{1.12} + \frac{\$10}{(1.12)^2} + \frac{\$10}{(1.12)^3}\right) + \\
& .25\left(\frac{\$15}{1.12} + \frac{\$15}{(1.12)^2} + \frac{\$15}{(1.12)^3}\right) \\
=\; & \$2.22.
\end{aligned}
$$

Since the NPEV is positive, a simple DCF analysis says that Nutrastuff should invest in the project.

A simple DCF analysis does not take into account every option Nutrastuff has relative to this project. That is, they are not really forced to either start it today or abandon it. For example, they could delay starting it one or two years while they test the market's response to the product. Or even if they do start it today, they could abandon it after one or two years if things do not go well. That is, managers could continue to make decisions concerning the project once it is started. This right to postpone starting a project or revise a strategy once a project is started is called a **managerial option**. It is also called a **real option** to distinguish it from options on financial assets such as stock.

Sercu and Uppal [1995] devised a toy decision problem that contains many of the features found in actual decisions concerning real options. We present that problem next, and we show how its solution can be modeled and solved using influence diagrams. Lander and Shenoy [1999] originally developed the influence diagram solution to the problem.

8.1   Solving Real Options Decision Problems

Sercu and Uppal [1995] developed a problem that concerns a U.S. firm. At the start of each year, that firm can produce one additional unit of some product, can liquidate the firm, or can postpone doing anything until the next year. The product is sold in the U.K., and there is uncertainty concerning the future exchange rate between the U.S. dollar and the pound Sterling. We show a sequence of examples concerning this scenario, which are increasingly complex. The first example has a one-year time frame, the second example has a two-year time frame, and the third example has a three-year time frame. In the first two examples, we first solve the problem using a decision tree to once again illustrate that decision trees and influence diagrams solve the same problems. In the third example, we do not do this because the decision tree would be forbiddingly large, while the influence diagram handles the problem readily.

Example 8.2 *Suppose a U.S. firm has a one-year investment opportunity in which the firm can choose to either produce one unit of a product at time $t = 0$ or liquidate. The product will be sold in the U.K. Specifically, the investment opportunity has the following properties:*

1. *The firm can produce one unit of the product.*

2. *It will take one year to produce the product and cost $1.80 (U.S. dollar) at the beginning of the year to produce it.*

3. *The product will be sold at the end of the year for £1.00 (pound Sterling).*

4. *Instead of producing the product, the firm could liquidate the plant for net liquidation value equal to $.20. However, if one unit is produced, the plant has no liquidation value.*

5. *The current spot exchange rate is $2/£1. That is, $2.00 is worth £1.00.*

6. *In one year the exchange rate will change by a multiplicative factor M. The possible values of M are $u = 1.2$ and $d = .8$. Furthermore,*

$$P(M = u) = .631$$

$$P(M = d) = .369.$$

7. *The U.S. risk-free rate r_f^{US} and the U.K. risk-free rate r_f^{UK} both equal .05.*

8. *Assume the risky discount rate is .105 for this investment opportunity.*

The firm has to make a decision D with the two alternatives: "Produce" and "Liquidate." The solved decision tree for this decision appears in Figure 8.1. The values in that decision tree are net present values. Next, we show how those values were computed. The value of the outcome $M = u$ is

$$\frac{\$2(1.2)}{1.105} - \$1.8 = \$.372,$$

while the value of the outcome $M = d$ is

$$\frac{\$2(.8)}{1.105} - \$1.8 = -\$.352.$$

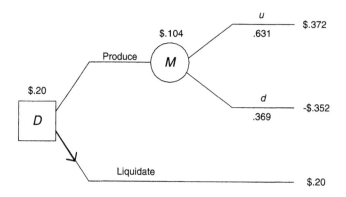

Figure 8.1: The solved decision tree for the decision in Example 8.2.

Therefore,

$$E(Produce) = E(M) = .631(\$.372) + .369(-\$.352) = \$.105$$

$$E(Liquidate) = \$.20.$$

So

$$E(D) = \max(\$.104, \$.20) = \$.20,$$

and the decision is to liquidate.

Figure 8.2 shows a solved influence diagram for this problem instance. The diagram was developed using the software package Netica. The expected values of the two decision alternatives appear in node D. That diagram also shows a table containing the value of U for each combination of values of D and M. A nice feature of influence diagram development packages such as Netica is that we do not need to hand-calculate the values of the utility node. Rather we can write a function which calculates them for us. In this case the Netica function is as follows:

U(D,M) =
    D == Liquidate ?
      .20:
      2*M/1.105 - 1.8.

The following is pseudocode for this function:

if $(D ==$ Liquidate) **then**
    $U(D, M) = 20$
else
    $U(D, M) = 2 \times M/1.105 - 1.8.$

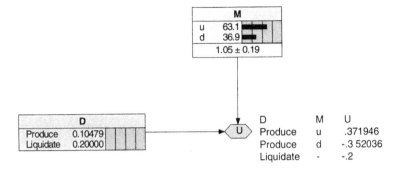

Figure 8.2: The solved influence diagram for the decision in Example 8.2.

Writing a function instead of hand-calculating the utility values is not much of a savings in this small problem instance. However, as we shall see in the examples that follow, it can be very useful in larger problem instances.

Notice in the previous example how our decision is affected by the uncertainty concerning the exchange rate. If we knew the rate was going up $(P(M = u) = 1)$, the discount rate would be the risk-free rate because there is no risk. So we would have that the value of the outcome $M = u$ is

$$\frac{2(1.2)}{1.05} - 1.8 = \$.486,$$

and

$$E(Produce) = E(M) = 1(\$.486) = \$.486.$$

So in this case we would choose to produce.

Next, we look at a decision in which we must look further into the future.

Example 8.3 Now suppose a U.S. firm is in the same situation as in Example 8.2, except that at time $t = 1$ the firm has the option to produce another unit of the product. Specifically, we make the following assumptions:

1. The firm can produce one unit of the product this year $(t = 0)$ and one unit next year $(t = 1)$.

2. For each time frame, it takes one year to produce the product, and it costs $1.80 (U.S. dollar) at the beginning of the year to produce it.

3. For each time frame, the product is sold at the end of the year for £1.00 (pound Sterling).

4. At time $t = 0$ the firm could temporarily exit the market at a cost of $.10. This means it does not produce a unit at $t = 0$. If the firm then wants to produce a unit at time $t = 1$, it will cost $.10 to re-enter the market.

5. Instead of producing at time $t = 0$ or $t = 1$, the firm could liquidate the plant for a net liquidation value equal to $.20. However, if a unit is produced at time $t = 1$, the plant has no liquidation value at time $t = 2$.

6. The current spot exchange rate is $2/£1. That is, $2.00 is worth £1.00.

7. Each year the exchange rate changes by a multiplicative factor M. The possible values of M are $u = 1.2$ and $d = .8$. Furthermore,

$$P(M = u) = .631$$

$$P(M = d) = .369.$$

8. The U.S. risk-free rate r_f^{US} and the U.K. risk-free rate r_f^{UK} both equal .05.

9. The risky discount rate is .105.

The firm has to make a decision D_0 with the three alternatives: "Produce," "Liquidate," and "Postpone." The last alternative is the decision to temporarily exit the market and postpone producing or liquidating. If it chooses to produce or postpone, it must make a second decision with two alternatives: "Produce" and "Liquidate." The solved decision tree for this decision appears in Figure 8.3. Next we show how some of the values in that decision tree were computed. If we choose to produce twice, $M_{01} = u$, and $M_{11} = u$, the present value of the outcome is equal to

$$\frac{\$2(1.2)}{1.105} - \$1.8 + \frac{\$2(1.2)^2}{(1.105)^2} - \frac{\$1.8}{1.105} = \$1.\,102.$$

If we choose to produce twice, $M_{01} = u$, and $M_{11} = d$, the present value of the outcome is equal to

$$\frac{\$2(1.2)}{1.105} - \$1.8 + \frac{\$2(1.2)(.8)}{(1.105)^2} - \frac{\$1.8}{1.105} = \$.315.$$

If we choose to postpone (temporarily exit) and then choose to produce, $M_{02} = u$, and $M_{13} = u$, the present value of the outcome is equal to

$$-\$.10 - \frac{\$.10}{1.105} + \frac{2(1.2)^2}{(1.105)^2} - \frac{1.8}{1.105} = \$.539.$$

If we choose to postpone and then choose to liquidate, the present value of the outcome is equal to

$$-\$.10 + \frac{\$.20}{1.105} = \$.081.$$

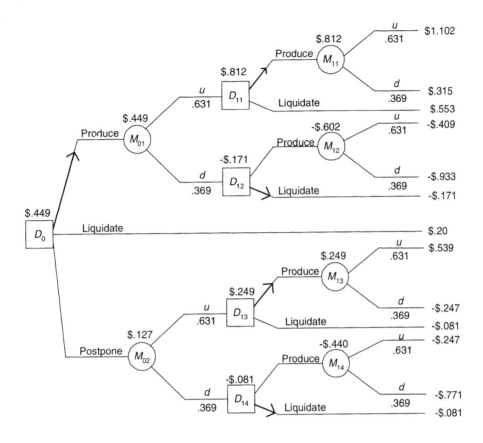

Figure 8.3: The solved decision tree for the decision in Example 8.3.

It is left as an exercise to compute the values of the other outcomes. After determining these values, we solve the decision tree by passing expected values to chance nodes and maximums to decision nodes. For example,

$$E(M_{11}) = .631(\$1.102) + .369(\$.315) = \$.812.$$

$$E(D) = \max(\$.812, \$.553) = \$.812.$$

Finally, we see that alternative "Produce" has the highest expected value ($.449); so that is our decision.

Figure 8.4 shows a solved influence diagram for this problem instance, along with a table showing the utility values. The Netica function that created this table is as follows:

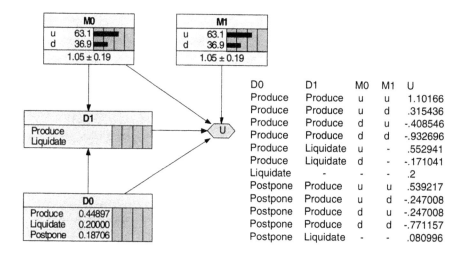

D0	D1	M0	M1	U
Produce	Produce	u	u	1.10166
Produce	Produce	u	d	.315436
Produce	Produce	d	u	-.408546
Produce	Produce	d	d	-.932696
Produce	Liquidate	u	-	.552941
Produce	Liquidate	d	-	-.171041
Liquidate	-	-	-	.2
Postpone	Produce	u	u	.539217
Postpone	Produce	u	d	-.247008
Postpone	Produce	d	u	-.247008
Postpone	Produce	d	d	-.771157
Postpone	Liquidate	-	-	.080996

Figure 8.4: The solved influence diagram for the decision in Example 8.3.

$U(D0,M0,D1,M1) =$

    $D0 == Liquidate ?$

      $.20:$

    $D0 == Produce \&\& D1 == Produce ?$

      $2*M0/1.105 - 1.8 + 2*M0*M1/1.105^2 - 1.8/1.105:$

    $D0 == Produce \&\& D1 == Liquidate ?$

      $2*M0/1.105 - 1.8 + .2/1.105:$

    $D0 == Postpone \&\& D1==Produce ?$

      $-.1 - .1/1.105 + 2*M0*M1/1.105^2 - 1.8/1.105:$

    $D0 == Postpone \&\& D1 == Liquidate ?$

      $-.1 + .2/1.105.$

Notice how much smaller the influence diagram is than the decision tree, and notice that the statement necessary to produce the utility values is small compared to the table it produces.

Next, we look at a decision in which we must look further into the future.

Example 8.4 *Now suppose a U.S. firm is in the same situation as in Example 8.2, except that at times $t = 1$ and $t = 2$ the firm has the option to produce another unit of the product. Specifically, we make the following assumptions:*

    1. *The firm can produce one unit of the product this year ($t = 0$), one unit next year ($t = 1$), and one unit the following year ($t = 2$).*

2. For each time frame, it takes one year to produce the product, and it costs $1.80 (U.S. dollar) at the beginning of the year to produce it.

3. For each time frame, the product is sold at the end of the year for £1.00 (pound Sterling).

4. At times $t = 0$ and $t = 1$ the firm could temporarily exit the market at a cost of $.10. If the firm exits at time $t = 0$, it will cost $.10 to re-enter at time $t = 1$ or $t = 2$. However, it does not cost anything to remain out at $t = 1$. If the firm exits at time $t = 1$, it will cost $.10 to re-enter at time $t = 2$.

5. Instead of producing at time $t = 0$, $t = 1$, or $t = 2$, the firm could liquidate the plant for a net liquidation value equal to $.20. However, if a unit is produced at time $t = 2$, the plant has no liquidation value at time $t = 3$.

6. The current spot exchange rate is $2/£1. That is, $2.00 is worth £1.00.

7. Each year the exchange rate changes by a multiplicative factor M. The possible values of M are $u = 1.2$, and $d = .8$. Furthermore,

$$P(M = u) = .631$$

$$P(M = d) = .369.$$

8. The U.S. risk-free rate r_f^{US} and the U.K. risk-free rate r_f^{UK} both equal .05.

9. The risky discount rate is .105.

The firm has to make a decision D_0 with three alternatives: "Produce," "Liquidate," and "Postpone." If it chooses to produce or postpone, it must make a second decision with three alternatives: "Produce," "Liquidate," and "Postpone." If the second decision is to produce or postpone, it must make a third decision with two alternatives: "Produce" and "Liquidate." We don't show a decision for this problem instance because it would be too large. However, we do show one calculation for the decision tree. If all three decisions are to produce, and M always takes the value u, the present value of the outcome is equal to

$$\frac{\$2(1.2)}{1.105} - \$1.8 + \frac{\$2(1.2)^2}{(1.105)^2} - \frac{\$1.8}{1.105} + \frac{\$2(1.2)^3}{(1.105)^3} - \frac{\$1.8}{(1.105)^2} = \$2.189.$$

The solved influence diagram for this decision appears in Figure 8.5. The top-level decision is to produce because that alternative has the largest expected value (.79457). Notice that the influence diagram is not very large. Rather, it has only increased linearly relative to the influence diagram in the previous example. We do not show the table containing the utility values because it is so large. (There are six edges into U.) However, the Netica function used to create the table is not that hard to write because it only has linear growth in terms of the number of decisions. It is as follows:

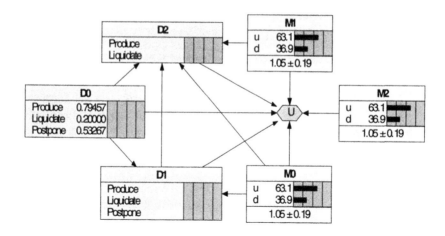

Figure 8.5: The solved influence diagram for the decision in Example 8.4.

$U(D0,M0,D1,M1,D2,M2) =$

    $D0 == Liquidate$?

        .20:

    $D0 == Produce$ && $D1==Produce$ && $D2==Produce$?

        $2*M0/1.105$ - $1.8 + 2*M0*M1/1.105^2$ - $1.8/1.105 +$

        $2*M0*M1*M2/1.105^3$ - $1.8/1.105^2:$

    $D0 == Produce$ && $D1 == Produce$ && $D2 == Liquidate$?

        $2*M0/1.105$ - $1.8 + 2*M0*M1/1.105^2$ - $1.8/1.105 + .20/1.105^2:$

    $D0 == Produce$ && $D1 == Liquidate$?

        $2*M0/1.105$ - $1.8 + .20/1.105:$

    $D0 == Produce$ && $D1 == Postpone$&& $D2 == Produce$?

        $2*M0/1.105$ - 1.8 -$.10/1.105$ - $.10/1.105^2 +$

        $2*M0*M1*M2/1.105^3$ - $1.8/1.105^2:$

    $D0 == Produce$ && $D1 == Postpone$ && $D2 == Liquidate$?

        $2*M0/1.105$ - 1.8 - $.10/1.105 + .20/1.105^2:$

    $D0 == Postpone$ && $D1 == Produce$ && $D2 == Produce$?

        $-.10$ - $.10/1.105 + 2*M0*M1/1.105^2$ - $1.8/1.105 +$

        $2*M0*M1*M2/1.105^3$ - $1.8/1.105^2:$

    $D0 == Postpone$ && $D1 == Produce$ && $D2 == Liquidate$?

        $-.10$ - $.10/1.105 + 2*M0*M1/1.105^2$ - $1.8/1.105 + .20/1.105^2:$

    $D0 == Postpone$&& $D1 == Liquidate$?

        $-.10 + .20/1.105:$

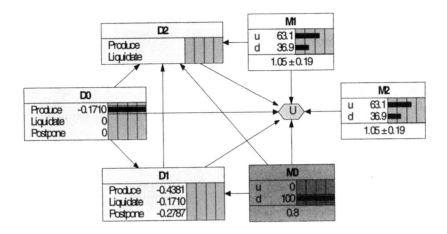

Figure 8.6: The solved influence diagram for the decision in Example 8.5.

$D0 == Postpone$ && $D1 == Postpone$ && $D2 == Produce$?

$-.10 - .10/1.105\hat{\ }2 + 2*M0*M1*M2/1.105\hat{\ }3 - 1.8/1.105\hat{\ }2:$

$D0 == Postpone$ && $D1 == Postpone$ && $D2 == Liquidate$?

$-.10 + .20/1.105\hat{\ }2.$

8.2 Making a Plan

The influence diagram not only informs the manager of the decision alternative to choose at the top level, but it also yields a plan. That is, it informs the manager of the decision alternative to choose in the future after new information is obtained. This is illustrated in the following examples.

Example 8.5 *Suppose we have the situation discussed in Example 8.4, we make the decision to produce at time $t = 0$, and $M0$ ends up taking the value d. What decision should we make at time $t = 1$? Figure 8.6 shows the influence diagram after D0 is instantiated to "Produce" and M0 is instantiated to d. The decision alternative that maximizes D1 is now "Liquidate" with value -0.1710. So we are informed that we should cut our losses and get out. Note that this is the opposite of what some decision makers tend to do. Some individuals feel they need to stay in the investment so that they have a chance to get their losses back.*

Example 8.6 *Suppose we have the situation discussed in Example 8.4, we make the decision to produce at times $t = 0$, and $M0$ ends up taking the value u. Then the recommended alternative at time $t = 1$ would be to produce again. Suppose we choose this alternative, and $M1$ ends up taking the value d. What*

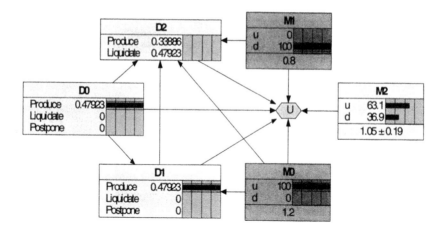

Figure 8.7: The solved influence diagram for the decision in Example 8.6.

decision should we make at time $t = 2$? Figure 8.7 shows the influence diagram after D0 and D1 are both instantiated to "Produce," M0 is instantiated to u, and M1 is instantiated to d. The decision alternative that maximizes D3 is now "Liquidate" with value .47923. So we are informed that we should get out with our current profit.

8.3   Sensitivity Analysis

The risky discount rate is difficult to estimate. So we should perform a sensitivity analysis concerning its value. It is left as an exercise to show that in Example 8.4 the decision will be to produce as long as that rate is $< .267$, which is more than twice our estimate.

The probability of the exchange rate going up or down each year is also difficult to ascertain. It is left as an exercise to show that the decision will be to produce as long as that rate is $> .315$, which is less than half of the estimate.

EXERCISES

Section 8.1

Exercise 8.1 Suppose company EyesRUs is considering developing a contact lens that can be worn for two weeks without removal. The cost of the project is $5 million. Decision makers at EyesRUs are uncertain as to the demand for this product. However, they estimate the product will generate cash flow for the next four years. If the demand is low, they will realize a cash flow of $500,000 in each of those years; if it is average, the cash flow will be $2 million each year; and if it is high, the cash flow will be $3 million each year. Finally, they estimate that the probabilities of the demand being low, average, and high are .2, .6, and .2, respectively, and that the risky discount rate for this project is .11. Determine the $NPEV$ for the project. Based on this analysis, does it seem reasonable to undertake the project?

Exercise 8.2 Modify Example 8.2 as follows:

- The product costs $1.90 (U.S. dollar) to produce.

- The firm could liquidate the plant for a net liquidation value equal to $.25.

- In one year the exchange rate will change by a multiplicative factor M. The possible values of M are $u = 1.1$ and $d = .9$. Furthermore,

$$P(M = u) = .4$$

$$P(M = d) = .6.$$

- The U.S. risk-free rate r_f^{US} and the U.K. risk-free rate r_f^{UK} both equal .06.

- The risky discount rate is .10 for this investment opportunity.

1. Represent the problem with a decision tree, and solve the decision.

2. Using Netica, represent and solve the problem using an influence diagram.

Exercise 8.3 Modify Example 8.3 with the modifications given in Exercise 8.2.

1. Represent the problem with a decision tree, and solve the decision.

2. Using Netica, represent and solve the problem using an influence diagram.

Exercise 8.4 Modify Example 8.4 with the modifications given in Exercise 8.2. Using Netica represent and solve the problem using an influence diagram.

Section 8.2

Exercise 8.5 Suppose we have the situation discussed in Exercise 8.4, we make the decision to produce at time $t = 0$, and $M0$ ends up taking the value d. What is the recommended decision at time $t = 1$?

Exercise 8.6 Suppose we have the situation discussed in Exercise 8.4, we make the decision to produce at times $t = 0$, and $M0$ ends up taking the value u.

1. What is the recommended decision at time $t = 1$?

2. Suppose at time $t = 1$ we choose to produce and $M1$ ends up taking the value u. What is the recommended decision at time $t = 2$?

Section 8.3

Exercise 8.7 Show that in Example 8.4 the decision will be to produce as long as the risky discount rate is $< .267$.

Exercise 8.8 Show that in Example 8.4 the decision will be to produce as long as the exchange rate is $> .315$.

Exercise 8.9 Do a sensitivity analysis on the risky discount rate in the problem described in Exercise 8.4.

Exercise 8.10 Do a sensitivity analysis on the exchange rate in the problem described in Exercise 8.4.

Chapter 9

Venture Capital Decision Making

Start-up companies often do not have access to sufficient capital, but, if they could obtain that capital, they may have the potential for good long-term growth. If a company is perceived as having such potential, investors can hope to obtain above-average returns by investing in such companies. Money provided by investors to start-up firms is called **venture capital (VC)**. Wealthy investors, investment banks, and other financial institutions typically provide venture capital funding. According to the National Venture Capital Association, the total venture capital invested in 1990 was $3.4 billion distributed among 1317 companies, while by 2000 it was $103.8 billion distributed among 5458 companies.

Venture capital investment can be very risky. A study in [Ruhnka et al., 1992]

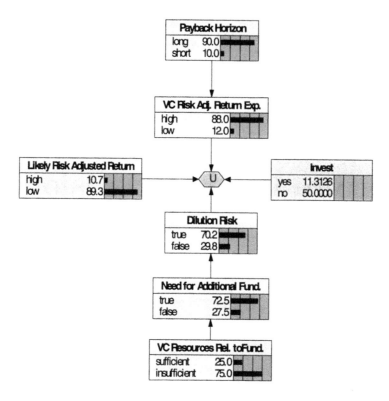

Figure 9.1: A simple influence diagram modeling the decision of whether to invest in a start-up firm.

indicates that 40% of backed ventures fail. Therefore, careful analysis of a new firm's prospects is warranted before deciding whether to back the firm. Venture capitalists are experts who analyze a firm's prospects. Kemmerer et al. [2006] performed an in-depth interview of an expert venture capitalist and elicited a causal (Bayesian) network from that expert. They then refined the network, and finally assessed the conditional probability distributions for the network with the help of the venture capitalist. As discussed in [Shepherd and Zacharakis, 2002], such models often outperform the venture capitalist, whose knowledge was used to create them. In Section 9.1 we present a simple influence diagram, based on their causal network, that models the decision of whether to invest in a given firm. This simple influence diagram is for the purpose of providing an accessible introduction. Section 9.2 shows a detailed influence diagram obtained from their causal network. In Section 9.3 we show the result of using the influence diagram to model a real decision.

9.1 A Simple VC Decision Model

A simple influence diagram modeling the decision of whether to invest in a start-up firm appears in Figure 9.1. The influence diagram was developed using Netica. The following report, produced by Netica, shows the conditional distributions and utilities in the diagram:

```
Dil_Risk:

true        false        Need_Add_Fund
0.95        0.05         true
0.05        0.95         false


Need_Add_Fund:

true        false        VC_Res_Fund
0.05        0.95         sufficient
0.95        0.05         insufficient


VC_Res_Fund:

sufficient  insufficient
0.25        0.75


VC_Risk_Adj:

high        low          Pay_Hor
0.95        0.05         long
0.25        0.75         short


Pay_Hor:

long        short
0.9         0.1


Likely_Ret:

high        low
0.107       0.893
```

U	Invest	Likely_Risk_Adj_Ret	VC_Risk_Adj_Ret_Exp	Dilution_Risk
85	yes	high	high	true
90	yes	high	high	false
87.5	yes	high	low	true
95	yes	high	low	false
1	yes	low	high	true
1	yes	low	high	false
10	yes	low	low	true
15	yes	low	low	false
50	no	high	high	true
50	no	high	high	false
50	no	high	low	true
50	no	high	low	false
50	no	low	high	true
50	no	low	high	false
50	no	low	low	true
50	no	low	low	false

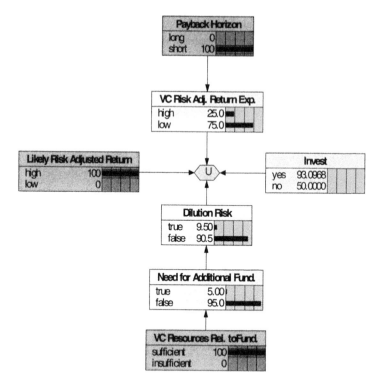

Figure 9.2: When the root nodes are instantiated to their most favorable values, the utility of the investment is over 93.

We see that only three variables directly affect the utility of the investment. These variables are the *Likely_Adjusted_Return*, *Dilution_Risk*, and *VC_Risk_Adj_Return_Exp*. The first of these variables concerns the likely return of the investment, the second concerns the risk inherent in the firm's ability to repay the loan, while the third concerns the risk-adjusted expectation concerning how soon the firm will repay the loan. The utilities are not actual returns. The utility can be thought of as a measure of the potential of the firm, with 0 being lowest and 1 being highest. An investment with a value of 1 can be thought of as the perfect investment. If *Likely_Risk_Adj_Return* is high, *VC_Risk_Adj_Return_Exp* is low, and *Dilution_Risk* is false, then the firm has the most possible potential the model can provide, and a utility of 95 is assigned. So even when the variables directly related to the value node have their most favorable values, the utility is still only 95. We compare the investment choice to a utility of 50. If the utility of the investment is less than 50, it would not be considered a good investment, and we would choose to not invest.

Notice in Figure 9.1 that the utility of the investment is only 11.3 when nothing is known about the firm. This indicates that, in the absence of any

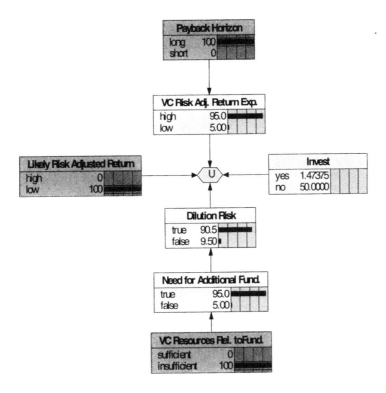

Figure 9.3: When the root nodes are instantiated to their least favorable values, the utility of the investment is about 1.47.

information about a firm, the firm would not be considered a good investment. Figure 9.2 shows the influence diagram with the root nodes instantiated to their most favorable values. We see then that the utility of the investment is over 93. Figure 9.3 shows the influence diagram with the root nodes instantiated to their least favorable values. We see then that the utility of the investment is about 1.47.

9.2   A Detailed VC Decision Model

The unrealistic aspect of the model just discussed is that ordinarily we would not be able to estimate whether the likely risk adjusted return is high or low. Rather there are many other variables that impact this variable, and we could estimate the values of many of them. Figure 9.4 shows an influence diagram based on the complete causal model obtained from the expert venture capitalist. The conditional probability distributions for this influence diagram appear in the appendix of this chapter. Notice that when nothing is known about a particular firm, the expected value of the firm's potential is still only 11.3.

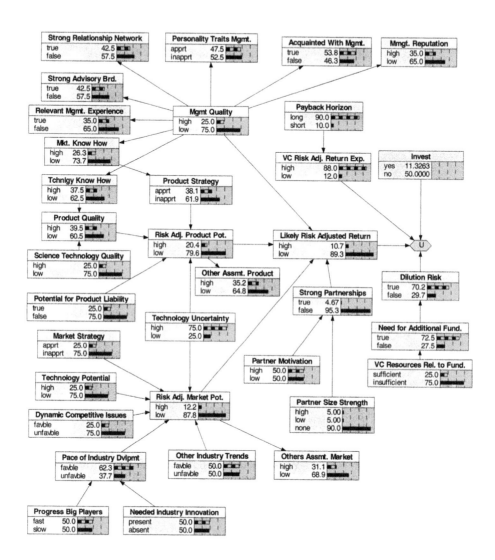

Figure 9.4: The detailed model of the venture capital funding decision.

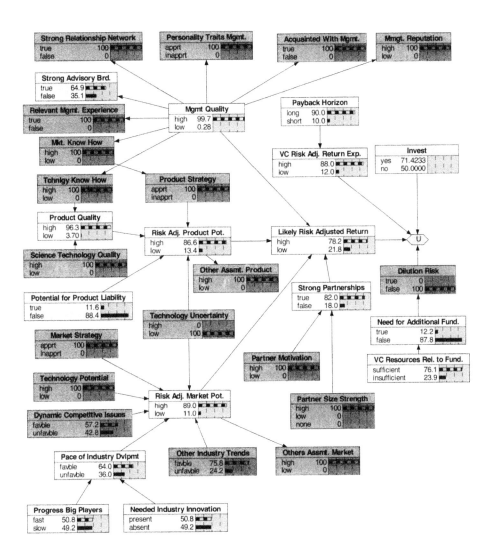

Figure 9.5: The influence diagram in Figure 9.4 with nodes instantiated to assessed values for a firm in the chemical industry.

9.3   Modeling Real Decisions

In order to test the system, Kemmerer et al. [2006] asked the venture capitalist to recall an example of an investment opportunity from the past which he chose and which ended up being successful. A firm in the chemical industry was chosen. Next, the venture capitalist was asked to assess values for as many variables as possible relative to this firm, but the assessed values should be the ones the venture capitalist would have assessed at the time the decision was made. Figure 9.5 shows the influence diagram in Figure 9.4 with nodes instantiated to these assessed values. Notice that the utility of deciding to invest is about 71.4, which is somewhat above the baseline value of 50.

Next, the venture capitalist was asked to recall an investment decision which had a disastrous outcome and to enter assessed values for variables at the time the decision was made. The utility of deciding to invest turned out to be 57.4, which is still above the baseline. However, when correct hindsight information regarding the market potential was entered, the utility turned out to be only 25.4, which is in line with the disastrous outcome. So the relatively high utility seemed to be due to poor assessment of the values of the variables.

Notice in Figure 9.5 that the instantiated variable *Other_Industry_Trends* is not instantiated to high or low. Rather, the probability of it having value favorable has changed from .5 to .758. Instead of instantiating a variable to a precise value, Netica allows the user to enter the new probability of the variable based on the user's evidence E. In this case, the capitalist entered

$$P(Other\_Industry\_Trends = favorable|E) = .75.$$

The probability then became .758 because the variable *Others_Assmt_Market* was set to high, and these variables have a dependency through the variable *Risk_Adj_Market_Pot*. Similarly, the probability of *Dynamic_Competitive _Issues* being favorable became .572 because the capitalist entered

$$P(Dynamic\_Competitive\_Issues = favorable|E) = .5.$$

EXERCISES

Section 9.1

Exercise 9.1 Using Netica, develop the influence diagram shown in Figure 9.1. Investigate the investment decision for various instantiations of the chance nodes.

Section 9.2

Exercise 9.2 Using, Netica develop the influence diagram shown in Figure 9.5.

1. Investigate the investment decision for various instantiations of the nodes.

2. Try to obtain information on an actual firm that tried to obtain venture capital. Enter as much information as you know for the firm. Compare the recommended decision to that which was actually made concerning the firm. If the firm obtained funding, compare the recommended decision to the outcome.

9.A   Appendix

The following report, produced by Netica, shows the conditional probability distributions and utilities in the influence diagram in Figure 9.4:

```
Dil_Risk:

true          false         Need_Add_Fund
0.95          0.05          true
0.05          0.95          false


VC_Res_Fund:

sufficient    insufficient
0.25          0.75


Mgmt_Rep:

high          low           Mgmt_Qual
0.8           0.2           high
0.2           0.8           low


Acq_Mgmt:

true          false         Mgmt_Qual
0.65          0.35          high
0.5           0.5           low


Strong_Rel_Net:

true          false         Mgmt_Qual
0.65          0.35          high
0.35          0.65          low


Strg_Adv:

true          false         Mgmt_Qual
0.65          0.35          high
0.35          0.65          low


Mgmt_Exp:

true          false         Mgmt_Qual
0.8           0.2           high
0.2           0.8           low


Other_Mrkt:

high          low           Risk_Mrkt
0.75          0.25          high
0.25          0.75          low


Part_Mot:

high          low
0.5           0.5


Tech_Pot:

high          low
0.25          0.75
```

Tch_Know:

high	low	Mgmt_Qual
0.75	0.25	high
0.25	0.75	low

Ind_Dvlpmt:

favble	unfavble	Prog_Big_Play	Nd_Ind_Innov
0.94	0.06	fast	present
0.75	0.25	fast	absent
0.75	0.25	slow	present
0.05	0.95	slow	absent

Ind_Trends:

favble	unfavble
0.5	0.5

Mkt_Know:

high	low	Mgmt_Qual
0.75	0.25	high
0.1	0.9	low

Mgmt_Qual:

high	low
0.25	0.75

Prod_Qlty:

high	low	Sci_Tech_Qlty	Tch_Know
0.93	0.07	high	high
0.83	0.17	high	low
0.55	0.45	low	high
0.05	0.95	low	low

Sci_Tech_Qlty:

high	low
0.25	0.75

Mrkt_Strgy:

apprt	inapprt
0.25	0.75

Pot_Prod_Lblty:

true	false
0.25	0.75

`Dyn_Comp`:

favble	unfavble
0.25	0.75

`Nd_Ind_Innov`:

present	absent
0.5	0.5

`Prog_Big_Play`:

fast	slow
0.5	0.5

`Other_Prod`:

high	low	Risk_Prod
0.75	0.25	high
0.25	0.75	low

`Tech_Uncrty`:

high	low
0.75	0.25

`VC_Risk_Adj`:

high	low	Pay_Hor
0.95	0.05	long
0.25	0.75	short

`Part_Size`:

high	low	none
0.05	0.05	0.9

`Strong_Part`:

true	false	Part_Mot	Part_Size
0.82	0.18	high	high
0.25	0.75	high	low
0	1	high	none
0.75	0.25	low	high
0.05	0.95	low	low
0	1	low	none

`Prod_Strat`:

apprt	inapprt	Mkt_Know
0.75	0.25	high
0.25	0.75	low

Pay_Hor:

long	short
0.9	0.1

Need_Add_Fund:

true	false	VC_Res_Fund
0.05	0.95	sufficient
0.95	0.05	insufficient

Risk_Prod:

high	low	Pot_Prod_Lblty	Prod_Qlty	Prod_Strat	Tech_Uncrty
0.05	0.95	true	high	apprt	high
0.05	0.95	true	high	apprt	low
0.05	0.95	true	high	inapprt	high
0.05	0.95	true	high	inapprt	low
0.05	0.95	true	low	apprt	high
0.05	0.95	true	low	apprt	low
0.05	0.95	true	low	inapprt	high
0.05	0.95	true	low	inapprt	low
0.9	0.1	false	high	apprt	high
0.95	0.05	false	high	apprt	low
0.25	0.75	false	high	inapprt	high
0.3	0.7	false	high	inapprt	low
0.1	0.9	false	low	apprt	high
0.15	0.85	false	low	apprt	low
0.05	0.95	false	low	inapprt	high
0.075	0.925	false	low	inapprt	low

Likely_Ret:

high	low	Mgmt_Qual	Risk_Prod	Risk_Mrkt	Strong_Part
0.9	0.1	high	high	high	true
0.9	0.1	high	high	high	false
0.35	0.65	high	high	low	true
0.3	0.7	high	high	low	false
0.45	0.55	high	low	high	true
0.4	0.6	high	low	high	false
0.3	0.7	high	low	low	true
0.25	0.75	high	low	low	false
0.5	0.5	low	high	high	true
0.25	0.75	low	high	high	false
0.2	0.8	low	high	low	true
0.1	0.9	low	high	low	false
0.2	0.8	low	low	high	true
0.1	0.9	low	low	high	false
0.1	0.9	low	low	low	true
0.01	0.99	low	low	low	false

Pers_Mgmt:

apprt	inapprt	Mgmt_Qual
0.55	0.45	high
0.45	0.55	low

Risk_Mrkt:

high	low	Dyn_Comp	Tech_Uncrty	Tech_Pot	Mrkt_Strgy	Ind_Dvlpmt	Ind_Trends
0.95	0.05	favble	high	high	apprt	favble	favble
0.95	0.05	favble	high	high	apprt	favble	unfavble
0.95	0.05	favble	high	high	apprt	unfavble	favble
0.92	0.08	favble	high	high	apprt	unfavble	unfavble
0.33	0.67	favble	high	high	inapprt	favble	favble
0.3	0.7	favble	high	high	inapprt	favble	unfavble
0.28	0.72	favble	high	high	inapprt	unfavble	favble
0.25	0.75	favble	high	high	inapprt	unfavble	unfavble
0.33	0.67	favble	high	low	apprt	favble	favble
0.3	0.7	favble	high	low	apprt	favble	unfavble
0.28	0.72	favble	high	low	apprt	unfavble	favble
0.25	0.75	favble	high	low	apprt	unfavble	unfavble
0.01	0.99	favble	high	low	inapprt	favble	favble
0.01	0.99	favble	high	low	inapprt	favble	unfavble
0.01	0.99	favble	high	low	inapprt	unfavble	favble
0.01	0.99	favble	high	low	inapprt	unfavble	unfavble
0.95	0.05	favble	low	high	apprt	favble	favble
0.92	0.08	favble	low	high	apprt	favble	unfavble
0.87	0.13	favble	low	high	apprt	unfavble	favble
0.78	0.22	favble	low	high	apprt	unfavble	unfavble
0.28	0.72	favble	low	high	inapprt	favble	favble
0.26	0.74	favble	low	high	inapprt	favble	unfavble
0.24	0.76	favble	low	high	inapprt	unfavble	favble
0.22	0.78	favble	low	high	inapprt	unfavble	unfavble
0.28	0.72	favble	low	low	apprt	favble	favble
0.26	0.74	favble	low	low	apprt	favble	unfavble
0.24	0.76	favble	low	low	apprt	unfavble	favble
0.22	0.78	favble	low	low	apprt	unfavble	unfavble
0.01	0.99	favble	low	low	inapprt	favble	favble
0.01	0.99	favble	low	low	inapprt	favble	unfavble
0.01	0.99	favble	low	low	inapprt	unfavble	favble
0.01	0.99	favble	low	low	inapprt	unfavble	unfavble
0.69	0.31	unfavble	high	high	apprt	favble	favble
0.63	0.37	unfavble	high	high	apprt	favble	unfavble
0.59	0.41	unfavble	high	high	apprt	unfavble	favble
0.53	0.47	unfavble	high	high	apprt	unfavble	unfavble
0.19	0.81	unfavble	high	high	inapprt	favble	favble
0.17	0.83	unfavble	high	high	inapprt	favble	unfavble
0.16	0.84	unfavble	high	high	inapprt	unfavble	favble
0.15	0.85	unfavble	high	high	inapprt	unfavble	unfavble
0.19	0.81	unfavble	high	low	apprt	favble	favble
0.17	0.83	unfavble	high	low	apprt	favble	unfavble
0.16	0.84	unfavble	high	low	apprt	unfavble	favble
0.15	0.85	unfavble	high	low	apprt	unfavble	unfavble
0.01	0.99	unfavble	high	low	inapprt	favble	favble
0.01	0.99	unfavble	high	low	inapprt	favble	unfavble
0.01	0.99	unfavble	high	low	inapprt	unfavble	favble
0.01	0.99	unfavble	high	low	inapprt	unfavble	unfavble
0.6	0.4	unfavble	low	high	apprt	favble	favble
0.54	0.46	unfavble	low	high	apprt	favble	unfavble
0.51	0.49	unfavble	low	high	apprt	unfavble	favble
0.46	0.54	unfavble	low	high	apprt	unfavble	unfavble
0.17	0.83	unfavble	low	high	inapprt	favble	favble
0.15	0.85	unfavble	low	high	inapprt	favble	unfavble
0.14	0.86	unfavble	low	high	inapprt	unfavble	favble
0.13	0.87	unfavble	low	high	inapprt	unfavble	unfavble
0.17	0.83	unfavble	low	low	apprt	favble	favble
0.15	0.85	unfavble	low	low	apprt	favble	unfavble
0.14	0.86	unfavble	low	low	apprt	unfavble	favble
0.13	0.87	unfavble	low	low	apprt	unfavble	unfavble
0.01	0.99	unfavble	low	low	inapprt	favble	favble
0.01	0.99	unfavble	low	low	inapprt	favble	unfavble
0.01	0.99	unfavble	low	low	inapprt	unfavble	favble
0.01	0.99	unfavble	low	low	inapprt	unfavble	unfavble

U	Invest	Likely_Ret	VC_Risk_Adj	Dil_Risk
85	yes	high	high	true
90	yes	high	high	false
87.5	yes	high	low	true
95	yes	high	low	false
1	yes	low	high	true
1	yes	low	high	false
10	yes	low	low	true
15	yes	low	low	false
50	no	high	high	true
50	no	high	high	false
50	no	high	low	true
50	no	high	low	false
50	no	low	high	true
50	no	low	high	false
50	no	low	low	true
50	no	low	low	false

Chapter 10

Bankruptcy Prediction

Predicting whether a firm will go bankrupt is quite important to the firm's auditors, creditors, and stockholders. However, even auditors, who should best know a firm's situation, often fail to see an impending bankruptcy [McKee, 2003]. So several models have been developed to predict the likelihood of bankruptcy from information about a firm. These include models based on simple univariate analysis [Beaver, 1966], logistic regression [Ohlson, 1980], neural networks [Tam and Kiang, 1992], genetic programming [McKee and Lensberg, 2002], and Bayesian networks [Sarkar and Sriram, 2001], [Sun and Shenoy, 2006]. We show the work in [Sun and Shenoy, 2006] as it carefully provides guidance in the selection of variables and the discretization of continuous variables.

First, we show the network and how it was developed. Then we discuss experiments testing its accuracy. Finally, we present possible improvements to the model.

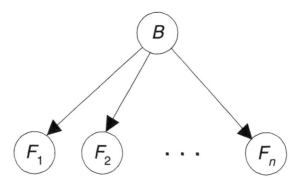

Figure 10.1: A naive Bayesian network.

10.1   A Bayesian Network for Predicting Bankruptcy

Naive Bayesian networks have been shown to perform well in bankruptcy prediction [Sarkar and Sriram, 2001]. The model presented here builds on these results and uses a naive Bayesian network. So first we discuss such networks, and then we develop a naive Bayesian network for bankruptcy prediction.

10.1.1   Naive Bayesian Networks

A **naive Bayesian network** is a Bayesian network with a single root, all other nodes are children of the root, and there are no edges between the other nodes. Figure 10.1 shows a naive Bayesian network. As is the case for any Bayesian network, the edges in a naive Bayesian network may or may not represent causal influence. Often, naive Bayesian networks are used to model classification problems. In such problems the values of the root are possible classes to which an entity could belong, while the leaves are features or predictors of the classes. In the current application, there are two classes: one consisting of companies that will go bankrupt and the other consisting of companies that will not. The features are variables whose values should often be different for companies in different classes. Since bankruptcy would not cause the features that would predict bankruptcy, in this application the edges are not causal.

10.1.2   Constructing the Bankruptcy Prediction Network

Next, we develop a naive Bayesian network for bankruptcy prediction. First, we determine the relevant variables. Then we learn the structure of the network. Finally, we obtain the parameters for the network.

	Construct	Var.	What the Variable Represents
Financial Factors	Size	TA	ln((Total Assets)/(GNP Implicit Price Deflator Index))
	Liquidity	CA	(Current Assets − Current Liabilities)/(Total Assets)
		CR	(Current Assets)/Current Liabilities
		OF	(Operating Cash Flow)/(Total Liabilities)
		LM	$(L+\mu)/\sigma$ L = Cash + short-term marketable securities μ, σ = Mean, standard deviation of quarter-to-quarter changes in L over prior 12 quarters
		CA	(Current Assets)/(Total Assets)
		CH	Cash/(Total Assets)
	Leverage	TL	(Total Liabilities)/(Total Assets)
		LTD	(Long-Term Debts)/(Total Assets)
	Turnover	S	Sales/(Total Assets)
		CS	(Current Assets)/Sales
	Profitability	E	(Earnings before Interest and Taxes)/(Total Assets)
		NT	(Net Income)/(Total Assets)
		IT	If Net Income over last 2 years < 0, = 1, else = 0
		RE	(Retained Earnings)/(Total Assets)
		CHN	(Net Income year t − Net Income year t-1)/ (Abs(Net Income year t) + Abs(Net Income year t-1))
Market Factors		M	ln(Firm's size relative to CRSP NYSE/AMEX/ /NASDAQ market capitalization index)
		R	Firm's stock return year t-1 minus value-weighted CRSP NYSE/AMEX/NASDAQ index return year t-1
Other Factors		AU	If Compustat code auditor's opinion is "1: unqualified," = 0, else = 1
		IFR	Industry failure rate calculated as avg. bankruptcy rate, where avg. bankruptcy rate is (# bankruptcies in 2-digit SIC ind.)/(# firms in ind.)

Table 10.1: Potential predictor variables for bankruptcy.

Determining Relevant Variables

Through their own analysis and reviewing past research, Sun and Shenoy [2006] identified the 20 variables in Table 10.1 to be potential predictors of bankruptcy.

Learning Structure

Next, we show how the structure of the network was learned from data. First, we discuss the sample from which the data were obtained, and then we present the method for learning from these data. It was predetermined that the network will be a naive Bayesian network. So by "learning structure" we mean learning which variables to include in the network.

The Sample Firms used in the study were publicly traded firms in the major stock exchanges across various industries during the period 1989-2002. Bankrupt firms were identified as follows: First, bankrupt firms were located through the Compustat and Lexis-Nexis Bankruptcy Report data bases. Then, bankruptcy filing dates were determined by searching the Lexis-Nexis Bankruptcy Report library, Lexis-Nexis News, and firms' Form 8-K reports. Firms without available bankruptcy filing dates were eliminated. For each bankrupt firm, the most recent annual report prior to its bankruptcy filing date was obtained. If the time between the fiscal year-end of the most recently filed annual report and the bankruptcy filing date was greater than two years, the firm was eliminated.[1] This procedure resulted in 890 bankrupt firms.

To identify non-bankrupt firms, first 500 firms were randomly selected from all active firms available in Compustat for each of the years between 1989 and 2002 inclusive. Once a non-bankrupt firm was selected for one year, it was excluded from selection for later years. This procedure resulted in 7000 active firms. Of these firms, 63 had missing data on all of the identified predictors, so they were eliminated. The final sample consisted of 6937 non-bankrupt firms. So the total size of the sample was $890 + 6937 = 7827$.

The Heuristic Learning Technique Our goal is to learn a Bayesian network containing a node for bankruptcy and a node for each of the relevant variables in Table 10.1. The most straightforward way to do this would be to learn a Bayesian network from the sample just described using one of the structure learning packages described in Chapter 4, Section 4.6. Alternatively, if for most firms we have data on all 20 variables, we could just learn a Markov blanket of the target variable (bankruptcy). The reason is that the variables in this set shield the target variable from the influence of all other variables. So if we know all their values, the values of the other variables are irrelevant to the conditional probability of the target variable. On the other hand, if we often have missing values in the Markov blanket, then we would want to learn a network containing the other variables also. Many of the learning packages (e.g., Tetrad) can also determine a Markov blanket from data.

Sun and Shenoy [2006] chose not to use any of the learning packages. Prakash Shenoy [private correspondence] said they made this decision for the following reasons:

> All learning packages assume that we have a random sample for the variables we are trying to learn the model from. We didn't have a random sample.

> The models that these learning packages learn usually need very large sample sizes. We didn't have a very large sample of bankrupt firms.

Instead, they developed a heuristic method to guide the selection of variables in a Markov blanket of the bankruptcy variable. Then they would use these

---

[1]This was done to ensure that the data used were current.

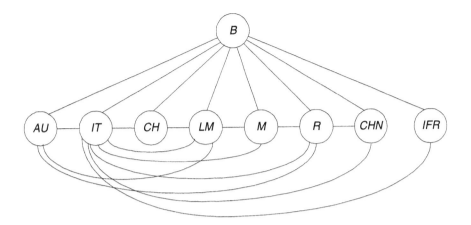

Figure 10.2: There is an edge between two variables if they are correlated.

variables as the children in a naive Bayesian network in which the bankruptcy variable was the root.

Their method proceeded as follows: They had 20 predictor variables and 1 target variable, namely bankruptcy status, making a total of 21 variables, and a sample consisting of 7827 records containing values of these variables. Based on this sample, they determined the correlation coefficient of each of the 21 variables with each of the other 21 variables. If the absolute value of the correlation coefficient was $\geq .1$, they judged the two variables to be correlated.[2] According to this criterion, only eight of the variables were correlated with bankruptcy. Figure 10.2 shows the relevant variables. There is an edge between two predictors if they were found to be correlated. It could be that, for example, CHN is correlated with B only through R. That is, we could have

$$I_P(CHN, B|R).$$

If this were the case, CHN would not be necessary to a Markov blanket. To determine this, all conditional correlation coefficients were computed. For these particular variables, it was found that

$$Corr(CHN, B|R) = -.22.$$

If we again use a cutoff of .1 for the absolute value of the correlation coefficient, this result indicates that CHN is not independent of B conditional on R, and so we should leave CHN in the network. A similar check of all relevant conditional correlation coefficients showed that none of the variables should be eliminated from the network.

The structure of the resultant naive Bayesian network appears in Figure 10.3. Note that there was no check to see if the conditional independencies

---

[2] Values of .05, .15, and .2 were also tried. Using a cutoff of .1 was found to yield a model with the best prediction capability while minimizing the number of variables.

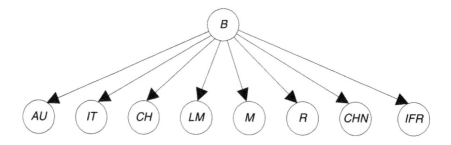

Figure 10.3: The structure of a naive Bayesian network for bankruptcy prediction.

entailed by the network were actually satisfied. For example, the network entails

$$I_P(AU, IT|B).$$

We know from Figure 10.2 that AU and IT are dependent. It could be that they remain dependent conditional on B. Prakash Shenoy [private correspondence] addresses this criticism as follows:

> The naïve Bayes model we used for the bankruptcy model is just that: a model. We realize that the predictors are not exactly conditionally independent given the bankruptcy variable. There is a trade-off between the complexity of the model and the number of parameters one needs to estimate from the data. The advantage of the naïve Bayes model is that the number of parameters is kept to a minimum. We made sure the naïve Bayes assumption was satisfied as much as possible by carefully selecting the subset of variables in the model. The procedure we describe for selecting the variables was motivated by the conditional independence assumption of a naïve Bayes model.

Learning Parameters

Most of the predictor variables are continuous; however, the target variable (bankruptcy) is discrete. In such cases we can sometimes get better predictive results by discretizing the continuous variable (see the example concerning cervical spinal-cord trauma in Chapter 4, Section 4.7.1). This was the path chosen for the current model. When we are trying to predict whether a company will go bankrupt, unusually low cash flow is indicative they will, while unusually high cash flow is indicative they won't [McKee and Lensberg, 2002]. Values in the middle are not indicative one way or the other. Recall from Chapter 3, Section 3.5.2, that in such cases we often obtain better results if we group the values in each tail together, and the Pearson-Tukey Method does this. This method replaces the range of a continuous random variable X by three discrete values. In the current model these values are *low*, *medium*, and *high*. Values of X below the 18.5 percentile of the cumulative distribution of X are replaced by the value *low*, values between the 18.5 and 81.5 percentile are replaced by the value *medium*, and values to the right of the 81.5 percentile are replaced by the value *high*. In this way,

$$\begin{aligned} P(low) &= .185 \\ P(medium) &= .63 \\ P(high) &= .185. \end{aligned}$$

Recall from Section 10.1.2 that a sample of 7827 records, containing data on 6937 non-bankrupt firms and 890 bankrupt firms, was developed. To apply the Pearson-Tukey Method, for each continuous variable X in Figure 10.3, points x_1 and x_2 were determined such that 18.5% of the values of X in the sample fell below x_1 and 81.5% fell below x_2. Then for each firm (record) in the sample, the following discrete values were assigned to X for that firm:

low if the value of X for the firm is $< x_1$;

medium if the value of X for the firm is in $[x_1, x_2]$;

high if the value of X for the firm is $> x_2$.

This procedure was used to determine the values of CH, LM, M, R, CHN, and IFR for each firm in the sample. The variables AU and IT were already binary variables with values 0 and 1. The variable B was already binary with values "yes" and "no."

Next, the values of the conditional probabilities for the Bayesian network in Figure 10.3 were learned from the sample. The sample proportion of bankruptcy was larger than the population proportion [McKee and Greenstein, 2000]. So rather than using either proportion for the prior probability of bankruptcy, a value of .5 was used. In this way if the probability of the data given bankruptcy is greater than the probability of the data given no bankruptcy (i.e., the likelihood ratio is > 1), the posterior probability of bankruptcy will be greater than .5, and it will be less than .5 otherwise.

The following report, produced by Netica, shows the probability distributions in the network:

```
B:
yes             no
0.5             0.5

AU:
zero            one             B
0.44            0.56            yes
0.75            0.25            no

CH:
low             medium          high            B
0.29            0.63            0.08            yes
0.19            0.63            0.18            no

LM:
low             medium          high            B
0.31            0.64            0.05            yes
0.15            0.63            0.22            no

M:
low             medium          high            B
0.44            0.54            0.02            yes
0.15            0.65            0.2             no

R:
low             medium          high            B
0.67            0.28            0.05            yes
0.12            0.68            0.2             no

CHN:
low             medium          high            B
0.35            0.53            0.12            yes
0.16            0.64            0.2             no

IFR:
low             medium          high            B
0.09            0.69            0.22            yes
0.19            0.75            0.06            no

IT:
zero            one             B
0.37            0.63            yes
0.76            0.24            no
```

The Bayesian network, displayed using Netica, appears in Figure 10.4 (a). In Figure 10.4 (b) all predictor variables are instantiated to their values which are most indicative of bankruptcy. In Figure 10.4 (c) all predictor variables are instantiated to their values which are least indicative of bankruptcy. We see that, according to the network, we can become very certain of bankruptcy or no bankruptcy based on the evidence.

10.2   Experiments

Next, we describe experiments testing the accuracy of the bankruptcy prediction Bayesian network.

10.2.1   Method

A 10-fold analysis was used to test the system. First, the entire sample was divided randomly into 10 equally sized subsets. Then nine subsets were randomly

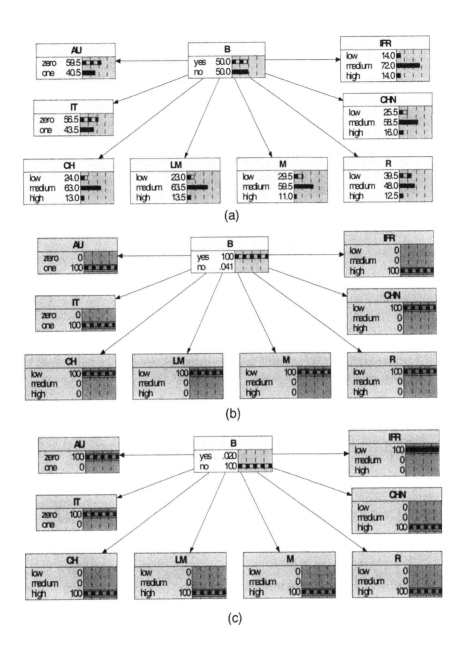

Figure 10.4: The bankruptcy Bayesian network appears in (a). In (b) all predictor variables are instantiated to their values which are most indicative of bankruptcy. In (c) all predictor variables are instantiated to their values which are least indicative of bankruptcy.

	9-Node Network		21-Node Network	
Trial	% Bkpt. Correct	% NonBkpt. Correct	% Bkpt. Correct	% NonBkpt. Correct
1	79.78	82.25	74.16	80.38
2	86.52	81.24	84.27	81.82
3	78.65	82.56	80.90	83.29
4	87.64	83.14	85.39	83.86
5	74.16	83.69	73.03	83.98
6	79.78	81.53	87.64	82.40
7	85.39	77.06	89.89	78.35
8	79.78	82.25	80.90	81.24
9	79.78	82.4	78.65	81.10
10	79.78	82.4	80.90	81.39
Avg.	81.12	81.85	81.57	81.78

Table 10.2: Prediction success for 9 node and 21 node naive Bayesian networks.

selected as the training set, and the remaining subset was the test set. The conditional probabilities for the network were learned from the training set. Next, the resultant network was used to determine the probability of bankruptcy for all firms in the test set. If the posterior probability of bankruptcy was $> .5$, the prediction was that the firm would go bankrupt; otherwise, the prediction was that the firm would not go bankrupt. This process was repeated 10 times. The conditional probabilities shown in the previous sections were the ones obtained in 1 of the 10 trials; however, those obtained in the other 9 trials were about the same.

The same analysis was also done using a naive Bayesian network that included all 20 predictor variables.

10.2.2   Results

We first show results of the experiment described above. Then we compare the results to results obtained using other versions of the network and using logistic regression.

Results of the Basic Experiment

Table 10.2 shows the results of the 10 trials and the averages over all trials. When we say, for example, that "% Bkpt. Correct" is 81.12, we mean that the system said that 81.12% of the bankrupt firms would go bankrupt. We see that the results are fairly good and that the 21-node network has essentially the same predictive accuracy as the 9-node network.

Comparison to the Use of Other Variable Ranges

The model used the Pearson-Tukey Method to discretize a continuous variable, which yields only three states for the variable. Perhaps better results could be

# States	% Bkpt. Correct	% NonBkpt. Correct
2	82.58	77.55
3	83.37	77.44
4	83.82	74.94
5	83.37	75.45
6	83.15	73.83
7	82.25	75.13
8	82.36	72.36
9	81.57	71.44
10	80.67	69.46
Continuous	83.6	77.51

Table 10.3: Effect of number of states on prediction accuracy.

obtained if more states were used. This conjecture was also tested. If n is the number of states, and bankrupt firms tend to have values in the extreme left tail of the distribution (which is the case for M, R, CH, L, and CHN), then the following points were used as the percentiles that determine the cutoffs:

$$\frac{1}{n+1}, \frac{2}{n+1}, \frac{3}{n+1}, \cdots \frac{n-1}{n+1}.$$

For example, if $n = 4$, the points were

$$\frac{1}{5}, \frac{2}{5}, \frac{3}{5}.$$

In the case of IFR, bankrupt firms tend to have values in the extreme right tail. So the following points were used for this variable:

$$\frac{2}{n+1}, \frac{3}{n+1}, \frac{4}{n+1}, \cdots \frac{n}{n+1}.$$

The average results appear in Table 10.3. Untabulated t-test results showed that there is no significant difference in the accuracy for two states and three states. Furthermore, the accuracy is about the same as it was when the Pearson-Tukey Method was used. However, when the number of states is four, the results for correctly predicting bankruptcy are statistically insignificant compared to the results for two and three states, but the results for correctly predicting non-bankruptcy are significantly worse ($p < .001$). After that, they continue to degrade as the number of states increases. One interpretation of these results is that bankrupt firms may have extreme values in one tail of each distribution, and non-bankrupt firms in the other tail. So two states would be enough to capture the distinction, and too many states result in overfitting.

Perhaps better performance could be obtained by not discretizing the variables at all. This was also tested by giving the variables normal distributions. The average results appear in the last row of Table 10.3. Untabulated t-test

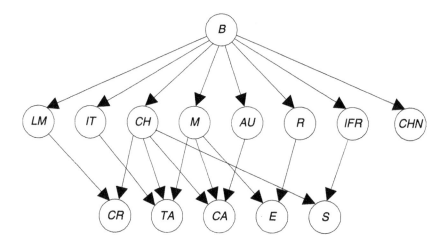

Figure 10.5: The structure of a cascaded naive Bayesian network for bankruptcy prediction.

results showed that the accuracy, when we use continuous variables for predicting bankruptcy, is statistically insignificant compared to the accuracy when we use the Pearson-Tukey Method; however, the accuracy for predicting non-bankruptcy is significantly worse ($p < .001$).

Comparison to a Cascaded Naive Bayesian Network

Many firms have missing data on one or more of the children nodes in Figure 10.4 (a). Specifically, 1678 firms have a missing value on IT, 2419 firms have a missing value on M, 1537 firms have a missing value on AU, 2331 firms have a missing value on R, 964 firms have a missing value on IFR, 5345 have a missing value on LM, 1086 have a missing value on CH, and 1679 have a missing value on CHN. Rather than entering no information for a node, we might obtain better results if we entered information relevant to the value of these nodes. For example, if the value of CR is probabilistically dependent on the value of LM, and the value of LM is missing for a given firm but the value of CR is not, then the value of CR would tell us something about the probable value of LM, which in turn should tell us something about the probable value of B.

To include predictors of predictors, a **cascaded naive Bayesian network** was developed. In such a network each child node (in the top-level naive Bayesian) is the root of a second-level naive Bayesian network. The nodes in this second-level network are predictors for that child. For each child in the network in Figure 10.4 (a), the relevant predictors were learned using the same technique used to learn the relevant predictors for bankruptcy. The result was the cascaded network in Figure 10.5.

The cascaded naive Bayesian network correctly predicted 81.12% of bankruptcies and 80.08% of non-bankruptcies. Recall that the analogous results for the naive network were 81.12% and 81.85%. Untabulated t-test results showed the difference in the accuracies for predicting bankruptcy to be insignificant; however the cascaded network performed significantly worse ($p < .05$) at predicting non-bankruptcy.

These results lend support for the use of the more parsimonious model in Figure 10.4 (a). However, if even more data were missing on a given firm, it could be that the cascaded model would perform better.

Comparison to Logistic Regression

Recall from Chapter 2, Section 2.5.3, that linear regression is often used to predict continuous target variables. A similar standard statistical technique, called **logistic regression**, is used when the target variable is discrete. Since logistic regression has been widely used as a bankruptcy prediction tool (see, e.g., [Ohlson, 1980]), the performance of the naive Bayesian network model in Figure 10.4 (a) was compared to that of logistic regression. Logistic regression cannot be used when there are missing data unless techniques are used to estimate the missing values. Therefore, the comparison was restricted to firms with complete information on all eight predictors. This reduced the sample to including 1435 non-bankrupt and 414 bankrupt firms.

The following logit was learned using logistic regression:

$$\ln\left(\frac{B}{1-B}\right) = -4.755 + .933AU + 1.098IT - 3.165CH - .056LM$$
$$-2.222R - .391CHN + .356IFR - .156M.$$

The naive Bayesian network model in Figure 10.4 (a) had an average predictive accuracy of 80.43% for bankruptcy and 80.00% for non-bankruptcy, while logistic regression had an average predictive accuracy of 79.48% for bankruptcy and 82.02% for non-bankruptcy. These results were found to be statistically insignificant.

Recall that the Bayesian network model had about the same predictive accuracy when we included firms with missing data. Perhaps this is the greatest argument for the use of such models. Namely, they readily handle missing data, and, at least in this study, did so with no degradation in performance.

10.2.3 Discussion

The parsimonious naive Bayesian network model in Figure 10.4 (a) performed as well or better than all other models considered. Indeed, the even more parsimonious model that discretized continuous variables into only two states performed just as well as the one in Figure 10.4 (a). The question remains of whether we could improve on the performance with other more complex models. Recall from Section 10.1.2 that Prakash Shenoy said, "We realize that the predictors are not exactly conditionally independent given the bankruptcy variable. There is a trade-off between the complexity of the model and number of parameters one needs to estimate from the data. The advantage of the naïve Bayes model is that the number of parameters is kept to a minimum." Perhaps we could obtain better performance if we modeled the dependencies between predictors. Furthermore, we could abandon the naive network approach completely and try to learn a general Bayesian network from data on all 21 variables. Another possible improvement, which would not increase the complexity of the network, would be to use the population proportion of bankruptcies as the prior probability of bankruptcy. Since this is a Bayesian model, the probabilities represent our beliefs, and, as such, they need not all come from the same source.

EXERCISES

Section 10.1

Exercise 10.1 In a naive Bayesian network are the children mutually independent? Are the children mutually independent conditional on the root?

Exercise 10.2 Using Netica, develop the Bayesian network in Figure 10.4 (a).

1. Instantiate AU to low. Have the probability distributions of the other nodes changed?

2. Instantiate B to yes. Now instantiate AU to low. Have the probability distributions of the other nodes changed further?

3. Investigate how various instantiations of the leaves affect the probability distribution of the bankruptcy node.

Exercise 10.3 Modify your network by using the sample proportion of bankrupt firms as the prior probability of bankruptcy. For various instantiations of the leaves, compare the posterior probability distribution of the bankruptcy node in this network to the one in the original network.

Section 10.2

Exercise 10.4 Obtain data on non-bankrupt and bankrupt firms from the Compustat and Lexis-Nexis Bankruptcy Report data bases. From these data learn a Bayesian network using a Bayesian network learning package such as Tetrad. Try both discretizing the data and using normal distributions. Compare the predictive results of your models to those of the network in Figure 10.4 (a).

Part III

Marketing Applications

Chapter 11

Collaborative Filtering

Suppose Joe decides to order books from an on-line store such as the one at Amazon.com. After Joe has made several purchases, we can learn something about Joe's interests and preferences regarding books from these purchases. A good marketing strategy would be to then recommend further books to him based on what we learned about his preferences. These recommendations would also supply useful information to Joe. One way to do this would be to recommend books that were purchased by individuals who had made similar purchases to those of Joe. This process of recommending items of interest to an individual, based on the interests of similar individuals, is called **collaborative filtering**. It is particularly effective for on-line stores, where a data set containing individuals' purchases can be automatically maintained.

There are two collaborative filtering techniques. The first technique, called **implicit voting**, interprets an individual's preferences from the individual's behavior. For example, if the individual purchased the text *War and Peace*, we may infer that the individual voted 1 for that text, while if the individual did not purchase it, we may infer that the individual voted 0. A technique related to collaborative filtering is **market basket analysis**, which attempts to model the relationships among items purchased. Market basket analysis has traditionally been used by retail stores to look at the items that people purchase at the same visit and then rearrange their stocking methods so that items that are often purchased together are stocked nearer or farther away from each other. However, this traditional application cannot take advantage of information concerning a current user's purchases. A more sophisticated application of market basket analysis is to develop a model that enables us to infer how likely an individual is to buy one or more items based on items the individual already purchased. So when the behavior investigated is whether or not an item is purchased, collaborative filtering with implicit voting is a market basket analysis application. The example just shown concerning Joe is such an application.

The second collaborative filtering technique, called **explicit voting**, learns an individual's preferences by asking the individual to rank the items on some scale, which is usually numeric. For example, suppose we ask five individuals to rank four books on a scale of 1 to 5, and we obtain the votes in the following table:

Person	Great Gatsby	Rumor of War	War and Peace	Vanity Fair
Gloria	1	4	5	4
Juan	5	1	1	2
Amit	4	1	2	1
Sam	2	5	4	5
Judy	1	5	5	5

Suppose further that Joe votes as follows for the first three books:

Person	Great Gatsby	Rumor of War	War and Peace
Joe	1	5	5

We want to estimate whether Joe would like *Vanity Fair*. Just by perusing the data, we see that Joe seems similar to Gloria, Sam, and Judy in his preferences. Since they liked *Vanity Fair*, we can guess that Joe would also.

Next, we discuss two formal methods for performing the type of analysis just illustrated. The first method is **memory-based**, while the second is **model-based**.

11.1 Memory-Based Methods

Henceforth we refer to an individual as a **user**, and the individual whose preferences we wish to predict as the **active user**. We denote the active user as $user_a$. Memory-based collaborative filtering algorithms predict the votes of the active user based on weights obtained from a data set of user votes. We start with a data set as follows:

Person	$item_1$	$item_2$	$\cdots$	$item_m$
$user_1$	v_{11}	v_{12}	$\cdots$	v_{1m}
$user_2$	v_{21}	v_{22}	$\cdots$	v_{2m}
$\vdots$	$\vdots$	$\vdots$	$\ddots$	$\vdots$
$user_n$	v_{n1}	v_{n2}	$\cdots$	v_{nm}

In memory-based algorithms, we first determine a measure of how similar the active user is to each of the other users. This measure is called a weight and is denoted $w(user_a, user_i)$. A straightforward measure to use for the weight is the correlation coefficient, which was introduced in Chapter 2, Section 2.5.2. Recall that the correlation coefficient ρ of two random variables X and Y is given by

$$\rho(X, Y) = \frac{E\left((X - \bar{X})(Y - \bar{Y})\right)}{\sigma_X \sigma_Y},$$

where E denotes expected value, $\bar{X} = E(X)$, and σ_X is the standard deviation of X.

If we use the correlation coefficient as our weight, we have

$$w(user_a, user_i) = \rho(V_a, V_i) \quad = \quad \frac{E\left((V_a - \bar{V}_a)(V_i - \bar{V}_i)\right)}{\sigma_{V_a} \sigma_{V_i}}$$

$$= \quad \frac{\sum_j \left(v_{aj} - \bar{V}_a\right)\left(v_{ij} - \bar{V}_i\right)}{\sqrt{\sum_j (v_{aj} - \bar{V}_a)^2}\sqrt{\sum_j (v_{ij} - \bar{V}_i)^2}},$$

where the sums are over all items on which $user_a$ and $user_i$ have both voted. Intuitively, if this weight is close to 1, the preferences of $user_i$ are very similar to those of $user_a$, while if it is close to -1, their preferences are very dissimilar.

Example 11.1 *Suppose we have the following data set of preferences:*

Person	$item_1$	$item_2$	$item_3$	$item_4$
$user_1$	1	4	5	4
$user_2$	5	1	1	2
$user_3$	4	1	2	1
$user_4$	2	5	4	5
$user_5$	1	5	5	5

Suppose further that $user_a$ has these preferences:

Person	$item_1$	$item_2$	$item_3$
$user_a$	1	5	5

Note that these are the same preferences we showed in the introduction when we discussed Joe, but now we use variables rather than mentioning individuals. Next, we determine $w(user_a, user_1)$. We have

$$\bar{V}_a = \frac{\sum_{j=1}^3 v_{aj}}{3} = \frac{1 + 5 + 5}{3} = 3.67$$

$$\bar{V}_1 = \frac{\sum_{j=1}^{3} v_{1j}}{3} = \frac{1+4+5}{3} = 3.33$$

$$
\begin{aligned}
w(user_a, user_1) &= \frac{\sum_{j=1}^{3}(v_{aj} - \bar{V}_a)(v_{3j} - \bar{V}_3)}{\sqrt{\sum_{j=1}^{3}(v_{aj} - \bar{V}_a)^2}\sqrt{\sum_{j=1}^{3}(v_{1j} - \bar{V}_1)^2}} \\
&= \frac{(1-3.67)(1-3.33)+(5-3.67)(4-3.33)+(5-3.67)(5-3.33)}{\sqrt{(1-3.67)^2+(5-3.67)^2+(5-3.67)^2}\sqrt{(1-3.33)^2+(4-3.33)^2+(5-3.33)^2}} \\
&= .971.
\end{aligned}
$$

Since this weight is close to 1, we conclude that the preferences of $user_1$ are very similar to those of $user_a$. It is left as an exercise to show

$$
\begin{aligned}
w(user_a, user_2) &= -.1.000 \\
w(user_a, user_3) &= -.945 \\
w(user_a, user_4) &= .945 \\
w(user_a, user_5) &= 1.000.
\end{aligned}
$$

Note that the preferences of $user_5$ are perfectly positively correlated with those of $user_a$ and they have the exact same preferences, while the preferences of $user_5$ are perfectly negatively correlated with $user_a$ and they have opposite preferences.

Once we have these weights, we can predict how $user_a$ might vote on a new item, $item_k$, based upon how the users in the data set voted on that item, using the following formula:

$$\hat{v}_{ak} = \bar{V}_a + \alpha \sum_{i=1}^{n} w(user_a, user_i)(v_{ik} - \bar{V}_i),$$

where α is a normalizing constant chosen so that the absolute values of the weights sum to 1, and $\hat{v}_{ak}$ is the predicted preference. The idea is that we estimate how much the active user's preference for $item_k$ will differ from that user's average preference based on how much the preferences for $item_k$ of users in the data set differ from their average preferences.

Example 11.2 Suppose we have the preferences in Example 11.1, and we want to predict how $user_a$ will vote on $item_4$. We have

$$\alpha = \frac{1}{.971 + |-1| + |-.945| + .945 + 1} = .206$$

$$
\begin{aligned}
\hat{v}_{a4} &= \bar{V}_a + \alpha \sum_{i=1}^{n} w(user_a, user_i)(v_{i4} - \bar{V}_i) \\
&= 3.67 + .206[.971(4 - 3.33) - 1(2 - 2.33) \\
&\quad -.945(1 - 2.33) + .945(5 - 3.67) + 1(5 - 3.67)] \\
&= 4.66.
\end{aligned}
$$

We predict $user_a$ should have a high preference for $item_4$. This is not surprising as the three users who are similar to $user_a$ like the item, while the two users who are dissimilar to $user_a$ do not like it.

An alternative measure to use for the weights is **vector similarity**, which is discussed in [Breese et al., 1998]. Extensions to memory-based algorithms include **default voting**, **inverse-user frequency**, and **case amplification**. These extensions are also discussed in [Breese et al., 1998].

11.2 Model-Based Methods

The method discussed in the previous section is a heuristic method. In Chapter 1, Section 1.2, we distinguished heuristic methods from model-based methods. Next, we present model-based methods for doing collaborative filtering.

11.2.1 Probabilistic Collaborative Filtering

In the framework of probability theory, we can use the expected value of the active user's ($user_a$) vote for $item_k$ as our prediction of that vote. That is, if the votes are integral values between 1 and r inclusive,

$$\hat{v}_{ak} = E(V_{ak}) = \sum_{i=1}^{r} i \times P(V_{ak} = i | \mathbf{V}_a),$$

where $\mathbf{V}_a$ is the set of votes so far cast by the active user. The probability distribution is learned from the data set containing the votes of the other users. Shortly, we discuss two methods for doing this probabilistic modeling, but first we give an example.

Example 11.3 *Suppose we have the data set and active user in Example 11.1. Then*

$$\mathbf{V}_a = \{v_{a1} = 1, v_{a2} = 5, v_{a3} = 5\}.$$

If from the data set we learn that

$$P(V_{a4} = 1 | \mathbf{V}_a) = .02$$
$$P(V_{a4} = 2 | \mathbf{V}_a) = .03$$
$$P(V_{a4} = 3 | \mathbf{V}_a) = .1$$
$$P(V_{a4} = 4 | \mathbf{V}_a) = .4$$
$$P(V_{a4} = 5 | \mathbf{V}_a) = .45,$$

then

$$
\begin{aligned}
\hat{v}_{a4} = E(V_{a4}) &= \sum_{i=1}^{r} i \times P(V_{a4} = i | \mathbf{V}_a) \\
&= 1 \times .02 + 2 \times .03 + 3 \times .1 + 4 \times .4 + 5 \times .45 \\
&= 4.23.
\end{aligned}
$$

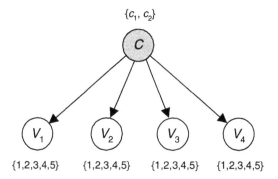

Figure 11.1: A cluster model we might learn from the data set in Example 11.1.

In the previous problem we simply made up reasonable conditional probabilities for the sake of illustration. Next, we show methods for learning these probabilities from a data set.

11.2.2   A Cluster Model

One application of Bayesian networks with hidden variables is to the **cluster learning problem**, which is as follows: Given a collection of unclassified entities and features of those entities, organize the entities into classes that in some sense maximize the similarities of the features in the same class. Autoclass [Cheeseman and Stutz, 1995] models this problem using a Bayesian network in which the root is a discrete hidden variable, and its possible values correspond to the underlying classes of entities. The features are all children of the root and may or may not have edges between them. Given a data set containing values of the features, Autoclass searches over variants of this model, including the number of possible values of the hidden variable, and it selects a variant so as to approximately maximize the posterior probability of the variant.

Applying this method to the collaborative filtering problem, the clusters (values of the root) are groups of users with common preferences, while the features (children of the root) are the votes on the items. There is one node for each item. It is reasonable in this application to assume these features are conditionally independent given the root (i.e., we have a naive Bayesian network), but this is not necessary. This assumption entails that if, for example, we know a user, say Joe, is in a given group, there is a probability that he will like *Rumor of War*, and finding out that he likes *War and Peace* does not change this probability. Intuitively, we already know he is in a group that likes war movies. So finding out he likes one war movie does not change the probability that he likes another one.

We learn the cluster model from a data set of user preferences using Autoclass. After that, we can do inference for a new user (the active user) using a Bayesian network inference algorithm.

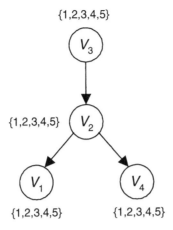

Figure 11.2: The structure of the Bayesian network model we might learn from the data set in Example 11.1.

Example 11.4 *Suppose we have the data set and active user in Example 11.1. Then from the data base we might learn a naive Bayesian network in which there are two clusters (i.e., the variable C has two possible values). This model is illustrated in Figure 11.1. Note that only the possible values of each node are shown. That is, the conditional probabilities of these values are not shown even though they too would have been learned from the data set. Once we have the model, we can compute $P(V_{a4} = i | V_a)$ for $1 \leq i \leq 5$, where, for example,*

$$V_a = \{v_{a1} = 1, v_{a2} = 5, v_{a3} = 5\}.$$

We can then use these probabilities to obtain a predicted vote for $item_4$ as illustrated in Example 11.3.

11.2.3   A Bayesian Network Model

Rather than assuming the existence of clusters, we can simply learn the probabilistic relationships among the votes using a Bayesian network learning algorithm. We can then use the model learned to do inference for an active user.

Example 11.5 *Suppose we have the data set and active user in Example 11.1. From this data set we might learn a Bayesian network that has the DAG structure shown in Figure 11.2. Again, no conditional probabilities are shown even though they too would be learned from the data set. Once this model is learned, we can do inference for the active user as illustrated in Example 11.4.*

11.3   Experiments

Breese et al. [1998] compared the predictive accuracy of the models we've discussed using three data sets. After describing the data sets, we explain their experimental method and results.

11.3.1   The Data Sets

The following three data sets were used in the study:

1. **MSWeb:** This data set contains users' visits to various areas of the Microsoft corporate web site. The voting here is implicit. That is, the vote is 1 if the area is visited and 0 if it is not visited.

2. **Neilsen:** This data set contains Neilsen network television viewing data for users during a two-week period in the summer of 1996. Again the vote is implicit, being 1 if the show is watched and 0 otherwise.

3. **EachMovie:** This data set contains explicit votes obtained from the EachMovie collaborative filtering site maintained by Digital Equipment Research Center. The votes are from the period 1995 through 1997. Votes range in value from 0 to 5.

The table that follows provides details about these data sets:

Data Set	# Users	# Items	Mean # votes per user
MSWeb	3453	294	3.95
Neilsen	1463	203	9.55
EachMovie	4119	1623	46.4

11.3.2   Method

Breese et al. [1998] divided each data set into a training set and a test set. The training set was used as the collaborative filtering data set for the memory-based algorithms and as the learning set for the probabilistic models. After the models were learned, each user in the test set was treated as an active user. The user's votes were divided into a set of votes that were treated as observed V_a and a set of votes which were to be predicted P_a. The algorithms were then used to predict the votes in P_a from the votes in V_a. The following protocols were used to choose the votes in P_a:

1. **All but 1:** A single vote was predicted (in P_a), and all other votes in the test set were treated as observed (in V_a).

2. **Given 2:** Two votes were treated as observed (in V_a), and all other votes were predicted (in P_a).

3. **Given 5:** Five votes were treated as observed (in V_a), and all other votes were predicted (in P_a).

4. **Given 10:** Ten votes were treated as observed (in V_a), and all other votes were predicted (in P_a).

Collaborative filtering applications use two approaches to presenting new items to the users. Next, we describe each approach and the method Breese et al. [1998] used to analyze the effectiveness of a model when the approach is used.

One-at-a-Time Approach

The first approach is to simply present new items to the user one at a time, along with ratings predicting the user's interest in the items. The GroupLens system [Resnick et al., 1994] works in this fashion. Each Netnews article has a bar chart indicating the prediction concerning the user's interest in the article. To analyze the effectiveness of a model when this approach is used, we simply compare the predicted values of the votes in P_a to their actual values. That is, for a given active user $user_a$, we let

$$Score_a = \frac{1}{k_a} \sum_{V_{aj} \in P_a} |\hat{v}_{aj} - v_{aj}|, \tag{11.1}$$

where $\hat{v}_{aj}$ is the predicted value of V_{aj}, v_{aj} is its actual value, and k_a is the number of predicted votes. We then take the average of $Score_a$ over all users in the test set. Lower scores are better. Evaluating a model according to how well it does according to this criterion is called **average absolute deviation scoring**.

Ordered-List Approach

The second approach is to provide the active users with an ordered list of recommended items. This list is called a ranked list of items, and evaluating a model according to how well it does when this approach is used is called **ranked scoring**. Examples of this type of system include PHOAKS [Terveen et al., 1997] and SiteSeer [Rucker and Polanco, 1997]. These systems present the user with a ranked list of recommended items (e.g., music recordings), where higher ranked items are predicted to be more preferred. Analyzing the effectiveness of a model in the case of ranked scoring is not straightforward. Breese et al. [1998] developed the analysis technique that we describe next.

For each user $user_a$ in the test set, we first order the items, whose votes are predicted, in decreasing order according to the values of those predictions. This is the ranked list of items. Next, we determine a default or neutral vote d_j for each item $item_j$. In the case of implicit voting, we set the default to 0, meaning the item was not chosen, while in the case of the explicit votes in the EachMovie data set, we set the default to the approximate middle value 3. For each actual vote on the items in the ranked list, we then determine how much more the actual vote is than the default vote. This difference is the utility of the item to the user. We then say that the expected utility of the ranked list is

Algorithm	All but 1	Given 2	Given 5	Given 10
CR	0.994	1.257	1.139	1.069
BC	1.103	1.127	1.144	1.138
BN	1.066	1.143	1.154	1.139
VSIM	2.136	2.113	2.177	2.235
REQ	0.043	0.022	0.023	0.025

Table 11.1: The results of computing the average absolute deviations for the EachMovie data set. Lower scores are better.

the sum over all items of the product of the utility of the item and probability of the user viewing that item. That is,

$$R_a \equiv EU(list) = \sum_{j=1}^{k_a} \max(v_{aj} - d_j, 0) P(item_j \text{ will be viewed}),$$

where k_a is the number of predicted votes for $user_a$, and the votes are sorted by index j according to declining values of $\hat{v}_{aj}$ (the predicted votes). We still need to derive the probability values P. We suppose that the probability of viewing items in the list declines exponentially as we go down the list and that the first item will be viewed for certain. If we let $item_\alpha$ be the item which has a .5 probability of being viewed, we then have

$$R_a \equiv \sum_{j=1}^{k_a} \max(v_{aj} - d_j, 0) \left(\frac{1}{2^{(j-1)/(\alpha-1)}} \right).$$

Breese et al. [1998] assumed $\alpha = 5$ in their analysis. (That is, $item_5$ has a .5 probability of being viewed.) However, they obtained the same results with $\alpha = 10$. Finally, the score of the ranked list for all active users in the test set is given by

$$R = 100 \frac{\sum_a R_a}{\sum_a R_a^{\max}}, \tag{11.2}$$

where $R_a^{\max}$ is the score obtained for a list sorted according to the user's actual votes. In this way, 100 is the highest possible score.

11.3.3 Results

Table 11.1 shows the results of average absolute deviation scoring (i.e., using Equality 11.1) for the EachMovie data set. Lower scores are better. The symbols in the left column represent the following algorithms:

- **CR:** Memory-based method using the correlation coefficient

- **BC:** Cluster method

- **BN:** General Bayesian network method

Algorithm	All but 1	Given 2	Given 5	Given 10
BN	66.89	59.95	59.84	53.92
CR+	63.59	60.64	57.89	51.47
VSIM	61.70	59.22	56.13	49.33
BC	59.42	57.03	54.83	47.83
POP	49.77	49.14	46.91	41.14
REQ	0.93	0.91	1.82	4.49

Table 11.2: The results of computing the rank scores for the MSWeb data set. Higher scores are better.

Algorithm	All but 1	Given 2	Given 5	Given 10
BN	44.92	34.90	42.24	47.39
CR+	39.49	39.44	43.23	43.47
VSIM	36.23	39.20	40.89	39.12
BC	16.48	19.55	18.85	22.51
POP	13.91	20.17	19.53	19.04
REQ	2.40	1.53	1.78	2.42

Table 11.3: The results of computing the rank scores for the Neilsen data set. Higher scores are better.

- **VSIM:** Memory-based method using the vector similarity method with inverse-user frequency

- **REQ:** The difference between any other two column values must be at least this big to be considered statistically significant at the 90% level

Memory-based methods that use the correlation coefficient with inverse-user frequency, default voting, or case amplification extensions are not shown as these extensions are not effective for this type of scoring. We see from Table 11.1 that in this comparison the correlation coefficient method did slightly better than the Bayesian network method, which did slightly better than the cluster method. The vector similarity method did much worse that the other three.

Tables 11.2, 11.3, and 11.4 show the results of rank scoring (i.e., using Equality 11.2) for the MSWeb, Neilsen, and EachMovie data sets, respectively. The symbols in the left column represent the same methods as in Table 11.1 with the following additions:

- **CR+:** Memory-based method using the correlation coefficient with inverse user frequency, default voting, and case amplification extensions

- **POP:** A baseline method that represents no collaborative filtering, in which we simply present the user with a list ordered according to the average votes over all users; that is, the items are ranked according to overall popularity

Algorithm	All but 1	Given 2	Given 5	Given 10
CR+	23.16	41.60	42.33	41.46
VSIM	22.07	42.45	42.12	40.15
BC	21.38	38.06	36.68	34.98
BN	23.49	28.64	30.50	33.16
POP	13.94	30.80	28.90	28.01
REQ	0.78	0.75	0.75	0.78

Table 11.4: The results of computing the rank scores for the EachMovie data set. Higher scores are better.

We see from these tables that the correlation coefficient method with extensions performed about as well as the Bayesian network method at the task of presenting users with a ranked list and that both of these methods performed better than the other methods.

Looking at the results of both average absolute deviation scoring and rank scoring, we see that the correlation coefficient method with extensions and the Bayesian network method were very close in performance. An advantage of the Bayesian network method (and model-based methods in general) over memory-based methods is that once the model is learned, only the Bayesian network must be stored and reasoned with in real time, whereas memory-based methods must store and reason with the entire data set in real time. Breese et al. [1998] note that the model-based algorithms are about four times as fast as the memory-based algorithms in generating recommendations.

EXERCISES

Section 11.1

Exercise 11.1 Suppose we have the following data set of preferences:

Person	$item_1$	$item_2$	$item_3$	$item_4$	$item_5$
$user_1$	2	3	5	4	1
$user_2$	3	4	4	3	2
$user_3$	5	1	1	1	4
$user_4$	4	2	2	3	5
$user_5$	3	3	4	3	3

Suppose further that $user_a$ has these preferences:

Person	$item_1$	$item_2$	$item_3$	$item_4$
$user_a$	3	4	5	3

1. Use the memory-based method with correlation coefficient to compute $w(user_a, user_i)$ for $1 \leq i \leq 5$.

2. Using these weights obtain an estimate of how $user_a$ will vote on $item_5$.

Section 11.2

Exercise 11.2 Suppose we have the data set and active user in Exercise 11.1. Then

$$\mathbf{V}_a = \{v_{a1} = 3, v_{a2} = 4, v_{a3} = 5, v_{a4} = 3\}.$$

Suppose further that from the data set we learn a Bayesian network, and, using that network, we compute that

$$P(V_{a5} = 1|\mathbf{V}_a) = .15$$

$$P(V_{a5} = 2|\mathbf{V}_a) = .4$$

$$P(V_{a5} = 3|\mathbf{V}_a) = .3$$

$$P(V_{a5} = 4|\mathbf{V}_a) = .1$$

$$P(V_{a5} = 5|\mathbf{V}_a) = .05.$$

Compute $E(V_{a5})$.

Exercise 11.3 Obtain a large data set such as MSWeb that contains information on implicit votes. Using a Bayesian network learning package such as Tetrad, learn a Bayesian network for vote prediction from this data set. Compare the predictive accuracy of this model with that of memory-based methods.

Exercise 11.4 Obtain a large data set such as EachMovie that contains information on explicit votes. Using a Bayesian network learning package such as Tetrad, learn a Bayesian network for vote prediction from this data set. Compare the predictive accuracy of this model with that of memory-based methods.

Chapter 12

Targeted Advertising

One way for a company to advertise its product would be to simply try to reach as many potential customers as possible. For example, they could just dump or mail flyers all over town. However, this could prove to be costly since there is no point in wasting an advertisement on someone who most probably will not buy the product regardless or, worst yet, who would be turned off by the advertisement. Another approach would be to try to identify those customers such that we can expect to increase our profit if we send them advertisements. **Targeted advertising** is the process of identifying such customers. We present a decision theoretic approach to targeted advertising developed by Chickering and Heckerman [2000]. The method uses class probability trees. So first we review such trees.

12.1 Class Probability Trees

Recall that Bayesian networks represent a large joint distribution of random variables succinctly. Furthermore, we can use a Bayesian network to compute the conditional probability of any variable(s) of interest given values of some other variables. This is most useful when there is no particular variable that is our target, and we want to use the model to determine conditional probabilities of different variables depending on the situation. For example, in the case of collaborative filtering (see Chapter 11) we want to be able to predict users' preferences for a number of different items based on their preferences for other items. However, in some applications there is a single target variable, and we are only concerned with the probability of the values of that variable given values of other variables. For example, in Chapter 2, Example 2.46, we were only interested in predicting the travel time from the miles driven and the deliveries. We were never interested in predicting the miles driven from the travel time and the deliveries. In situations like this we can certainly model the problem using a Bayesian network. Indeed, this is precisely what we did in Chapter 7, Section 7.3, and Chapter 10, Section 10.1. However, there are other techniques available for modeling problems that concentrate on a target variable. As mentioned in Chapter 10, Section 10.2.2, standard techniques include linear and logistic regression. The machine learning community developed a quite different method, called **class probability trees**, for handling discrete target variables. This method, which we discuss next, makes no special assumptions about the probability distribution.

Suppose we are interested in whether an individual might buy some particular product. Suppose further that income, sex, and whether the individual is mailed a flyer all have an influence on whether the individual buys, and we articulate three ranges of income, namely low, medium, and high. Then our variables and their values are as follows:

Target Variable	Values
Buy	{*no, yes*}

Predictor Variables	Values
Income	{*low, medium, high*}
Sex	{*male, female*}
Mailed	{*no, yes*}

There are $3 \times 2 \times 2 = 12$ combinations of values of the predictor variables. We are interested in the conditional probability of *Buy* given each of these combinations of values. For example, we are interested in

$$P(Buy = yes | Income = low, Sex = female, Mailed = yes).$$

We can store these 12 conditional probabilities using a class probability tree. A **complete class probability tree** has stored at its root one of the predictors (say *Income*). There is one branch from the root for each value of the variable stored at the root. The nodes at level 1 in the tree each store the same second

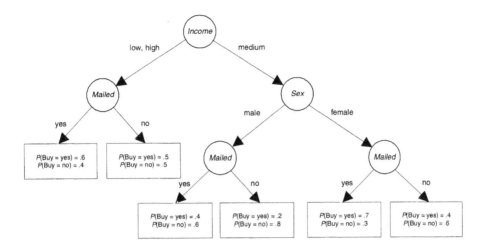

Figure 12.1: A class probability tree.

predictor (say *Sex*). There is one branch from each of those nodes for each value of the variable stored at the node. We continue down the tree until all predictors are stored. The leaves of the tree store the target variable along with the conditional probability of the target variable given the values of the predictors in the path leading to the leaf. In our current example there are 12 leaves, one for each combination of values of the predictors. If there are many predictors, each with quite a few values, a complete class probability tree can become quite large. Aside from the storage problem, it might be hard to learn a large tree from data. However, some conditional probabilities may be the same for two or more combinations of values of the predictors. For example, the following four conditional probabilities may all be the same:

$$P(Buy = yes | Income = low, Sex = female, Mailed = yes)$$

$$P(Buy = yes | Income = low, Sex = male, Mailed = yes)$$

$$P(Buy = yes | Income = high, Sex = female, Mailed = yes)$$

$$P(Buy = yes | Income = high, Sex = male, Mailed = yes).$$

In this case we can represent the conditional probabilities more succinctly using the class probability tree in Figure 12.1. There is just one branch from the *Income* node for the two values *low* and *high* because the conditional probability is the same for these two values. Furthermore, this branch leads directly to a *Mailed* node because the value of *Sex* does not matter when *Income* is *low* or *high*.

We call the set of edges emanating from an internal node in a class probability tree a **split**, and we say there is a split on the variable stored at the node. For example, the root in the tree in Figure 12.1 is a split on the variable *Income*.

To retrieve a conditional probability from a class probability tree we start at the root and proceed to a leaf following the branches that have the values on which we are conditioning. For example, to retrieve $P(Buy = yes|Income = medium, Sex = female, Mailed = yes)$ from the tree in Figure 12.1, we proceed to the right from the *Income* node, then to right from the *Sex* node, and, finally, to the left from the *Mailed* node. We then retrieve the conditional probability value .7. In this way, we obtain

$$P(Buy = yes|Income = medium, Sex = female, Mailed = yes) = .7.$$

The purpose of a class probability tree learning algorithm is to learn from data a tree that best represents the conditional probabilities of interest. This is called "**growing the tree**." Trees are grown using a greedy one-ply lookahead search strategy and a scoring criterion to evaluate how good the tree appears based on the data. The classical text in this area is [Breiman et al., 1984]. Buntine [1993] presents a tree learning algorithm that uses a Bayesian scoring criterion. This algorithm is used in the IND Tree Package [Buntine, 2002], which comes with source code and a manual.

12.2   Application to Targeted Advertising

Suppose we have some population of potential customers such as all individuals living in a certain geographical area or all individuals who are registered users of Microsoft Windows. In targeted advertising we want to mail (or present in some way) an advertisement to a given subpopulation of this population only if we can expect to increase our profit by so doing. For example, if we learn we can expect to increase our profit by mailing to all males over the age of 30 in some population, we should do so. On the other hand, if we cannot expect to increase our profit by mailing to females under 30, we should not do so. First, we obtain an equality which tell us whether we can expect to increase our profit by mailing an advertisement to members of some subpopulation. Then we show how to use this inequality along with class probability trees to identify potentially profitable subpopulations. Finally, we show some experimental results.

12.2.1   Calculating Expected Lift in Profit

We can distinguish the following four types of potential customers:

Customer Type	Does Not Receive Ad	Receives Ad
Never-Buy	Won't buy	Won't buy
Persuadable	Won't buy	Will buy
Anti-Persuadable	Will buy	Won't buy
Always-Buy	Will buy	Will buy

A never-buy customer will not buy no matter what; a persuadable customer will buy only if an advertisement is received; an anti-persuadable customer will buy only if an advertisement is not received (such a customer is perhaps aggravated by the advertisement, causing the customer to reverse the decision to buy); and an always-buy customer will buy regardless of whether an advertisement is received. An advertisement is wasted on never-buy and always-buy customers, while it has a negative effect on an anti-persuadable customer. So the ideal mailing would send the advertisement only to the persuadable customers. However, in general, it is not possible to identify exactly this subset of customers. So we make an effort to come as close as possible to an ideal mailing.

In some subpopulation let N_{Never}, N_{Pers}, N_{Anti}, and N_{Always} be the number of people with behaviors never-buy, persuadable, anti-persuadable, and always-buy, and let N be the total number of people in the subpopulation. Furthermore, let c denote the cost of mailing (or in some way delivering) the advertisement to a given person, let r_u denote the profit obtained from a sale to an unsolicited customer, and let r_s denote the profit from a sale to a solicited customer. The reason r_s may be different from r_u is that we may offer some discount in our advertisement. Now suppose we pick a person at random from the subpopulation and mail that person the advertisement. We have

$$P(Buy = yes | Mailed = yes) = \frac{N_{Pers} + N_{Always}}{N}, \qquad (12.1)$$

where we are also conditioning on the fact that the person is in the given subpopulation without explicitly denoting this.

So the expected amount of profit received from each person who is mailed the advertisement is

$$\left(\frac{N_{Pers} + N_{Always}}{N}\right) \times (r_s - c) + \left(1 - \frac{N_{Pers} + N_{Always}}{N}\right) \times (-c)$$

$$= \frac{N_{Pers} + N_{Always}}{N} \times r_s - c.$$

Similarly,

$$P(Buy = yes | Mailed = no) = \frac{N_{Anti} + N_{Always}}{N}, \qquad (12.2)$$

and the expected amount of profit received from each person who is not mailed the advertisement is

$$\frac{N_{Anti} + N_{Always}}{N} \times r_u.$$

We should mail the advertisement to members of this subpopulation only when the expected profit received from a person who is mailed the advertisement exceeds the expected profit received from a person who is not mailed the advertisement. So we should mail the advertisement to members of the subpopulation only if

$$\frac{N_{Pers} + N_{Always}}{N} \times r_s - c > r_u \times \frac{N_{Anti} + N_{Always}}{N},$$

or equivalently

$$\frac{N_{Pers} + N_{Always}}{N} \times r_s - \frac{N_{Anti} + N_{Always}}{N} \times r_u - c > 0.$$

We call the left side of this last inequality the **expected lift in profit (ELP)** that we obtain by mailing a person the advertisement. Owing to Equality 12.1, we can obtain an estimate of $(N_{Pers} + N_{Always})/N$ by mailing many people in the subpopulation the advertisement and seeing what fraction buy. Similarly, owing to Equality 12.2, we can obtain an estimate of $(N_{Anti} + N_{Always})/N$ by not mailing the advertisement to a set of people and see what fraction of these people buy.

We developed the theory using N_{Never}, N_{Pers}, N_{Anti}, and N_{Always} to show the behavior in the subpopulation that results in the probabilities. The relevant formula in deciding whether or not to mail to a person in a subpopulation is simply

$$ELP = P(Buy = yes|Mailed = yes)r_s - P(Buy = yes|Mailed = no)r_u - c,$$
$$(12.3)$$

where the conditional probabilities are obtained from data or from data along with prior belief. We mail to everyone in the subpopulation if $ELP > 0$, and we mail to no one in the subpopulation if $ELP \leq 0$.

12.2.2   Identifying Subpopulations with Positive ELPs

Next, we show how to use class probability trees to identify subpopulations that have positive ELPs.

First, we take a large sample of individuals from the entire population and obtain data on the target variable Buy, the indicator variable $Mailed$, and n other indicator variables $X_1, X_2, \ldots, X_n$ for members of this sample. The n other indicator variables are attributes such as income, sex, salary, etc. Next, from these data we grow a class probability tree using a tree growing algorithm as discussed at the end of Section 12.1. For example, suppose $n = 2$, $X_1 = Sex$, and $X_2 = Income$. We might learn the class probability tree in Figure 12.1.

Once we learn the tree, we can calculate the ELP for every subpopulation. The next examples illustrate this.

Example 12.1 *Suppose*

$$
\begin{aligned}
c &= .5 \\
r_s &= 8 \\
r_u &= 10,
\end{aligned}
$$

and we are investigating the subpopulation consisting of individuals with medium income who are male using the tree in Figure 12.1. We follow the right branch from the root of the tree and then the left branch from the Sex node to arrive at the Mailed node corresponding to this subpopulation. We then find for this subpopulation that

$$
\begin{aligned}
ELP &= P(Buy = yes|Mailed = yes)r_s - P(Buy = yes|Mailed = no)r_u - c \\
&= .4 \times 8 - .2 \times 10 - .5 = .7.
\end{aligned}
$$

Since the ELP is positive we mail to this subpopulation.

Example 12.2 *Suppose the same values of c, r_s, and r_u as in the previous example, and we are investigating the subpopulation consisting of individuals with low income who are female. We follow the left branch from the root to arrive at the Mailed node corresponding to this subpopulation. We then find for this subpopulation that*

$$ELP = P(Buy = yes|Mailed = yes)r_s - P(Buy = yes|Mailed = no)r_u - c$$
$$= .6 \times 8 - .5 \times 10 - .5 = -.7.$$

Since the ELP is negative we do not mail to this subpopulation. Notice that for members of this subpopulation there is an increased probability of buying if we mail them the advertisement (.6 versus .5). However, owing to the cost and profits involved, the ELP is negative.

In general, the goal of a tree growing algorithm is to identify a tree that best represents the conditional probability of interest. That is, the scoring criterion evaluates the predictive accuracy of the tree. So a classical tree growing algorithm would attempt to maximize the accuracy of our estimate of $P(Buy|Mailed, X_1, X_2, \ldots, X_n)$ for all values of the variables. However, in this application we want to maximize expected profit, which is determined by the ELP. Notice in Equality 12.3 that the ELP will be 0 if $P(Buy = yes|Mailed = yes) = P(Buy = yes|Mailed = no)$. So if in some path from the root to a leaf there is no split on $Mailed$, the value of the ELP at the leaf will be 0. It seems a useful heuristic would be to modify the tree growing algorithm to force a split on $Mailed$ on every path from the root to a leaf. In this way for every subpopulation we will determine that either the ELP is positive (and therefore we should definitely mail), or the ELP is negative (and therefore we should definitely not mail). One way to implement this heuristic would be to modify the algorithm so that the last split is always on the $Mailed$ node. Chickering and Heckerman [2000] developed a tree growing algorithm with this modification.

12.2.3 Experiments

Chickering and Heckerman [2000] performed an experiment applying the technique just described. The experiment concerned deciding which subpopulations of Windows 95 registrants to mail an advertisement soliciting for an MSN subscription. For each of these registrants they had data on 15 variables from their registration forms. Examples of these variables included gender and amount of memory in the individual's computer. After discussing the method used in the experiment, we provide the results.

Method

A sample of about 110,000 registrants was obtained, and 90% of them were mailed the advertisement. The advertisement did not offer a special price. So $r_s = r_u$ in this experiment. After a specified period of time the experiment ended, and it was recorded whether each individual subscribed. So besides the

15 variables from the registration form, they now had values of *Mailed* and *Buy* (subscribe) for each individual in the sample. The sample was divided into a training set consisting of 70% of the sample and a testing set consisting of the remaining 30%. Trees were grown using the training test, and the resultant trees were then evaluated using the testing set. Two algorithms were used to grow the trees. The first, called *Normal*, was a standard algorithm that simply tried to maximize the scoring criterion, while the second, called *Force*, forced the last split to be on *Mailed*. A Bayesian scoring criterion was used to evaluate candidate trees.

After a tree was created, for each record in the test set the *ELP* was evaluated using Equality 12.3. An individual was marked for mailing if and only if the *ELP* was positive; otherwise, the individual was marked for not mailing. An individual was removed from the test set if the marked value did not match what was actually done for the individual. For example, if Joe Smith was mailed an advertisement but the *ELP* for Joe Smith was negative, the record for Joe Smith was removed from the test set. So the final test set consisted only of individuals who were mailed advertisements and had positive *ELP*s and ones who were not mailed advertisements and had negative *ELP*s. Total revenue was computed using this final test set as follows: If an individual were mailed an advertisement and subscribed, $r_s - c$ was added to the total; if an individual were mailed an advertisement and did not subscribe, $-c$ was added to the total; if an individual were not mailed an advertisement and subscribed, r_u (which in this case equals r_s) was added to the total; and if an individual were not mailed an advertisement and did not subscribe, 0 was added to the total. Finally, the total revenue was divided by the number of individuals in the final data set to obtain **revenue per person**.

The revenue per person was also calculated using the simple strategy of mailing to everyone. This value is obtained quite easily. For each individual in the test set, we simply add $r_s - c$ to the total revenue if the individual subscribed and $-c$ otherwise.

Let rev_{Normal}, rev_{Force}, and rev_{All} be the revenue per person obtained using the *Normal* algorithm, the *Force* algorithm, and the mail-to-all strategy, and set

$$rev\_inc_{Normal} = rev_{Normal} - rev_{All}$$
$$rev\_inc_{Force} = rev_{Force} - rev_{All}.$$

The variables rev_inc_{Normal} and rev_inc_{Force} tell us how much increase in per person revenue we can expect to obtain by doing the mailing determined by the respective algorithm, rather than simply mailing to everyone.

Results

Figure 12.2 shows the results for algorithm *Force* using $c = .42$ and varying the subscription rate $r_s = r_u$ from 1 to 15. The value of $r_s = r_u$ is plotted on the x-axis, while the value of rev_inc_{Force} is plotted on the y-axis. The results for algorithm *Normal* were just slightly worse than those for algorithm *Force*,

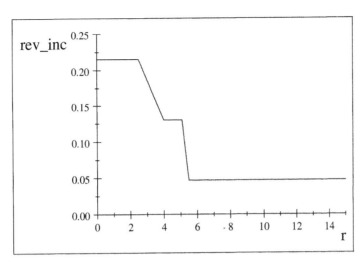

Figure 12.2: Expected increase in average revenue by doing targeted advertising using algorithm *Force* as a function of subscription rate.

so we do not bother to show them. We see that the improvement decreases significantly as the subscription rate increases to about 5.5. However, after that it remains fixed at about .047. Notice that, regardless of the subscription rate, we can expect to benefit by doing targeted advertising according to the method described in this chapter rather than simply mailing to everyone. The reason is that the method does seem to identify persuadable individuals who would buy only if they are sent an advertisement and anti-persuadable individuals who are turned off by the advertisement.

EXERCISES

Section 12.1

Exercise 12.1 Retrieve the following probability from the tree in Figure 12.1:

$$P(Buy = yes | Income = medium, Sex = male, Mailed = no).$$

Exercise 12.2 Retrieve the following probability from the tree in Figure 12.1:

$$P(Buy = yes | Income = low, Mailed = yes).$$

Section 12.2

Exercise 12.3 Suppose

$$
\begin{aligned}
c &= .6 \\
r_s &= 7 \\
r_u &= 9.
\end{aligned}
$$

1. Compute the ELP for the subpopulation consisting of individuals with medium income who are male using the tree in Figure 12.1. Should we mail to this subpopulation?

2. Compute the ELP for the subpopulation consisting of individuals with medium income who are female using the tree in Figure 12.1. Should we mail to this subpopulation?

3. Compute the ELP for the subpopulation consisting of individuals with low income using the tree in Figure 12.1. Should we mail to this subpopulation?

Bibliography

[Aczel and Sounderpandian, 2002]   Aczel, A., and J. Sounderpandian, *Complete Business Statistics*, McGraw-Hill, New York, 2002.

[Anderson et al., 2005]   Anderson, D., D. Sweeny, and T. Williams, *Statistics for Business and Economics*, Southwestern, Mason, Ohio, 2005.

[Basye et al., 1993]   Basye, K., T. Dean, J. Kirman, and M. Lejter, "A Decision-Theoretic Approach to Planning, Perception and Control," *IEEE Expert*, Vol. 7, No. 4, 1993.

[Beaver, 1966]   Beaver, W.H., "Financial Ratios as Predictors of Failure," *Journal of Accounting Research,* Vol. 4 (supplement), 1966.

[Beinlich and Herskovits, 1990]   Beinlich, I.A., and E. H. Herskovits, "A Graphical Environment for Constructing Bayesian Belief Networks," in Henrion, M., R.D. Shachter, L.N. Kanal, and J.F. Lemmer (Eds.): *Uncertainty in Artificial Intelligence 5*, North-Holland, Amsterdam, 1990.

[Bell, 1982]   Bell, D.E., "Regret in Decision Making Under Uncertainty," Operations Research, Vol. 30, 1982.

[Berry, 1996]   Berry, D.A., *Statistics: A Bayesian Perspective*, Wadsworth, Belmont, California, 1996.

[Berry et al., 1988]   Berry, M.A., E. Burmeister, and M.B. McElroy, "Sorting Out Risk Using Known APT Factors," *Financial Analyst Journal*, Vol. 44, No. 4, 1988.

[Bodie et al., 2004] Bodie, Z., A. Kane, and A. Marcus, *Investments*, McGraw-Hill, New York, 2004.

[Breese et al., 1998] Breese, J., D. Heckerman, and C. Kadie, "Empirical Analysis of Predictive Algorithms for Collaborative Filtering," in Cooper, G.F., and S. Moral (Eds.): *Uncertainty in Artificial Intelligence; Proceedings of the Fourteenth Conference*, Morgan Kaufmann, San Mateo, California, 1998.

[Breiman et al., 1984] Breiman, L., J. Friedman, R. Olshen, and C. Stone, *Classification and Regression Trees*, Wadsworth and Brooks, Monterey, California, 1984.

[Buntine, 1993] Buntine, W., "Learning Classification Trees," in Hand, D.J. (Ed.): *Artificial Intelligence Frontiers in Statistics: AI and Statistics III*, Chapman and Hall, London, 1993.

[Buntine, 2002] Buntine, W., "Tree Classification Software," *Proceedings of the Third National Technology Transfer Conference and Exposition*, Baltimore, Maryland, 2002.

[Burmeister et al., 1994] Burmeister, E., R. Roll, and S. Ross, "A Practioner's Guide to Arbitrage Pricing Theory," in *A Practitioner's Guide to Factor Models*, Research Foundation of the Institute of Chartered Financial Analysts, Charlottesville, Virginia, 1994; "Using Macroeconomic Factors to Control Portfolio Risk," March 9, 2003, is a revision of this paper and can be obtained at http://birr.com/methodology.cfm.

[Campbell, 1999] Campbell, J., "Asset Prices, Consumption, and the Business Cycle," in Taylor, J., and M. Woodward (Eds.): *Handbook of Macroeconomics*, Vol. 1, North-Holland, Amsterdam, 1999.

[Cheeseman and Stutz, 1995] Cheeseman, P., and J. Stutz, "Bayesian Classification (Autoclass): Theory and Results," in Fayyad, D., G. Piatesky-Shapiro, P. Smyth, and R. Uthurusamy (Eds.): *Advances in Knowledge Discovery and Data*

Mining, AAAI Press, Menlo Park, California, 1995.

[Chickering, 1996]

Chickering, D., "Learning Bayesian Networks Is NP-Complete," in Fisher, D., and H. Lenz (Eds.): *Learning From Data*, Springer-Verlag, New York, 1996.

[Chickering and Heckerman, 2000]

Chickering, D., and D. Heckerman, "A Decision-Theoretic Approach to Targeted Advertising," in Boutilier, C., and M. Goldszmidt (Eds.): *Uncertainty in Artificial Intelligence; Proceedings of the Sixteenth Conference*, Morgan Kaufmann, San Mateo, California, 2000.

[Clemen, 1996]

Clemen, R.T., *Making Hard Decisions*, PWS-KENT, Boston, Massachusetts, 1996.

[Cooper, 1990]

Cooper, G.F., "The Computational Complexity of Probabilistic Inference Using Bayesian Belief Networks," *Artificial Intelligence*, Vol. 33, 1990.

[Dagum and Chavez, 1993]

Dagum, P., and R.M. Chavez, "Approximate Probabilistic Inference in Bayesian Belief Networks," *IEEE Transactions on Pattern Analysis and Machine Intelligence*, Vol. 15, No. 3, 1993.

[Dagum and Luby, 1993]

Dagum, P., and M. Luby, "Approximate Probabilistic Inference in Bayesian Belief Networks Is NP-hard," *Artificial Intelligence*, Vol. 60, No. 1, 1993.

[Dean and Wellman, 1991]

Dean, T., and M. Wellman, *Planning and Control*, Morgan Kaufmann, San Mateo, California, 1991.

[de Finetti, 1937]

de Finetti, B., "La prévision: See Lois Logiques, ses Sources Subjectives," *Annales de l'Institut Henri Poincaré*, Vol. 7, 1937.

[Demirer et al., 2007]

Demirer, R., R. Mau, and C. Shenoy, "Bayesian Networks: A Decision Tool to Improve Security Analysis," to appear in *Journal of Applied Science*, 2007.

[Diez and Druzdzel, 2006]   Diez, F.J., and M.J. Druzdzel, "Canonical
                                   Probabilistic Models for Knowledge Engi-
                                   neering," submitted for publication, 2006.

[Druzdzel and Glymour, 1999]   Druzdzel, M.J., and C. Glymour, "Causal
                                   Inferences from Databases: Why Universi-
                                   ties Lose Students," in Glymour, C., and
                                   G.F. Cooper (Eds.): *Computation, Causa-
                                   tion, and Discovery*, AAAI Press, Menlo
                                   Park, California, 1999.

[Eells, 1991]   Eells, E., *Probabilistic Causality*, Cam-
                                   bridge University Press, London, 1991.

[Fama and MacBeth, 1973]   Fama, E., and J. MacBeth, "Risk, Return,
                                   and Equilibrium: Empirical Tests," *Jour-
                                   nal of Political Economy*, Vol. 81, No. 3,
                                   1973.

[Feller, 1968]   Feller, W., *An Introduction to Probability
                                   Theory and Its Applications*, Wiley, New
                                   York, 1968.

[Flanders et al., 1996]   Flanders, A.E., C.M. Spettell, L.M.
                                   Tartaglino, D.P. Friedman, and G.J. Her-
                                   bison, "Forecasting Motor Recovery af-
                                   ter Cervical Spinal Cord Injury: Value of
                                   MRI," *Radiology*, Vol. 201, 1996.

[Fung and Chang, 1990]   Fung, R., and K. Chang, "Weighing and
                                   Integrating Evidence for Stochastic Simu-
                                   lation in Bayesian Networks," in Henrion,
                                   M., R.D. Shachter, L.N. Kanal, and J.F.
                                   Lemmer (Eds.): *Uncertainty in Artificial
                                   Intelligence 5*, North Holland, Amsterdam,
                                   1990.

[Graham and Zweig, 2003]   Graham, B., and J. Zweig, *The Intelligent
                                   Investor: The Definitive Book on Value
                                   Investing, Revised Edition*, HarperCollins,
                                   New York, 2003.

[Heckerman, 1996]   Heckerman, D., "A Tutorial on Learn-
                                   ing with Bayesian Networks," Technical
                                   Report # MSR-TR-95-06, Microsoft Re-
                                   search, Redmond, Washington, 1996.

[Heckerman et al., 1999]   Heckerman, D., C. Meek, and G. Cooper,
                                   "A Bayesian Approach to Causal Discov-
                                   ery," in Glymour, C., and G.F. Cooper

(Eds.): *Computation, Causation, and Discovery*, AAAI Press, Menlo Park, California, 1999.

[Heckerman and Meek, 1997]   Heckerman, D., and C. Meek, "Embedded Bayesian Network Classifiers," Technical Report MSR-TR-97-06, Microsoft Research, Redmond, Washington, 1997.

[Herskovits and Dagher, 1997]   Herskovits, E.H., and A.P. Dagher, "Applications of Bayesian Networks to Health Care," Technical Report NSI-TR-1997-02, Noetic Systems Incorporated, Baltimore, Maryland, 1997.

[Hogg and Craig, 1972]   Hogg, R.V., and A.T. Craig, *Introduction to Mathematical Statistics*, Macmillan, New York, 1972.

[Huang et al., 1994]   Huang, T., D. Koller, J. Malik, G. Ogasawara, B. Rao, S. Russell, and J. Weber, "Automatic Symbolic Traffic Scene Analysis Using Belief Networks," *Proceedings of the Twelfth National Conference on Artificial Intelligence (AAAI94)*, AAAI Press, Seattle, Washington, 1994.

[Hume, 1748]   Hume, D., *An Inquiry Concerning Human Understanding*, Prometheus, Amhurst, New York, 1988 (originally published in 1748).

[Ingersoll, 1987]   Ingersoll, J., *Theory of Financial Decision Making*, Rowman & Littlefield, Lanham, Maryland, 1987.

[Iversen et al., 1971]   Iversen, G.R., W.H. Longcor, F. Mosteller, J.P. Gilbert, and C. Youtz, "Bias and Runs in Dice Throwing and Recording: A Few Million Throws," *Psychometrika*, Vol. 36, 1971.

[Jensen et al., 1990]   Jensen, F.V., S. L. Lauritzen, and K.G. Olesen, "Bayesian Updating in Causal Probabilistic Networks by Local Computation," *Computational Statistical Quarterly*, Vol. 4, 1990.

[Jensen, 2001]   Jensen, F.V., *Bayesian Networks and Decision Graphs*, Springer-Verlag, New York, 2001.

[Kahneman and Tversky, 1979]   Kahneman, D., and A. Tversky, "Prospect Theory: An Analysis of Decision Under Risk," *Econometrica*, Vol. 47, 1979.

[Keefer, 1983]   Keefer, D.L., "3-Point Approximations for Continuous Random Variables," *Management Science*, Vol. 29, 1983.

[Kemmerer et al., 2006]   Kemmerer, B., Mishra, S., and P. Shenoy, "Bayesian Causal Maps as Decision Aids in Venture Capital Decision Making," submitted to the *Entrepreneurship Division (ENT)*, 2006.

[Kennett et al., 2001]   Kennett, R., K. Korb, and A. Nicholson, "Seabreeze Prediction Using Bayesian Networks: A Case Study," *Proceedings of the 5th Pacific-Asia Conference on Advances in Knowledge Discovery and Data Mining-PAKDD*, Springer-Verlag, New York, 2001.

[Kerrich, 1946]   Kerrich, J.E., *An Experimental Introduction to the Theory of Probability*, Einer Munksgaard, Copenhagen, 1946.

[Lander and Shenoy, 1999]   Lander, D.M., and P. Shenoy, "Modeling and Valuing Real Options Using Influence Diagrams," School of Business Working Paper No. 283, University of Kansas, Lawrence, Kansas, 1999.

[Lauritzen and Spiegelhalter, 1988]   Lauritzen, S.L., and D.J. Spiegelhalter, "Local Computation with Probabilities in Graphical Structures and Their Applications to Expert Systems," *Journal of the Royal Statistical Society B*, Vol. 50, No. 2, 1988.

[Li and D'Ambrosio, 1994]   Li, Z., and B. D'Ambrosio, "Efficient Inference in Bayes' Networks as a Combinatorial Optimization Problem," *International Journal of Approximate Inference*, Vol. 11, 1994.

[Lindley, 1985]   Lindley, D.V., *Introduction to Probability and Statistics from a Bayesian Viewpoint*, Cambridge University Press, London, 1985.

[Luenberger, 1998]   Luenberger, D., *Investment Science*, Ox-
                            ford, New York, 1998.

[Lugg et al., 1995]   Lugg, J.A., J. Raifer, and C.N.F. González,
                            "Dehydrotestosterone is the Active Andro-
                            gen in the Maintenance of Nitric Oxide-
                            Mediated Penile Erection in the Rat," *En-
                            docrinology*, Vol. 136, No. 4, 1995.

[Lynch and Rothchild, 2000] Lynch, P., and J. Rothchild, *One Up On
                            Wall Street: How to Use What You Already
                            Know to Make Money in the Market*, Fire-
                            side, New York, 2000.

[Mani et al., 1997]   Mani, S., S. McDermott, and M. Val-
                            torta, "MENTOR: A Bayesian Model for
                            Prediction of Mental Retardation in New-
                            borns," *Research in Developmental Disabil-
                            ities*, Vol. 8, No. 5, 1997.

[McClennan and Markham, 1999] McClennan, K.J., and A. Markham, "Fi-
                            nasteride: A Review of Its Use in Male
                            Pattern Baldness," *Drugs*, Vol. 57, No. 1,
                            1999.

[McElroy and Burmeister, 1988] McElroy, M., and E. Burmeister, "Arbi-
                            trage Pricing Theory as a Restricted Non-
                            linear Multivariate Regression Model: It-
                            erated Nonlinear Seemingly Unrelated Re-
                            gression Estimates," *Journal of Business
                            and Economic Statistics*, Vol. 6, No. 1,
                            1988.

[McKee, 2003]   Mckee, T.E., "Rough Sets Bankruptcy Pre-
                            diction Models Versus Auditor Signaling
                            Rates," *Journal of Forecasting*, Vol. 22,
                            2003.

[McKee and Greenstein, 2000] McKee, T.E., and M. Greenstein, "Predict-
                            ing Bankruptcy Using Recursive Partition-
                            ing and a Realistically Proportioned Data
                            Set," *Journal of Forecasting*, Vol. 19, 2000.

[McKee and Lensberg, 2002] McKee, T.E., and T. Lensberg, "Genetic
                            Programming and Rough Sets: A Hybrid
                            Approach to Bankruptcy Classification,"
                            Journal of Operational Research, Vol. 138,
                            2002.

[McLachlan and Krishnan, 1997]   McLachlan, G.J., and T. Krishnan, *The EM Algorithm and Its Extensions*, Wiley, New York, 1997.

[Meek, 1995]   Meek, C., "Strong Completeness and Faithfulness in Bayesian Networks," in Besnard, P., and S. Hanks (Eds.): *Uncertainty in Artificial Intelligence; Proceedings of the Eleventh Conference*, Morgan Kaufmann, San Mateo, California, 1995.

[Neal, 1992]   Neal, R., "Connectionist Learning of Belief Networks," *Artificial Intelligence*, Vol. 56, 1992.

[Neapolitan, 1990]   Neapolitan, R.E., *Probabilistic Reasoning in Expert Systems*, Wiley, New York, 1990.

[Neapolitan, 1992]   Neapolitan, R.E., "A Survey of Uncertain and Approximate Inference," in Zadeh, L., and J. Kacprzyk (Eds.): *Fuzzy Logic for the Management of Uncertainty*, Wiley, New York, 1992.

[Neapolitan, 1996]   Neapolitan, R.E., "Is Higher-Order Uncertainty Needed?" in *IEEE Transactions on Systems, Man, and Cybernetics Part A: Systems and Humans*, Vol. 26, No. 3, 1996.

[Neapolitan, 2004]   Neapolitan, R.E., *Learning Bayesian Networks*, Prentice Hall, Upper Saddle River, New Jersey, 2004.

[Neapolitan and Morris, 2002]   Neapolitan, R.E., and S. Morris, "Probabilistic Modeling Using Bayesian Networks," in D. Kaplan (Ed.): *Handbook of Quantitative Methodology in the Social Sciences*, Sage, Thousand Oaks, California, 2002.

[Nease and Owens, 1997]   Nease, R.F., and D.K. Owens, "Use of Influence Diagrams to Structure Medical Decisions," *Medical Decision Making*, Vol. 17, 1997.

[Nefian et al., 2002]   Nefian, A.F., L. H. Liang, X.X. Liu, X. Pi. and K. Murphy, "Dynamic Bayesian Networks for Audio-Visual Speech Recognition," *Journal of Applied Signal Processing, Special Issue on Joint Audio Visual Speech Processing*, Vol. 11, 2002.

[Nicholson, 1996]   Nicholson, A.E., "Fall Diagnosis Using Dynamic Belief Networks," *Proceedings of the 4th Pacific Rim International Conference on Artificial Intelligence (PRICAI-96)*, Cairns, Australia, 1996.

[Ohlson, 1980]   Ohlson, J.A., "Financial Ratios and the Probabilistic Prediction of Bankruptcy," *Journal of Accounting Research*, Vol. 19, 1980.

[Olesen et al., 1992]   Olesen, K.G., S.L. Lauritzen, and F.V. Jensen, "aHUGIN: A System Creating Adaptive Causal Probabilistic Networks," in Dubois, D., M.P. Wellman, B. D'Ambrosio, and P. Smets (Eds.): *Uncertainty in Artificial Intelligence; Proceedings of the Eighth Conference*, Morgan Kaufmann, San Mateo, California, 1992.

[Olmsted, 1983]   Olmsted, S.M., "On Representing and Solving Influence Diagrams," Ph.D. Thesis, Dept. of Engineering-Economic Systems, Stanford University, California, 1983.

[Pearl, 1986]   Pearl, J., "Fusion, Propagation, and Structuring in Belief Networks," *Artificial Intelligence*, Vol. 29, 1986.

[Pearl, 1988]   Pearl, J., *Probabilistic Reasoning in Intelligent Systems*, Morgan Kaufmann, San Mateo, California, 1988.

[Pearl, 2000]   Pearl, J., *Causality: Models, Reasoning, and Inference*, Cambridge University Press, Cambridge, U.K., 2000.

[Piaget, 1966]   Piaget, J., *The Child's Conception of Physical Causality*, Routledge and Kegan Paul, London, 1966.

[Pradham and Dagum, 1996]   Pradham, M., and P. Dagum, "Optimal Monte Carlo Estimation of Belief Network Inference," in Horvitz, E., and F. Jensen (Eds.): *Uncertainty in Artificial Intelligence; Proceedings of the Twelfth Conference*, Morgan Kaufmann, San Mateo, California, 1996.

[Resnick et al., 1994]   Resnick, P., N. Iacovou, M. Suchak, P. Bergstrom, and J. Riedl, "Grouplens: An Open Architecture for Collaborative Filtering of Netnews," *Proceedings of the ACM 1994 Conference on Computer Supported Cooperative Work*, New York, 1994.

[Robinson, 1977]   Robinson, R.W., "Counting Unlabeled Acyclic Digraphs," in Little, C.H.C. (Ed.): *Lecture Notes in Mathematics, 622: Combinatorial Mathematics V*, Springer-Verlag, New York, 1977.

[Rucker and Polanco, 1997]   Rucker, J., and M.J. Polanco, "Siteseer: Personalized Navigation of the Web," *Communications of the ACM*, Vol. 40, No. 3, 1997.

[Ruhnka et al., 1992]   Ruhnka, J.C., H.D. Feldman, and T.J. Dean, "The 'Living Dead' Phenomena in Venture Capital Investments," *Journal of Business Venturing*, Vol. 7, No. 2, 1992.

[Russell and Norvig, 1995]   Russell, S., and P. Norvig, *Artificial Intelligence: A Modern Approach*, Prentice Hall, Upper Saddle River, New Jersey, 1995.

[Salmon, 1997]   Salmon, W., *Causality and Explanation*, Oxford University Press, New York, 1997.

[Sarkar and Sriram, 2001]   Sarkar, S., and R.S. Sriram, "Bayesian Models for Early Warning of Bank Failure," *Management Science*, Vol. 47, 2001.

[Savage, 1954]   Savage, L.J., *Foundations of Statistics*, Wiley, New York, 1954.

[Scarville et al., 1996]   Scarville J., S.B. Button, J.E. Edwards, A.R. Lancaster, and T.W. Elig, "Armed Forces 1996 Equal Opportunity Survey," DMDC Report No. 97-0279, Defense Manpower Data Center, Arlington, VA., 1996.

[Scheines et al., 1994]   Scheines, R., P. Spirtes, C. Glymour, and C. Meek, *Tetrad II: User Manual*, Lawrence Erlbaum, Hillsdale, New Jersey, 1994.

[Sercu and Uppal, 1995]   Sercu, P., and R. Uppal, *International Financial Markets and the Firm*, Southwestern College, Cincinnati, Ohio, 1995.

[Shachter, 1986]   Shachter, R.D., "Evaluating Influence Di-
                               agrams," *Operations Research*, Vol. 34,
                               1986.

[Shachter and Peot, 1990]   Shachter, R.D., and M. Peot, "Simulation
                               Approaches to General Probabilistic Infer-
                               ence in Bayesian Networks," in Henrion,
                               M., R.D. Shachter, L.N. Kanal, and J.F.
                               Lemmer (Eds.): *Uncertainty in Artificial
                               Intelligence 5*, North-Holland, Amsterdam,
                               1990.

[Shepherd and Zacharakis, 2002] Shepherd, D.A., and A. Zacharakis, "Ven-
                               ture Capitalists' Expertise: A Call for Re-
                               search into Decision Aids and Cognitive
                               Feedback," *Journal of Business Venturing*,
                               Vol. 17, 2002.

[Singh and Valtorta, 1995]   Singh, M., and M. Valtorta, "Construc-
                               tion of Bayesian Network Structures from
                               Data: A Brief Survey and an Efficient Al-
                               gorithm," *International Journal of Approx-
                               imate Reasoning*, Vol. 12, 1995.

[Spirtes et al., 1993, 2000]   Spirtes, P., C. Glymour, and R. Scheines,
                               Causation, Prediction, and Search,
                               Springer-Verlag, New York, 1993; 2nd ed.:
                               MIT Press, Cambridge, Massachusetts,
                               2000.

[Srinivas, 1993]   Srinivas, S., "A Generalization of the Noisy
                               OR Model," in Heckerman, D., and A.
                               Mamdani (Eds.): *Uncertainty in Artificial
                               Intelligence; Proceedings of the Ninth Con-
                               ference*, Morgan Kaufmann, San Mateo,
                               California, 1993.

[Stangor et al., 2002]   Stangor, C., J.K. Swim, K.L. Van Allen,
                               and G.B. Sechrist, "Reporting Discrimina-
                               tion in Public and Private Contexts," *Jour-
                               nal of Personality and Social Psychology*,
                               Vol. 82, 2002.

[Sun and Shenoy, 2006]   Sun, L., and P. Shenoy, "Using Bayesian
                               Networks for Bankruptcy Prediction:
                               Some Methodological Issues," School of
                               Business Working Paper No. 302, Uni-
                               versity of Kansas, Lawrence, Kansas,
                               2006.

[Tam and Kiang, 1992]   Tam, K.Y., and M.Y. Kiang, "Manager-
                               ial Applications of Neural Networks: The
                               Case of Bank Failure Predictions," *Man-
                               agement Science*, Vol. 38, 1992.

[Tatman and Shachter, 1990]   Tatman, J.A., and R.D. Shachter, "Dy-
                               namic Programming and Influence Dia-
                               grams," *IEEE Transactions on Systems,
                               Man, and Cybernetics*, Vol. 20, 1990.

[Terveen et al., 1997]   Terveen, L., W. Hill, B. Amento, D. Mc-
                               Donald, and J. Creter, "PHOAKS: A Sys-
                               tem for Sharing Recommendations," *Com-
                               munications of the ACM*, Vol. 40, No. 3,
                               1997.

[Tversky and Kahneman, 1981]   Tversky, A., and D. Kahneman, "The
                               Framing of Decisions and the Psychology
                               of Choice," *Science*, Vol. 211, 1981.

[van Lambalgen, 1987]   van Lambalgen, M., "Random Sequences,"
                               Ph.D. Thesis, University of Amsterdam,
                               1987.

[von Mises, 1919]   von Mises, R., "Grundlagen der Wahr-
                               scheinlichkeitsrechnung," *Mathematische
                               Zeitschrift*, Vol. 5, 1919.

[Wallace and Korb, 1999]   Wallace, C.S., and K. Korb, "Learning Lin-
                               ear Causal Models by MML Sampling," in
                               Gammerman, A. (Ed.): *Causal Models and
                               Intelligent Data Mining*, Springer-Verlag,
                               New York, 1999.

[Zadeh, 1995]   Zadeh, L., "Probability Theory and Fuzzy
                               Logic Are Complementary Rather Than
                               Competitive," *Technometrics*, Vol. 37,
                               1995.

Index

Printed and bound by CPI Group (UK) Ltd, Croydon, CR0 4YY

03/10/2024

01040313-0014